# Causal Asymmetries

Causation is asymmetrical in many different ways. Causes precede effects. Explanations cite causes not effects. Agents use causes to manipulate their effects; they do not use effects to manipulate their causes. Effects of a common cause are correlated; causes of a common effect are not.

This book by an eminent contemporary philosopher of science offers the first comprehensive treatment of causal asymmetries. It explains why a relationship that is asymmetrical in one of these regards is asymmetrical in others. Hausman discovers surprising hidden logical connections between apparently unrelated aspects of causation and traces them all to an asymmetry of independence, which he argues is constitutive of the causal relation.

This is a major book for metaphysicians and philosophers of science that will also prove stimulating to statisticians and scientists.

Daniel M. Hausman is Professor of Philosophy at the University of Wisconsin–Madison.

**Cambridge Studies in Probability, Induction, and Decision Theory**

*General Editor:* Brian Skyrms

Ellery Eells, *Probabilistic Causality*
Richard Jeffrey, *Probability and the Art of Judgment*
Robert C. Koons, *Paradoxes of Belief and Strategic Rationality*
Cristina Bicchieri and Maria Luisa Dalla Chiara (eds.), *Knowledge, Belief, and Strategic Interaction*
Patrick Maher, *Betting on Theories*
Cristina Bicchieri, *Rationality and Coordination*
J. Howard Sobel, *Taking Chances*
Jan von Plato, *Creating Modern Probability: Its Mathematics, Physics, and Philosophy in Historical Perspective*
Ellery Eells and Brian Skyrms (eds.), *Probability and Conditionals*
Cristina Bicchieri, Richard Jeffrey, and Brian Skyrms (eds.), *The Dynamics of Norms*
Patrick Suppes and Mario Zanotti, *Foundations of Probability with Applications*
Paul Weirich, *Equilibrium and Rationality*

# Causal Asymmetries

Daniel M. Hausman

CAMBRIDGE UNIVERSITY PRESS
Cambridge, New York, Melbourne, Madrid, Cape Town, Singapore, São Paulo

Cambridge University Press
The Edinburgh Building, Cambridge CB2 8RU, UK

Published in the United States of America by Cambridge University Press, New York

www.cambridge.org
Information on this title: www.cambridge.org/9780521622899



First published 1998
This digitally printed version 2008

*A catalogue record for this publication is available from the British Library*

*Library of Congress Cataloguing in Publication data*

Hausman, Daniel M., 1947–

Causal asymmetries / Daniel M. Hausman.

p. cm. – (Cambridge studies in probability, induction, and decision theory)

Includes bibliographical references and index.

ISBN 0-521-62289-1 (hardcover)

1. Causation. 2. Science – Philosophy. I. Title. II. Series.
BD531.H38 1998
122 – dc21 97-42437
CIP

ISBN 978-0-521-62289-9 hardback
ISBN 978-0-521-05242-9 paperback

# Contents

# List of Figures

# Acknowledgments

I have been thinking about causal asymmetries for nearly twenty years, and my intellectual debts have piled up. Despite the many acknowledgments that follow, I've probably forgotten some of the help I've had, and I apologize for inadvertent omissions.

My ideas on causal asymmetry took shape originally during the 1980–1 academic year while I was on leave from teaching with a grant from the National Science Foundation. I was inspired by Herbert Simon's essays (particularly "Causal Order and Identifiability" (1953)) and by long conversations with Douglas Ehring, who was completing a dissertation on causation. The similarities between Ehring's ideas and mine are evident from a comparison between Ehring's essay "Causal Asymmetry" (1982) and my "Causal Priority" (1984). I discuss Ehring's views briefly in footnote 7 of chapter 4*. I should have been inspired by Hans Reichenbach's work, especially his *The Direction of Time* (1956), but I did not read Reichenbach until later.

Versions of the ideas developed in this manuscript have appeared in several essays. "Causal Priority" (1984) is a first stab at the central ideas developed here in chapter 4. An early version of the ideas concerning causation and agency developed in chapter 5 appear in "Causation and Experimentation" (1986), and a much more recent version appears in "Causation, Agency, and Independence" (1998a). A large part of chapter 6 appears in "Causation and Counterfactual Dependence Reconsidered" (1996). One can trace the slow development of the ideas concerning the relations between causation and explanation that are presented in chapter 8 in "Causal and Explanatory Asymmetry" (1982), "The Insufficiency of Nomological Explanation" (1989), "Why Don't Causes Explain Their Effects?" (1993b), and "Linking Causal and Explanatory Asymmetries" (1993a). Many of the ideas on probabilistic causality presented in chapter 9 appear in "Deterministic Causation of Probabilities" (1998b).

During the many years before I began work on this manuscript, dozens of philosophers and economists helped my thoughts germinate. I want to thank Martin Bunzl, Ellery Eells, Malcolm Forster, Clark Glymour, Peter Menzies, Huw Price, Alexander Rosenberg, David Sanford, Herbert Simon, Elliott Sober, Paul Thagard, Judith Thomson, and Peter van Inwagen. An invitation from Peter Slezak to deliver a series of five lectures at the University of New South Wales in August of 1989 provided an occasion for a first attempt to pull my ideas on causality together.

A first rough draft of this manuscript was completed during the 1994–5 academic year in Paris, when I had a sabbatical leave from the University of Wisconsin with an additional grant from the Wisconsin Alumni Research Foundation. I couldn't have done this work without the friendship, encouragement, and stimulation provided by the philosophers and economists at the CREA (Centre de Recherche en Epistemologie Appliquée) and by the economists belonging to the THEMA research group at the University of Cergy Pontoise. Audiences in Paris, Cergy Pontoise, and at the London School of Economics helped enormously. Special thanks to Nancy Cartwright, Mary Morgan, David Papineau, and John Worrall at the LSE, Jean-Pierre Dupuy, Max Kistler, Philippe Mongin, Peter Railton, and Bernard Walliser at the CREA, and Jean-François Laslier at University of Cergy Pontoise. Ellery Eells, Richard Scheines, Robert Stalnaker, and James Woodward read portions of the 1994–5 version of the manuscript and helped me to improve it. In addition to those mentioned above, I would like to thank Alain Beraud, Marc Fleurbaey, Nicolas Gravel, and Alain Trannoy of the THEMA group. The office I was given in Cergy Pontoise provided an indispensable refuge from the temptations of Paris. Robert Card at the University of Wisconsin helped track down sources.

During the Fall of 1995, Martin Barrett, Berent Enç, Malcolm Forster, and Elliott Sober met with me weekly to read and criticize chapters of the first draft. I am deeply indebted to them for their criticisms. At the same time Nancy Cartwright and David Papineau discussed parts of the manuscript with a seminar at the London School of Economics and found further flaws. I also benefited from conversations in Florence in August 1995 with Paul Humphreys, Peter Menzies, and Huw Price. During the Spring of 1996 members of a graduate seminar at the University of Wisconsin worked through most of the manuscript and offered searching criticisms. I owe a large debt to this talented group of students: Gordon Barnes, Jinhee Choi, Peter DeSmidt, Leslie Graves, Rick Hefko, Sungsu Kim, Gregory Mougin, Collin O'Neil, Doug Smith, Chris Stevens, Richard Teng, and Eric Wolf. During the Spring of 1997 as I was preparing the final version, Richard Scheines and a group of his colleagues at Carnegie Mellon worked through most of the book and caught some of the errors that survived all the previous criticism. James Woodward and Nancy Cartwright also saved me from serious errors in chapter 11.

Finally I would like to thank the readers for Cambridge University Press, who provided still more much-needed criticism. Christopher Hitchcock identified himself as one of these readers, and I am glad to be able to thank him for his extensive help. The errors that remain are a testament to my thick-headedness.

I have had a great deal of financial support for this research. In addition to the grant in 1980–1 from the NSF and the support during the 1994–5 academic year from the University of Wisconsin, I received support from the University of Maryland General Research Board during the Spring of 1983, from the National Endowment for the Humanities during the Summer of 1986, and from the Wisconsin Alumni Research Foundation during the Spring of 1991.

# Introduction

# Causation and its Asymmetries

Causation apparently has several different asymmetrical features. In this book I shall say what these features are and how they are related to one another. Here is a list of many of these purported asymmetries:

*Time order*: Effects do not come before their causes (chapter 3).

*Probabilistic Independence*: Causes of a given effect are probabilistically independent of one another, while effects of a given cause are probabilistically dependent on one another (chapters 4, 12).

*Agency or manipulability*: Causes can be used to manipulate their effects, but effects cannot be used to manipulate their causes, and effects of a common cause cannot be used to manipulate one another (chapters 5, 7).

*Counterfactual dependence*: Effects counterfactually depend on their causes, while causes do not counterfactually depend on their effects and effects of a common cause do not counterfactually depend on one another (chapters 6, 7).

*Overdetermination*: Effects overdetermine their causes, while causes rarely overdetermine their effects (chapter 6).

*Explanation*: Causes can be cited to explain their effects, but effects cannot be cited to explain their causes and effects of a common cause cannot be cited to explain one another (chapter 8).

*Invariance*: If the dependent variables in an equation system are effects of the independent variables, then if one intervenes and changes the value of an independent variable and substitutes the new value in the equations, one has the best prediction of new values for the dependent variables. If on the other hand the independent variables causally depend on the dependent variables and one substitutes new values for the independent variables, then the values one calculates for the dependent variables will be incorrect (chapters 8, 11).

*Screening-off*: Causes screen off their effects – i.e., controlling for causes makes the probabilistic dependence among effects disappear – while effects do not screen off their causes and effects of a common cause do not screen off one another (chapter 10).

*Robustness*: The relationship between cause and effect is invariant with respect to the frequency of the cause or with respect to how the cause comes about but not with respect to the frequency of the effect or with respect to how the effect comes about (chapter 11).

*Fixity*: Causes are "fixed" no later than their effects (chapter 7).

*Connection dependence*: If one were to break the connection between cause and effect, only the effect would be affected (chapter 6).

Many of these claims are, at best, approximate, yet even those that break down help one to understand causation. One understands causal asymmetry only when one knows whether these claims are true and how they are related to one another. I shall place particular emphasis on an asymmetry of *causal independence*, which is not in the list, but which is linked to the asymmetry of probabilistic independence. The asymmetry of causal independence is implicated in most of the others, and it has a special role in linking them. I shall argue that causal independence is the central thread in causal asymmetry, even though it is not the whole cloth.

The first two chapters set the stage. Chapter 1 clarifies my presuppositions, and chapter 2 says what the causal relation relates. The remaining chapters, except for the last, discuss the asymmetries listed above and explore the relations among them. The last chapter takes up complications concerning event fusions, overdetermination, and preemption, and in it I state my conclusions. Whenever possible, I prove the assertions I make about relations among the various asymmetries. Unfortunately, these proofs and the precise statements of conditions they depend on are tedious. To make the book readable I have accordingly confined digression, subsidiary arguments, and most of the proofs to separate chapters with asterisks following their numbers.[1] Readers who are not interested in the by-roads and the proofs should skip the starred chapters. Along the way (though mainly in the starred chapters) I formulate explicit conditions and propositions, which I label with capital boldfaced letters. For easy reference, these are listed in alphabetical order in Appendix A. When I prove a proposition that is of interest or to which I shall want to refer, I call it a "theorem" and number it, beginning with the chapter number. For example, theorem 4.1 is the first theorem in chapter 4*. For convenient reference I list the theorems in numerical order in Appendix B. "Theorem" is a grandiose name for these humble results, but I needed some way to make convenient references to the propositions I prove.

Although most of this book explores the precise relations among these asymmetries, some general conclusions emerge: Human beings single out some lawful relations as causal, and they distinguish causes from effects. The reason is ultimately practical. When one factor can be independently manipulated without breaking its nomological links to others, then the factor that can be manipulated can be used to control whatever continues to be linked to it. The possibility of manipulation and control and the related possibility of giving a specifically causal explanation obtain when there is a certain pattern of independence within nomological relations. There are causal relations exactly when these patterns of independence hold. The asymmetries of independence are "objective," but their significance depends

[1] I am borrowing this expository technique from Amartya Sen's masterful *Collective Choice and Social Welfare* (1970).

on human interests. Their centrality to causation and explanation are explained by human interests and are manifest in the intricate relations between the asymmetry of independence and the other asymmetries listed above.

As the first chapter explains, I began with quite a different picture in mind, and these conclusions only emerged painfully out of the detailed arguments in this book. What these sketchy remarks mean and why one should believe them cannot be explained here in the introduction. The book as a whole is devoted to that task. I call this book *Causal Asymmetries*, because it deals with many asymmetric aspects of the causal relation. Articulating them and clarifying the relations among them tells one a great deal about causation.

# 1

# Metaphysical Pictures and Wishes

This chapter tells you where I started, explains how my initial hunches led to difficulties, and lays bare a fundamental ambivalence that nags the discussions in this book. It exposes unclarities in what I was looking for in a theory of causation.

## 1.1 Metaphysical Theories

In this century, metaphysics has been in ill repute. The word "metaphysics" conjures up an image of a philosopher envisioning "the essential features of reality" and then confidently dictating to artists, moralists, and scientists. Does metaphysics have any genuine content? Are there metaphysical questions that do not collapse into either empirical questions subject to scientific inquiry or semantic questions subject to conceptual analysis?

I sympathize with this "positivistic" skepticism concerning the possibility of conjuring substantive knowledge out of pure contemplation, and there is, I hope, no conjuring in this book. Although questions such as "What are the differences between causes and effects?" are more general than the questions scientists ask, they are not of a different kind. Metaphysical questions concerning causation are continuous with scientific questions. Although more abstract than theories in geology or sociology, the account of causation I defend in this book is intended as an empirical theory (jointly of nature and of human explanatory practices). This theory, like scientific theories generally, is acceptable only if it helps our beliefs and practices to fit together coherently.

## 1.2 The Question

One might say, "The explosion was caused by the foreman's striking a match," or "Margaret's hitting the tomato with a hammer smashed it." The second claim does not use the word "cause," but smashing is a kind of causing. To say, "The hammer blow smashed the tomato," appears to state physical relations between the hammer and the tomato and causal relations between the hammer blow and the smashing. Causation appears to be a *rela-*

*tion* between *event tokens*. This appearance may be misleading. Philosophers have questioned whether causation is a *relation*, whether if it is a relation, its relata are *tokens*, and whether if it relates tokens, these tokens are *events*. I shall grapple with these questions in chapter 2, which concludes that appearances are not misleading: Causation is a relation among token events. In this book I shall ask what this relation is. In particular I want to understand its asymmetry.

## 1.3 How to Begin

More needs to be said about how to proceed. Bear with me in these discussions, for they are abstract and have little immediate payoff. Beginnings are the hardest part (though, if truth be told, there aren't any easy parts!).

In any inquiry one needs criteria upon which to judge the adequacy of one's claims, lest one stab in the dark at one-knows-not-what. Often these criteria are settled. It is obvious what counts as a good answer to the question, "Did Cassius Clay change his name to 'Muhammed Ali'?" and it is obvious how one goes about finding the answer. It is less obvious how one assesses answers to the question, "Is Cassius Clay the same person as Muhammed Ali?" And I think it is less obvious still how one assesses theories of causation. On what basis can one judge one theory of causation superior to another?

Rather than listing a refined set of standards, which I couldn't have enunciated when I began and which have been shaped by the theorizing they are supposed to evaluate, let me begin by describing what I sought, even though I could not find it. I reached some of my conclusions reluctantly ("kicking and screaming" would be more like the truth), and I would like you to understand both why I wound up where I did and why I wanted to arrive somewhere else. In that way you may understand better some of the particular formulations, and you may be able to share the frustration, struggle, and excitement out of which emerged the propositions and theorems listed in the appendices. Stripped of the motivation, the false turns, and the unfulfilled hopes, my conclusions are frozen images of themselves. I shall explain how arguments drove me from "initial" hunches to articulated conclusions.[1]

Criteria to judge answers to metaphysical questions are like descriptions of ideal lovers. They are vague, conflicting, and subject to rewriting as one explores real alternatives. My commitment to the following picture is now

[1] What were my "initial" predispositions? Did I have them when I first thought about causality at age six, or when I studied science in high school and college, or when I first seriously began to wonder about causation, when I was finishing a dissertation in philosophy of economics? My "initial" predispositions are unavoidably my current reconstruction of the hunches that have kept me struggling with this difficult material.

cautious, qualified, and conflicted, but I still feel its pull, and my conclusions may be more intelligible in relation to this picture.

### 1.4 The Initial Picture

Here is the picture of causation I began with:

*Causation is "objective." It is a relation "in the objects." "Out there" are causal relations among events.* Most of these relations would obtain if there were no one to think about them. Without humans to notice them, large meteorites would still make craters in the Earth's surface, though meteorites would not be called "meteorites," and the results of their collisions would not be described as "craters." Nonhuman cognizers might describe meteorites and collisions in radically different ways. But substances such as meteorites, events such as collisions, and (*pace* Earman 1976) causal relations such as that holding between a collision and the creation of a crater do not depend on minds, whether they be human or Martian. Idealism and phenomenalism, which deny that anything is independent of mind, are false. There are substances and events that are independent of mind. If this general realism is untenable, it remains the case that events and causal relations are no *more* mind dependent than are substances.

*Events resemble substances. Like substances, the same event may be picked out by many different descriptions.* World War I is the war that began with the assassination of Archduke Ferdinand. The properties of events and substances are not exhausted by the descriptions we offer of them. Just as there are many different kinds of substances – blocks, stomachs, beetles, and principles – so there are many different kinds of events – lives, performances, conquests, and conversions. Some substances, such as cats and paper clips, have natural boundaries, while others, such as the aluminum comprising the bottom half of a beer can, are separated from other substances only by our descriptions. Similarly, some events, such as the great Chicago fire, have natural boundaries, while other events, such as the third hour of the D-Day landing, do not. Events enter into many different relations with one another. Some of these are the same as relations among substances. Events may, like substances, be part of one another, and events typically have spatial relations. Like Chicago itself, the great Chicago fire was located in Illinois. Events may be larger or smaller than other events. Like substances, events may have aesthetic qualities, and they may be the objects of psychological states.

*But events are not substances, and they enter into relations that substances do not enter into.* One event can be a temporal part of another (Mellor 1995, pp. 122–3). Pickett's charge is a temporal as well as a spatial part of the Battle of Gettysburg. The second day of the battle is a purely temporal part of it. A substance, on the other hand, cannot be a temporal part of another substance. Brian's-cat-during-the-third-day-of-its-life is either not a substance, or it is identical with Brian's cat. Substances, unlike events, are wholly present in single instants. Although some substances may *last* longer than others, substances cannot *be* temporally longer or shorter than one another. (To talk about how long a substance is in existence is arguably to talk about the length of an event.) Events, unlike substances, have temporal *dimensions,* not just temporal locations. When some people live longer than others, it is their lives, not the people themselves that have different temporal "sizes."

*Finally, and crucially, events enter into causal relations with one another*. These relations are apparently irreflexive, transitive, and asymmetric. Causes apparently never occur after their effects, and perhaps always precede them. Effects seem to be counterfactually dependent on their causes. Causation seems like a "glue" attaching events and like a "force" making things happen. Causation is reflected in regularities of at least a probabilistic form. Causation seems connected to intervention and manipulation: One can use causes to "wiggle" their effects. Causal knowledge seems crucial in decision making.

There are reasons to feel queasy about this sketch, because it says nothing about the close connection between causation and explanation. To cite a cause of an event is to explain the event. Since explanation is a human activity linked to human interests, this intimate bond between causation and explanation threatens the objectivity of causation. The picture of mind-independent causal relations among concrete events nevertheless motivates my work. This sketch is not particularly idiosyncratic, though not everyone finds it attractive. I no longer think it is tenable, and the detailed portrait of causality this book draws differs considerably from the initial sketch. But without the picture I wouldn't have had any idea what questions to ask or what sorts of answers to look for. The arguments I make and the strategies I employ should be understood against the background of this initial picture.

### 1.5 Wishes

The picture gave shape to my questions, but why bother asking these questions, and how can one judge purported answers? Hopes or wishes concerning what a satisfactory theory of causation ought to achieve also drive this inquiry. In a moment I will codify these aspirations as criteria of adequacy, but these are misleading, since I didn't know precisely what I was looking for when I began and I am still unsure about what a good theory of causation ought to do. One of the most difficult parts of metaphysics is to determine what metaphysical theories ought to achieve.

As I conceive it, my task is not to *analyze* causality or to define the term "causes." My job is instead to formulate general truths concerning the causal relation and what it has to do with other relations among events. A theory of causality ought to clarify what role causality plays in human practices such as explaining or making decisions and to explain why people (including scientists) hold the beliefs about causality that they do. One would like a theory of causality to make sense of the methods people use to determine what causes what and perhaps even to improve them. I sought a theory of causality that would link together beliefs about causality and practices of causal inference, and I hoped that a theory of causality would help answer questions such as why people know so much more about the past than the future. A theory of causality that analyzed causal relations in terms of simpler notions would be an exciting achievement, but I never hoped for such a reduction. Since causal claims are so fundamental and pervasive, such a reduction has always seemed unlikely.

According to John Mackie, theories of causality determine what causal claims *mean,* and they specify what causality *is* "in the objects" (1980, p. 1). One can go about the first task in two radically different ways. First, like an anthropologist or linguist, one might ask what people in fact mean when they make causal claims. One would regard a question such as, "Are causes regarded as necessary conditions for their effects?" as an empirical question about what people believe. When these beliefs are part of general linguistic competence, such questions can be answered by consulting linguistic intuitions. But they are still empirical questions about usage and everyday belief. Mackie argues, incidentally, that the meaning of causal claims is partly counterfactual: When we say that *a* is a cause of *b*, we mean that *a* and *b* occurred and, in the circumstances, if *a* had not occurred then *b* would not have occurred. No theory of causality can avoid the task of asking about the meaning of causal claims, because there are limits to how far a theory of causality can diverge from what people take causation to be. No acceptable theory of causation can make the meaning of causal claims bizarre and inexplicable. Causal language may be full of mistakes, but if a theory of causation finds only mistakes or if it makes the mistakes it finds inexplicable, then it undercuts its claim to be a theory of *causation.*

Analyzing the meaning of causal claims need not be understood as linguistic description. It can instead be an exercise in applied logic or conceptual analysis. If one begins with constraints that adequate definitions of concepts must satisfy, like those imposed by empiricists such as Hume or Mach, one can offer analytical reconstructions that revise and correct the faulty definitions people actually accept. In this way Bridgman proposed substituting operational definitions of scientific concepts for unsatisfactory nonoperational definitions – regardless of how generally accepted everyday nonoperational definitions might be (1938). Rather than discovering what beliefs are implicit in causal claims, one might explore the "logic" of causal claims and of their relations to other sorts of claims. A conceptual analysis of causation, just like a theory of what causation is in the objects, must explain the relevant linguistic phenomena, but it is not limited to providing a description of them.

I think it is futile and misleading to draw a line between stating the *meaning* of causal concepts and using those concepts to make substantive assertions about the world. Attempts to provide a correct *analysis* of causality have implications for what causality is "in the objects," and theories of what causality is have implications for the analysis of causal concepts. My skepticism about the value of separating claims about meaning and claims about the world originally derived from Quine's critique of the distinction between analytical claims that are true by virtue of the meanings of the terms they contain and synthetic claims that are true by virtue of the way the world is (1953, 1960). But it is ultimately independent of concerns about semantics. Like Putnam (1962), my point is methodological. I find it futile rather than impossible to demarcate statements that give the meaning of theoretical notions from statements that use those notions to make synthetic claims. Whether or not one is persuaded by Putnam's general argument against classifying claims as analytic or synthetic, there are special grounds for skepticism about distinguishing the analysis of causation from theories of what the causal relation is. Causation is both so abstract and so basic to human thought that claims about causation, whether intended as analytic or synthetic, will have fundamental implications for our beliefs and meanings. Considering the meaning of causal claims involves a consideration of the beliefs of those who make them. Such consideration is an unavoidable part of this book, because those beliefs may be true, and they roughly pick out the relation that I seek to understand. The aim is to provide true assertions about the causal relation.

I have boiled down the various desiderata into the following five criteria for evaluating theories of causality:

1. *Intuitive fit*: A theory of causation should fit our "intuitions."
2. *Empirical adequacy*: A good theory of causation should "fit the facts" and permit a coherent construal of human practices, including especially scientific practices.
3. *Epistemic access*: A good theory of causation should explain how to find out what causes what.
4. *Superseding competitors*: A good theory of causation should be better than competing theories and help explain why they succeed and fail.
5. *Metaphysical fecundity*: A good theory of causation should clarify the links between causation and other relations, such as temporal relations, and it should help one to answer other metaphysical questions.

*1. Intuitive fit*: A theory of causation should imply that paradigm causal claims are true. If a theory of causation implied that Gertrude's drinking poison did not cause her to die, that the Titanic's striking an iceberg did not cause its sinking, and that Darwin's publication of *The Origin of Species* did not cause people to reject the theory of divine creation, then that theory of causation would probably be mistaken. Causal language and causal beliefs are *data*. One might be willing to entertain a theory of causation that counted the above claims as false, if it offered a satisfactory explanation for how people could be so persistently in error, but even then causal beliefs remain crucial data.

A theory of causation should also make possible explanations of how people come to use causal language as they do, how they come to have the causal beliefs that they have, and why their causal claims mean what they do. If causation turns out to be a highly esoteric notion, then one must explain how, nevertheless, people acquire causal concepts and make the causal claims that they do. A theory could not be both correct and a theory of *causation* otherwise. A theory of causation fits our intuitions if it permits one to explain why we believe what we do about causal relations.

*2. Empirical adequacy*: This is a tricky requirement. One cannot test theories of causality in the laboratory. They are too abstract for that, and they are theories of human practices as well as of relations among events. "Testing" will inevitably be indirect and controversial. Nevertheless some theories of causality make for a neater, more flexible, more fruitful, more usable body of theories than others. A theory of causation should also illuminate the role of causal notions in other theories. The account of causation developed here explains and justifies features of experimental, explanatory, and inferential practices.

The discovery of causes has traditionally been thought to be a central task for the sciences, and satisfactory accounts of features of science such as explanation or confirmation will bear upon and be influenced by accounts of causation. The study of causation belongs to philosophy of science as well as to metaphysics. Causal language is as prevalent in the laboratory as in the kitchen. The criterion of empirical adequacy is accordingly analogous to the criterion of intuitive fit. Just as intuitive fit demands congruence with everyday causal beliefs, so empirical adequacy requires that an adequate theory of causation explain why commitment to particular scientific theories leads scientists to their causal beliefs. The task is more complicated in the case of the sciences, because scientists read philosophy and attempt to regiment their language to fit their philosophical commitments. Since these commitments are in many cases inconsistent with scientific practice, theories of causation will have to call either for a reform of scientific practices or for a revision in the philosophical positions many scientists adopt. In my view, scientists are better at science than at philosophy, and their practices are generally sounder than their philosophical reflections.

*3. Epistemic access*: A theory of causation should help explain how people can find out what causes what. Since I shall take empiricism for granted, I take the criterion of epistemic access to require that a theory of causation explain how people can learn about causal relations by means of observation and experiment. What counts as "learning by means of observation and experiment" is not clear, and there is a danger of construing this requirement so loosely that every theory of causation automatically satisfies it or of interpreting it so strictly that it cannot be satisfied. Acceptable theories of causation must explain how common methods of causal inference work, and one would like them to suggest improvements.

4. *Superseding competitors*: There are many theories of causation in the philosophical literature, and apart from relatives of the authors, nobody is fully satisfied with them. Yet these theories are not complete failures either. A good theory of causation ought to explain what is right about other theories of causation, why some philosophers have thought they were correct, and why these theories fail when they do. It is not essential that a theory satisfy this desiderata, but doing so counts in its favor.

One virtue of the theory in this book is that it illuminates the connections among other views of causation. It comes close to substituting one intricate unified mystery for many separate mysteries.

*5. Metaphysical fecundity*: Causation is intertwined with other metaphysical mysteries. For example, what is an event? Do events exist? Do these questions about events need to be resolved before one can begin theorizing about causality (as Thomson 1985 argues), or can one use facts about causality to constrain theorizing about events? Furthermore, there are other important asymmetries that appear to be connected to the asymmetry of causation. Time "moves" from past to future only. People know more about the past than about the future. The future is "open," while the past is "fixed." People makes decisions about the future, not the past; and they are frightened by the future much more than by the past. A good theory of causation should help us to understand these other asymmetries. I ran out of space and energy before exploring the fecundity of the theory of causation defended in this book.

These five criteria of adequacy may lead to conflicting rankings. It could be the case that one theory of causation provides the best intuitive fit, while another is most empirically adequate, and a third makes causal claims more empirically accessible. Such conflicts may suggest that there is more than one kind of causal relation, or they may demand that one weight these criteria. Let us proceed in the hope that problematic conflicts will not arise.

## 1.6 Problems with the Picture

One can say both that the melting of these three ice cubes caused the bourbon and soda in the glass to get colder and that the bourbon and soda in the glass caused the ice cubes to melt. Physically there is a single process whereby kinetic energy is transferred from the free water and alcohol molecules to those in the ice cubes. It is tempting to conclude that the iced bourbon and soda example constitutes "strong evidence that we must not conceive of causation as an ontological category, but rather must see it as an explanatory concept" (Dieks 1981, p. 106). If causing is a physical relation – some general sort of "forcing" – then how can the ice cool the drink while at the same time the bourbon and soda melts the ice? It seems that "causing" has to be something that omissions and recipients as well as donors of energy can do.

Reichenbach recognized that causation has an epistemic face, which he attempted to capture by taking causes as increasing the probability of their effects. Reichenbach believed that a physical characterization of the relationship in terms of an ability to transmit "marks" is coextensive with a characterization of the relationship in terms of an increased probability of the effect (see Otte 1986, pp. 59–61). But these characterizations are not coextensive, and one cannot identify causation, on the one hand with some variety of explanation and, on the other, with transmission of marks or transference of physical quantities. In Chapter 1*, I sketch some reasons to reject such physical reductions. Later, especially in chapter 8, I shall also argue against construing the explanatory aspect of causation in terms of an increase in probability.

To anticipate a long story, I shall argue that causal relations obtain among events when explanatory relations obtain among aspects of those events, and that explanation must be understood in relation to human interests. Yet I shall argue that causal relations are nevertheless "objective." Consider an analogy: Without human beings and their interests, plants would not be classified as "edible" or "nonedible." Whether a plant is edible or not depends on its chemical constitution, but since edibility does not match up neatly with any chemical classification, the term "edible" and the reference back to human digestive structure cannot be dropped. I shall argue that in the case of causation, in contrast, the relation among events that obtains when there are explanatory relations among aspects of the events coincides with a structure that can be cleanly specified without a reference back to human interests. Like Reichenbach I am going to argue that a single coin has both an explanatory and a nonexplanatory face, but I shall take issue with his characterization of both faces.

# 1*

# Transfer Theories

To provide a better sense of my picture and wishes, I shall discuss one recent theory of causation that fits my picture without satisfying my wishes. This is the so-called transference theory of causation presented and defended in the works of Aronson (1971a,b), Braddon-Mitchell (1993), Byerly (1979), Castañeda (1980, 1984), Dowe (1992a,b, 1995, 1996), Fair (1979), and Salmon (1994). In Jerrold Aronson's formulation, causes transfer some quantity ("e.g. velocity, momentum, kinetic energy, heat, etc.") to their effects through contact (1971a, p. 422). In David Fair's formulation, causation "is a physically-specifiable relation of energy-momentum flow from the objects comprising cause to those comprising effect" (1979, p. 220). According to Hector-Neri Castañeda, something in need of specification, which he calls "causity," is transferred. More recently Phil Dowe and Wesley Salmon describe causal interactions as intersections of causal processes that involve an exchange of some conserved quantity. Thus, for example, rolling billiard balls are causal processes, and their collisions (which involve an exchange of momentum) are causal interactions. I shall focus on Salmon and Dowe's version of a transfer theory.

Transfer theories fill in my picture of causation. They portray causation as fully objective. "Out there" world lines intersect, and there are exchanges of conserved quantities in some of these intersections. The study of causation, as the study of exchanges of conserved quantities, is a part of physics. In developing the notions of causal processes and causal interactions, Salmon and Dowe modify my naive ontology of substances and events; but they preserve the basic idea that causes and effects are things extended in time. Like me, Salmon and Dowe aim to say what causation in fact *is* rather than to provide an *analysis* of the notion.

Central to Salmon and Dowe's theory are the notions of a causal process and of a causal intersection. These notions replace an ontology of events. Salmon argues that to view cause and effect as events leads to intractable difficulties (1985, pp. 138–9, 155, 183). An entity such as a baseball flying through the air is not made up of events that are causally connected to one another. Instead the baseball is a causal process (Salmon 1985, p. 139).

Causal processes transmit causal influences. To transmit a causal influence is to manifest relevant properties continuously at every point in time.

Not all processes are causal processes capable of transmitting causal influences. Salmon gives the memorable example of the moving spot of light along the inside walls of a circular building, which results from the rotation of a spotlight set in the middle (1985, pp. 140–1). Such a moving spot of light is, in Salmon's terminology, a "pseudo-process," and any "mark" or modification made in the moving spot at one moment in time is not transmitted to later stages of the process. The distinction between causal processes and pseudo-processes has great physical significance, because a pseudo-process such as a moving spot of light, unlike a physical process, can travel faster than the speed of light.

In his 1985 book, Salmon distinguishes causal processes from pseudo-processes in terms of their capacity to carry "marks." This mark-transmission theory of causal processes is problematic, and Salmon has abandoned it. Instead, he now holds that a causal process is something extended through time that possesses a nonzero amount of some conserved quantity at each spacetime point it occupies without any spatiotemporal intersections that involve an exchange of the particular conserved quantity.[1] Spatiotemporal intersections among causal processes that involve an exchange of conserved quantities are "causal interactions."

Salmon's theory of causation consists of little more than these definitions of causal processes and causal interactions (plus the physical determination of which quantities are conserved). In particular he does not say how to evaluate causal claims such as "*a* causes *b*." Dowe makes some sketchy remarks in this regard (1992a, pp. 210–12), but much remains to be done. Furthermore, Salmon's theory says nothing about the asymmetry of causation. Dowe (1992a) attempts to fill this gap by arguing that the fact that causal interactions satisfy conservation laws means that the causal processes involved in any interaction can always be classified into two groups, "incoming" and "outgoing." In order to figure out which is which, one needs a reason to classify a process in one group as "prior" to a process in

[1] This combines, simplifies, and in one regard modifies Salmon's definitions $2_e$ and 3 (1994, p. 308). I think this is a fair reconstruction, but I should note that this reading is strikingly different from Dowe's (1995). Dowe omits the last clause and sees Salmon as attempting to coax an asymmetry out of the claim that causal processes "transmit" rather than merely "possess" a nonzero amount of a conserved quantity. Salmon argues that causal processes should be defined in terms of relativistically *invariant* (rather than conserved) quantities, because whether or not something is a causal process should not depend on one's frame of reference (*pace* Fair 1979, p. 240). The argument is dubious. Whether *c* causes *e* should not depend on one's frame of reference, but does it matter whether the magnitude of the quantity transferred is invariant? Conserved quantities are a more plausible choice, since these are precisely the quantities that do not change unless there is some interaction (Hitchcock 1995b, pp. 315–16).

the other. If causal interactions are all linked together into a single "net" in which there are no "loops," then one will be able to give a direction to all causal relations. One can find reasons to classify some processes as causally prior to others in the direction of entropy increase and the time irreversibility of K meson decay.[2] Dowe does not, however, say what causal priority is (and perhaps he would deny that there is such a thing as causal priority). He does not say why the physically asymmetric relations between some members of group one and some members of group two imply that the former are *causes* and the latter *effects*.[3]

Let us then consider how well the theory sketched in the paragraphs above satisfies the criteria of adequacy. With respect to *intuitive fit* and *empirical adequacy*, the theory of causal processes and interactions has no claim to be a theory of causality unless it vindicates many causal beliefs (at least of physicists) and explains how people (or at least physicists) go astray when they do.[4] There are four reasons to doubt that transfer theories can meet these demands.

1. Salmon and Dowe's theory faces counterexamples. Here is one proposed by Christopher Hitchcock (1995b, pp. 314–15). Consider a shadow of a constant shape moving over a uniformly charged metal plate. The shadowed surface possesses a constant quantity of electric charge, which is a conserved quantity. Dowe might respond that the surface lying within the shadow is "spatio-temporal junk," rather than a genuine object (1995, pp. 326–31), but to do so, he needs to defend a theory of objects. (And it is questionable whether one can distinguish objects from connected "junk" without relying on causal facts.)
2. Salmon and Dowe's theory apparently provides no account of the asymmetry of causation.
3. Salmon and Dowe's theory does not yet have clear implications concerning which causal claims are true and which are false. Some causes and effects are apparently neither causal processes nor causal interactions. The motions and expressions of a particular shadow "rabbit" on a wall are pseudo-processes, yet they are caused by a magician's clever hand movements and themselves cause a child to laugh.
4. People count absences and nonoccurrences as causes. A fire engine's flat tire may be one of the things that caused a house to burn down (Otte 1986, p. 61), but there is no causal process linking the flat tire to the house burning down. Transfer theorists could say that the house kept burning because of its physical properties. They could

[2] This way of theorizing about causal asymmetry is reminiscent of Reichenbach's (1956). For further relevant discussion, see Healey (1983).

[3] Recently (1996), Dowe interprets his own theory as linking the direction of causation to the direction of open "forks" (see §10.1). Such theories of causal asymmetry are discussed in chapter 10.

[4] In unpublished work, Sungsu Kim has argued convincingly that there is a natural "fit" between process theories and Ellery Eells's theory of probabilistic token causation (1991, ch. 6). Eells's account helps to map claims about processes and interactions into claims about causes and effects. See also Menzies (1989a) and especially Hitchcock (1996b, pp. 99f).

maintain (but see Dowe 1992b, p. 214) that the explanation in terms of the flat tire refers not to any actual cause of the burning, but to a merely possible cause of the fire's extinction. But these claims, whatever their merits, do not respond to the objection (see Dieks 1981). It seems that the notion of causation is in large part an explanatory and epistemic notion.

In response to the third and fourth objections, Salmon and Dowe could maintain that they are providing an account of only one kind of causation – physical causation – and that they are committed only to the thesis that causal claims *in physics* are translatable into claims about causal processes and causal interactions. They could thus deny that the flat tire is a specifically *physical* cause of the house burning down. Indeed, Salmon and Dowe might deny that intuitive fit is an appropriate criterion of evaluation. They might insist that they are only concerned to explicate the usage of physicists. But can they dismiss all concern with intuitive fit? Even if causation in physics is significantly different from causation in everyday life, one still wants to know how it differs and why causes in physics are called "causes."

Is there nothing in common between claims about physical causes and claims about the effects of shadows and omissions? Salmon and Dowe might invoke Fair's argument that one can treat causal claims involving omissions as *extensions* of causal concepts beyond the paradigm cases (1979, pp. 246–8). But without some other notion of causation that would explicate its links to explanations and predictions, such extensions of causal language would be mysterious. Alternatively, Salmon and Dowe might claim that they do not need to explain how omissions can be causes, because causal accounts in physics do not cite omissions (Otte 1986, pp. 63–4). But physicists do invoke absences. For example, the fact that one end of a laser is not coated with a fully reflective material is crucial to the explanation of how a laser works. An experiment may fail because a particular electromagnet was not turned on, or because there was a hole in a tank. In defense of a transfer theory, one might say that the fact that absences are important in *explanations* does not show that they are *causes*. But one cannot say this if one accepts a theory of explanation such as Salmon's, which maintains that explanation involves citing causes. Furthermore, even if one could defend the claim that omissions and pseudo-processes cannot be *physical* causes, one would still like some account of what sort of causes they are. Salmon and Dowe's theory does not do well with respect to the criteria of empirical adequacy and, especially, intuitive fit.

Nor is the theory a success with respect to the other criteria of evaluation. With regard to *epistemic access*, Salmon and Dowe have not yet said how their theory bears on questions concerning how people generally or physicists specifically learn about causal relations. With respect to the criterion of *superseding competitors*, there is equally little to say. With respect to *metaphysical coherence and fecundity*, Salmon and Dowe's project of unifying the theory of causation with physics is exciting, but it remains to be seen how well it will succeed and how fecund it will be.

The failures with respect to empirical adequacy and intuitive fit are especially serious. Since my concern is with the asymmetry of causation, the problems in accounting for the asymmetries of causation are decisive. These problems with transfer theories point to inadequacies in my initial picture. An account of causation in terms of physical relations "out there" cannot avoid addressing the links between causation and explanation.

# 2

# Is Causation a Relation Among Events?

In my initial picture, events are located in space and stretch through time. Causation links them. Our ability to infer that some events have occurred or will occur – when we know that other events occurred – rests on objective causal relations among events. This chapter will articulate and modify this picture. I shall argue that causation relates events in virtue of explanatory links between simple *tropes*. Tropes are located values of relevant variables or located instantiations of relevant properties. Claims about causal relations among events are true in virtue of the relations that obtain among simple tropes, and so it turns out that the theory of causation is largely independent of questions about the metaphysics of events and facts.

Some philosophers take tropes to be properties that are also particulars and specifically distinguish tropes from property exemplifications (Ehring 1997), and some take tropes to be fundamental and attempt to replace an ontology of substances and properties with an ontology of tropes (Williams 1953). My use of tropes is less metaphysically ambitious. I regard a trope – intuitively something like the whiteness of a particular pebble at a particular time – as a located instantiation of a property. I do not regard tropes as particularized properties, and I do not think they are ontologically fundamental, but I also think that this chapter's claims about the relata of the causal relation are independent of these questions concerning the status of tropes. The reason why I speak of "simple tropes" is that events can be conceived of as instantiations of exceedingly complicated properties and thus as themselves tropes.[1]

In this chapter I shall be talking about the relata of causal relations that hold among particular tokens. There are type-level causal relations, too. People make token-level claims such as, "Beverly Sill's debut performance at the Metropolitan Opera commanded a standing ovation," and they make

[1] Bennett argues that events are instantiations of very complicated properties and that corresponding to every event is a "companion fact" stating that this complicated property is instantiated at the relevant place and time (1988, pp. 126–9).

type-level claims such as, "Drinking hemlock causes death." Token causation appears to relate events. Type causation in contrast apparently relates properties or kinds. Some authors maintain that type and token causation are largely independent of one another. In chapter 5* I shall argue that type-level claims are generalizations of token-level claims. No independent theory of "type-level" causes or effects is needed.

The relation with which I am concerned is that of one item being *a token cause or causal condition* of another. I shall not distinguish between causes and causal conditions, and I shall not be concerned with the criteria by which some cause (in this sense) is picked out as "*the* cause" (Hart and Honoré 1959; Gorovitz 1965). In everyday talk, in contrast, people distinguish causes from causal conditions and debate which of the many causes is *the* cause. For example, in 1973 during the Middle East oil embargo, there were long lines at gasoline stations in the United States. People disagreed about whether the cause of these lines was the embargo, company conspiracy, or government price-setting. No one regarded the fact that automobiles have internal combustion engines as the cause of the long lines. Yet nobody denies that the fact that automobiles in 1973 required gasoline was one of the things causally responsible for the crisis. The fact that automobiles needed gasoline was, in my sense of "cause," one of the causes. In using the term "cause" this way, I am obviously departing from ordinary usage.

People also distinguish sharply between causes and preventatives. The chili Harriet ate for dinner last night and the antacid she took before bed may both bear a causal relationship to the extent of her stomach upset this morning. But we would say that the chili caused the upset, while the antacid kept it from getting worse. There will be occasions when I shall distinguish causes from preventatives, but in general "causes" in this book are any causally relevant factors – whether causal conditions or truly "causes," whether salient or uninteresting, whether positive or negative.

As the paragraph before last shows, a theory of causation that is faithful to everyday language apparently must take the relata of the causal relation to include both facts and events (and according to Byerly 1979, substances, too). But a theory of causation need not be this faithful to usage. All that is required is that one be able to explain why people talk as they do. What then does the causal relation relate?

### 2.1 Events

Philosophers agree on little concerning events. About all that most would grant are, first, that events are located in space and time and are thus not abstract entities. Second, most would agree that two events cannot have all the same causes and effects. If one takes "same causes and effects" to mean "same events causing them and caused by them," this claim may not be very useful, but it does say *something* about events.[2] Finally, most philosophers would grant that both the relational and nonrelational properties of events are not exhausted by the properties mentioned in the definite descriptions by which people refer to events. For example, the phrase "Jeremy and Jessica's wedding" refers to an event with many intrinsic properties besides being a wedding.

The disagreements concerning events are more substantial. First, some philosophers take events to be *changes* in substances. Others also count unchanging states of affairs as monotonous events. Still others deny that events must be linked to substances at all. Whether or not they count as events, unchanging states of affairs count as causes and effects in both science and everyday life. For example: "Because of the weaker gravity of the moon, Neil Armstrong weighed only one-sixth as much there as on Earth."

Second, philosophers disagree about the individuation of events. Some, such as W. V. O. Quine (1985), E. J. Lemmon (1967), and recently Donald Davidson (1985), defend a "coarse-grained" view of events as regions of space and time. On a coarse-grained view, events, like substances, have a limitless number of nonrelational properties. On this view, if events are tropes, the properties they instantiate must be boundlessly complicated. For example, when Hamlet stabs Claudius, he perforates his uncle's abdomen, assaults the King of Denmark, and avenges his father. How many actions does he perform? How many events occur? If events are regions of space and time, one needs to determine whether the stabbing, the perforating, the assault, and the avenging have the same spatiotemporal boundaries. The boundaries of the stabbing, the perforating, and the assault coincide, and so these are one and the same event. It is not obvious where and when the avenging is located. The coarse-grained view links up nicely to relativistic physics, and it captures the intuition that in speaking of the stabbing, assaulting, and the avenging, people are describing the same occurrence in different ways.[3]

According to Jaegwon Kim (1969, 1973, 1980) and Alvin Goldman

[2] This claim is challenged by Judith Thomson (1977, p. 70). See also Bennett 1988, pp. 163–4.

[3] It also permits one to count as literally true claims such as "That insult was an assault on my character." As Bennett points out (1988, p. 78), claims such as this last one – which do not directly invoke our shaky intuitions concerning event identity – may nevertheless require some sort of coarse-grained view. If in the context in which the particular insult occurred there was only one assault, and that insult was an assault, then the insult must be the assault.

(1970), distinct "constitutive" properties imply distinct events. One specifies an event as a triple consisting of a substance, a property, and a time.[4] On Kim and Goldman's view, the stabbing, the perforating, and the avenging are three different events, because they involve three different properties. Their theory accounts for the fact that what explains and is explained by the stabbing differs from what explains and is explained by the avenging, and it conforms to a common inclination to define human actions in a fine-grained way.

Neither of these opposing views fits all of our intuitions. Suppose (to use an example of Jonathan Bennett's (1988, p. 125)) two chess players simultaneously play two games of mental chess by means of signals that specify jointly their moves in both games. (For example, "125" might mean "Advance king's pawn one to king–3 in game one and advance king's knight to bishop–3 in game 2." ) Suppose also that it so happens that both games end at the same move, though one ends in a victory and the other in a stalemate. Contrary to the coarse-grained account, it seems that two games have been played and hence that two events have taken place at the same time and place. On the other hand, it is implausible to maintain that Hamlet's murdering or assaulting Claudius are different actions from his stabbing Claudius.

At first glance, neither the coarse- nor the fine-grained view of events seems satisfactory for the sciences or for human practices. On the coarse-grained view, causation is a connection between regions of space and time. Knowledge of such connections is often unsophisticated and uninformative. People want to know what the properties are in virtue of which events are causally linked. Kim's metaphysics apparently solves this problem, but it has its own drawbacks. It does not permit one to recognize causal relations until one identifies properties in virtue of which the causal relations hold, and it creates puzzles concerning the relations among the multitudes of events that share the same places and times.

Philosophers have defended notions of events that are less fine-grained than Kim's and more fine-grained than Quine's. None of these seems to me satisfactory.[5] Even if there were a promising theory of events, one should pause before erecting a theory of causation on the shaky foundations of a theory of events. Might there not be some way to avoid these problems?

[4] Bennett argues that one should take Kim and Goldman to be identifying events with tropes consisting of instantiations of the constitutive property by a substance at a time (1988, pp. 91–2). Ordered triples, unlike events, exist whenever all three of their constituents do, and they cannot be literally said to occur. But it is questionable whether Kim's events can be such tropes, since they have intrinsic properties that they merely "exemplify" in addition to their constitutive properties (Kim 1980, p. 170).

[5] For one fine example see Lombard (1986). For further discussions of the nature of events – and other proposals – see Brand (1977), Thomson (1977), Horgan (1978, 1980), and Bradie (1984).

## 2.2 Facts

Perhaps one should instead take *facts* to be the relata of the causal relation. Facts differ from events in several ways. They are not located in space or time. The fact that I had a big toe in 1993, unlike my big toe, did not travel with me. The fact did not come into existence and will never go out of existence. Facts are also abstract in a second way. There is nothing more to be said about the nonrelational features of facts than what is asserted in the sentences expressing them. As Jonathan Bennett puts it, "What you see is what you get" (1988, p. 9). The fact that I stubbed my toe has no intrinsic properties other than that I stubbed my toe. When one asks, "Did it hurt?" or "Was it swollen?" one is seeking further information about the event or about the toe, not about the fact.

One can only *refer to* an event, while one can both *state* and *refer to* a fact. The sentence "automobiles in 1973 required gasoline" states the fact that one can refer to by placing the words "the fact that" in front of the sentence. Facts, unlike events, are propositional. Facts are true propositions, and they are consequently in general referentially opaque. Propositions are identical if and only if they are true in all the same logically possible worlds. For example, even though "Jacqueline Onassis" and "President Kennedy's wife" refer to the same person, there are possible worlds in which "Jacqueline Onassis died in 1994" is true while "President Kennedy's wife died in 1994" is false. So these two sentences express different propositions and hence different facts. Davidson has argued (1980, pp. 152–3) that propositions cannot be referentially transparent, or else all true sentences would express the same proposition.

Given these features of facts, it seems that they cannot be causal relata. Here are four arguments: First, some facts cannot enter into causal relations. What could be the causes or effects of the fact that two is an integer? Second, causal relata are located in space and time, but facts are not. Third, causal relata are concrete, while facts are abstract. One can learn more about causal relata but not about facts. Fourth, statements referring to facts, unlike statements referring to causal relata, are referentially opaque. If Jacqueline Onassis is President Kennedy's wife, then Jacqueline Onassis's actions are the actions of President Kennedy's wife and have the same causes and effects.

These arguments are not decisive, and their conclusion seems paradoxical. At the beginning of this chapter, I wrote, "nobody denies that the fact that most automobiles in 1973 required gasoline was one of the things causally responsible for the crisis," yet here it seems that I am doing just this. One might attempt to rebut the above four arguments as follows: (1) Even though some facts cannot enter into causal relations, it does not follow that none can. (2) Causal relations might obtain between "zone facts" that are "associated" with the locations of what they concern (Bennett 1988, p. 86; Mellor 1995, p. 9), so the fact that facts are not located in space and time might not be a decisive objection. (3) One can deny that causal relata are concrete and (4) one can attempt to show how relations between facts of a particular form can permit substitution of co-referential terms (see, for example Mellor 1995, chs. 9–12).

I maintain, however, that because facts are not located in space and time, they cannot be causes or effects. In order for the fact that a match was struck to explain the match's lighting, the times and locations of the striking and the lighting must bear the right relations to one another. But facts don't have times and locations. At most they *concern* times and locations. Purported causal relations among facts presuppose spatial and temporal relations among ____ – and what could go in the blank except events? If causal relations are distinct from explanatory relations and relate entities located in time and space, then they don't relate facts. Those unhappy with this denial of "fact causation" should, I think, deny that there is a causal relation among tokens rather than arguing that facts can be causes or effects.

Statements apparently asserting the existence of causal relations among facts should be regarded as fragments of causal explanations (see Davidson 1980, pp. 161–2). In insisting that only events stand in causal relations, I am not denying that there are explanatory relations among facts and between facts and events; and I agree with those who maintain that explanatory relations among facts are of more scientific importance than are causal relations among events. It may even be plausible to maintain that explanatory relations among facts should be the primary object of philosophical inquiry.

### 2.3 Is Causation a Relation?

Some philosophers have argued that causation is not a relation at all, or that its relata are *aspects* of facts or events (Dretske 1977). Peter Achinstein argues as follows that causation is not a relation:

> We begin by supposing that
>
> 1. Socrates' *drinking hemlock* at dusk caused his death
>
> is true. . . . Let us assume that "Socrates' drinking hemlock at dusk" refers to a particular event, and that it refers to the same event no matter which words, if any, are emphasized within it. That is,
>
> 2. Socrates' drinking hemlock at dusk = Socrates' *drinking hemlock* at dusk = Socrates' drinking hemlock *at dusk*.
>
> If singular causal sentences are referentially transparent in cause positions, then from (1) and (2) we may infer

3. Socrates' drinking hemlock *at dusk* caused his death.

> But (3) is false, since it falsely selects the time of the drinking as causally efficacious; it states that the event's being at dusk is what causes Socrates to die. . . . I conclude that singular causal sentences are referentially opaque, i. e., not transparent, in the cause position. . . Hence singular causal sentences such as. . . (1) are not relational (1983, p. 194 [sentence numbering changed]).

Nothing rests on whether one regards "Socrates' drinking hemlock at dusk" as referring to an event or fact. The referential opacity Achinstein points out is an opacity within causal sentences. By hypothesis, in both (3) and (1) the very same entity is said to cause Socrates' death, yet (1) is true and (3) is false. If that entity causes Socrates' death, both should be true; and if it doesn't, then both should be false. Thus causation cannot be a relation between this entity and Socrates' death or a relation between entities at all, whether these be facts or events.

Emphasis or highlighting selects an *aspect* of events or facts as causally efficacious. Contrast (1) and (3) with

4. Socrates' drinking hemlock at dusk caused his death.

I would maintain that all three *refer to* the same cause. The difference is that the emphasis in (1) correctly implies that drinking hemlock was causally efficacious in bringing about Socrates' death, while (3) incorrectly implies that the time of day was causally efficacious. Emphasized references to facts or events imply assertions about the causal efficacy of particular aspects. (1) is true, while (3) is false because the implication of (1) is true and the implication of (3) is false. So the fact that (1) is true and (3) is false does not imply that causal contexts are opaque or that causation is not a relation.

Achinstein dismisses proposals such as this one with the following argument:

> (4) contains no emphasis at all. If so, then, I suggest, there are two possible claims we might make about it. One is that it is ambiguous and therefore without unique truth-value. Depending on where emphasis is placed (or understood) in the cause-term, (4) becomes true or false, as the case may be. . . . The second possible claim is that (4) without emphasis is not ambiguous but is to be understood as if *all* the words in the cause term are emphasized; it is to be understood as implying that all aspects of the event (implicit in its description) were causally operative. (1983, p. 206 [sentence numbering changed])

If the only alternatives were that (4) is ambiguous or that (4) emphasizes all its constituents, then my proposal would fail. On the first alternative, (4) has no truth value, while on the second (4) is false. But (4) is true.

There is, however, a third interpretation of unemphasized claims, which Achinstein does not consider. One can regard unemphasized causal claims such as (4) as implying that *some* aspect of the cause is causally relevant to

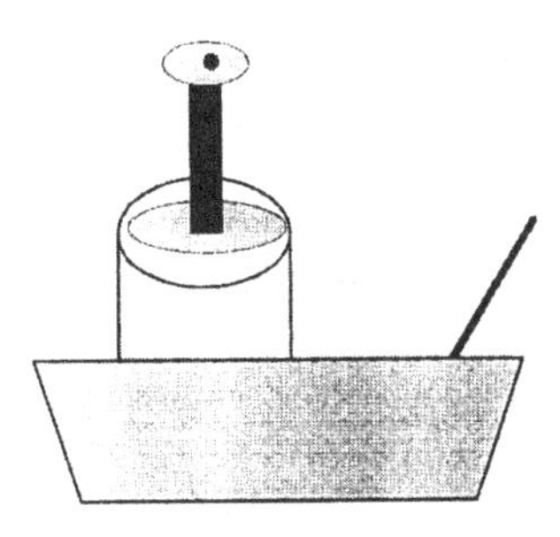

**Figure 2.1:** A causal set-up

*some* aspect of the effect. In causal claims relating events, the relevant aspects need not be mentioned, while in explanatory claims relating facts they must be. The semantic influences of highlighting are in this way consistent with taking causation to relate events and with granting that facts that do not confine themselves to causally relevant aspects can still be explanatory.

The phenomena of emphasis show that one cannot do justice to the concern to identify the causally relevant features of cause and effect by taking facts or fine-grained events to be causal relata, because both go only part of the way toward zeroing in on what is causally relevant. One can say truly, "The fact that Socrates drank hemlock at dusk explains his death," even though the time was irrelevant. What makes causal claims true are the relations that obtain between *aspects* of cause and effect.

## 2.4 Causation and Simple Tropes

In the apparatus shown in figure 2.1, a container of air with initial volume $v_0$* and pressure $p_0$* is placed in a large hot-water bath that is at temperature $t_0$*. When the temperature of the water is increased, the volume increases, and when the weights on top of the piston are increased, the volume decreases. The *events* of changing the temperature or the pressure are among the causes of the change in volume. The *fact* that the container of air was placed in a large hot-water bath at temperature $t$* helps explain the new volume. But given the physics of gases (which I here take to be captured by the ideal gas law, $PV = kT$), only the pressure and temperature changes are causally relevant to the volume changes. The definite descriptions of the events and facts mention both causally relevant and irrelevant properties. It does not matter whether the gas is placed in a bath of hot water or hot oil. There is no *mistake* in the claims that the events cited cause the volume increase and that the facts referred to explain it, but the descriptions do not zero in on *only* causally relevant features.

In this example, as in Achinstein's, one wants to know which aspects of the cause are responsible for which aspects of the effects. A satisfactory scientific theory implies relations among simple *variables* or *properties*. It might thus appear that one should take causation in the first instance to relate variables or properties and then to rely on the theory of variable or property causation to say something about causation as a relation among events. But variables and properties cannot be causal relata for the same reason that facts cannot be causal relata: They are not particulars located in space and time. Notice also that the relations among variables or properties may vary depending on the circumstances. In figure 2.1, the volume depends on the temperature and pressure. In an apparatus with a sealed cylinder in a hot-water bath, in contrast, the pressure depends on the volume and temperature.

Unlike variables, properties, and facts, *tropes* are particulars located in space and time. A trope is a located value of a variable or an instantiation of a property at a place and time. Like substances and events, tropes have locations, but, unlike substances and events, tropes have no nonrelational properties that are distinct from those in terms of which they are identified. The causally relevant aspects of facts and events are tropes. Causal relations among events and explanatory relations among facts obtain in virtue of the relations that obtain among simple tropes (compare Bennett 1988, pp. 135f).

Whether tropes are causally relevant to one another depends on the properties they instantiate. Causal relevance is nomological: An event *c* causes another event *e* only if there are aspects (tropes) *a* of *c* and *b* of *e*, which are related to one another by a law of nature. The tropes *a* and *b* are related to one another by a law of nature only if the properties *A* and *B* instantiated by *a* and *b* are related to one another by a law of nature. In saying that the tropes *a* and *b* are the causally relevant aspects of the cause *c* and the effect *e*, one implies that there is a law of nature linking *A* and *B*. Similarly one fact explains another only if there are lawful relations among properties mentioned or implied by the two facts. These are only necessary conditions. They cannot be sufficient, because causal relevance, unlike causation or explanation, is a symmetric relation.

"Causally relevant" tropes are as it were *doppelgänger* for laws, and one may wonder why such *doppelgänger* are needed. Why aren't laws enough? The reason is that the *asymmetries* that are my concern do not exist apart from the particular spatiotemporal configurations that arise when properties are instantiated. This is the reason why causal generalizations must always be relativized to some set of circumstances (see §3.2 and §5.2* and the beginning of chapter 5). For example, the gas law is symmetrical. Knowing that law will never tell one whether a particular pressure change in a gas causally depends on a particular temperature change. One needs to know whether one has a system like that in figure 2.1 or some other kind of system. Tropes enable one to zero in on what is causally relevant about cause and effect without losing the particularity which is essential to causation.

The relation between causally relevant aspects is better conceived of as a kind of causal explanation, than as a kind of causation. Like causal explanations, causal claims concerning relevant tropes zero in on particular properties and are not extensional. The circulation of blood is explained by the possession of a heart, not by the possession of a liver, even if all the same things possess hearts and livers. Nonextensionality at the trope level does not, however, imply that causal relations among *events* are not extensional. For to claim that one event causes another implies only that *some* aspect of the cause is causally relevant to some aspect of the effect (and that the purported cause is causally prior).

An account of the explanatory relations among tropes is part of rather than an alternative to a theory of causal relations among events. There are explanatory relations among relevant aspects if and only if there are causal relations among events and explanatory relations among facts. An event *c* causes an event *e* only if some aspect *a* of *c* is causally relevant to some aspect *b* of *e*, whether or not the aspect is mentioned. A fact **P** causally explains a fact **Q** only if **P** and **Q** entail the existence of tropes *a* and *b* and *a* is causally relevant to *b*.[6] Notice that *a* causes *b* and *b* causes *c* implies *a* causes *c* only if the properties *A*, *B*, and *C* are the relevant aspects of the events *a*, *b*, and *c*. For example, suppose *a* is the event of Mr. D putting potassium salts in a fireplace at $t_1$, *b* is the event of a purple fire burning in that fireplace a few minutes later at $t_2$, and *c* is the event of the hearth rug catching fire a little after $t_2$. *a* causes *b* and *b* causes *c*, yet *a* does not cause *c* (Ehring 1997, p. 76; for a more complicated example, see Hausman 1992). Transitivity fails here because the aspect in virtue of which *b* is an effect of *a* is different from the aspect of *b* in virtue of which it is a cause of *c*. When I speak of causal relations between events *a* and *b*, I shall assume that the definite description of the events refers to their relevant aspects and thus that there is a law relating properties *A* and *B*.

As I shall use the term "variable," variables are features of the world represented by symbols such as "*x*" and "*y*," not the symbols that represent them. Variables and their values are properties, and variables are to their values as determinables (such as color) are to determinants (such as turquoise). Statements specifying the values of variables thus state facts. Explanatory relations among tropes are in contrast relations among dated particulars, not relations among facts. Variables may be quantitative or qualitative. *v*, a car's speed, may be 60 miles per hour at time *t;* *w*, its color,

[6] Complex facts and event "fusions" create complications here. See §2.5 and chapter 13.

may be crimson; and *x*, its structural integrity, may be "1" (okay). In this way all tropes can be regarded as located values of variables, although in nonquantitative cases, it is equivalent and usually simpler to take tropes to be property instantiations.

Some readers may object that to say that causation depends upon relations among located values of variables is just a complicated way of saying that causation is a relation among "zone facts" or (more plausibly, since facts have no location) among events as conceived of by Kim and Goldman. Why all the rigmarole? Why not settle for Kim's and Goldman's view of events?

There are two reasons. First, if one takes causation and causal explanation as depending on the relations that obtain among relevant aspects, one can offer a unified account of event causation and of causal explanation involving facts. Causal relations among events and explanatory relations among facts obtain in virtue of causal-explanatory relations among relevant tropes. Second, the account here avoids defending any particular theory of events and avoids having to contend with the implausible features of analyses of events (see Bennett 1988).

To take causation as deriving from relations among tropes does compete with taking causation to be a relation among events. Consider an analogy: To say that the density of a pine log is less than the density of a beech log is to explicate a relation between the logs rather than to deny that such a relation obtains. Taking causation as deriving from relations among tropes attempts to say what causal relations among events are. Rather than providing an ontology of causal relata, it reifies aspects of events and *evades* ontological questions. I am dodging rather than answering questions about the entities whose aspects causation relates.

### 2.5 Artificial Events and Omissions

Sometimes causal relata seem to be omissions or "negative." For example, when a climber's rope breaks, but he nevertheless holds on to a rock outcropping, one might say, "Don does not die, because he does not fall" (Mellor 1995, p. 132). Suppose one rephrases and claims that Don's holding-on causes his survival. D. H. Mellor argues that the holding-on, like the not-falling, is not an event, and neither is Don's survival. The conclusion Mellor draws is that cause and effect here (and, Mellor would add, generally) are facts, not events.

> But now suppose. . . that "Don does not die" is made true by a single *negative event*, Don's survival, which exists just when Don is not dying. To make "Don does not die" entail both "Don does not die quickly" and "Don does not die slowly," Don's survival will have to be both quick and slow; but it cannot be both, so it does not exist. (1995, p. 134)

This argument shows that there is no event of "Don's not dying." Don's not falling is not an event that causes his not dying, and if Don's hanging on is identical to his not falling and Don's survival is identical to his not dying, then his hanging on is not an event that causes his survival. But none of this shows that there are causal relations between anything other than token events. Don's hanging on to that rock for 10 minutes is an event (which is not identical to the fact of his not falling), and it could perfectly well be a cause of the helicopter rescuing him, which is another event. To say that Don doesn't die because he doesn't fall does not single out the events that make this explanatory claim true, but there is no argument here against regarding causation as a relation among events (see Kistler, unpublished). Bennett's protests against regarding omissions as acts (1988, pp. 140–1) fail in the same way to establish the claim that causation does not relate events.

The notion of an event is vague, and there apparently are events corresponding to fusions of other events. Similarly, facts may consist of odd compounds of facts. Such complex events and facts pose difficulties. Does the event consisting of my pressing at time *t* the letter "r" on my computer keyboard and my son sneezing at school at *t* cause the letter "r" to appear on the display screen at *t′*? One might answer "yes," since part of the compound event is causally relevant, and event-causes always contain irrelevant parts. Yet the combination of the irrelevancy of the sneeze and the spatial discontinuity of this compound event might lead one to answer "no." For another example, suppose *a* causes *b* and *b* causes *c*. How is *b* related to the compound (and discontinuous) event *d* consisting of the fusion of *a* and *c*? There is no end to the problems, and, moreover, I believe that at a certain point there will no longer be any good reason to say whether one bizarre complex is or is not a cause of another.

I shall resolutely sidestep these problems until chapter 13 by stipulating that all the events, facts, properties, and variables I shall consider be "natural."[7] A property or variable is *natural* if it is not an arbitrary combination of properties and variables that are (relative to our ways of categorizing) "simple." Events and facts are natural if they are not arbitrary amalgams of tropes or of other events or facts. As our explanatory theories and our ways of classifying change, we may decide that some of those properties, variables, facts, and events we took to be natural are artificial, and some of those we took to be artificial are natural. The temperature of a gas, though not a simple property (since it is identical with the mean kinetic energy of the gas), is nevertheless a natural property. Neither the phlogiston content of the gas nor the combination of its temperature, location, and place of origin are natural properties. My typing the letter "r" and my son sneezing at school at the same time is not a natural event.

[7] I am indebted here to suggestions from Martin Barrett.

| | |
|---|---|
| properties | $A, B, C\ldots$ |
| variables | $x, y, z\ldots$ |
| values of variables | $x^*, y^*, z^*$ |
| times | $t, t_1, t'$, etc. |
| events and tropes | $a, b, c\ldots$ |
| variables ranging over events | $a, b, c\ldots$ |
| an event at place $s$ and time $t$ | $a_{s,t}$ |
| kinds of events | **a**, **b**, **c**,. . . |
| propositions and facts stating that events occur | **A**, **B**, **C**,. . . |
| probability that an event of kind **a** occurs | Pr(**a**) or Pr(A) |
| probability that property $A$ is instantiated | Pr(A) |

**Figure 2.2:** Notation

To confess the obvious: all this is imprecise. In my view, a theory of explanatory relations among simple tropes captures the fundamental truths upon which causal relations among events depend. To match our usage this theory will have to be supplemented by principles specifying what causal relations among nonnatural events follow from causal relations among natural events. I shall sketch some of these principles in chapter 13.

## 2.6 Formal Preliminaries

### 2.6.1 Notation and Terminology

Token causal relata – both events and simple tropes – will be denoted by italicized lowercase letters from the beginning of the alphabet, *a*, *b*, *c*, etc. I shall also sometimes use *a*, *b*, etc. as variables ranging over token causal relata.

The properties corresponding to the causally relevant aspects of events *a*, *b*, *c* will be denoted by the corresponding italicized capital letters *A*, *B*, *C*. This notation does not imply *A* is predicated of *a*. For example, *A* might be the property of drinking hemlock and *a* the event of Socrates drinking hemlock. It is of course Socrates, not the event, that drinks hemlock.

Italicized lowercase letters from near the end of the alphabet, *x*, *y*, *z*, etc. denote variables and lowercase letters with asterisks are the values of variables. So $x = x^*$ says that the variable $x$ has the value $x^*$. At one point in this project, I hoped to confine the word "causes" to relations among tokens only, and to say that variables, properties, or kinds "influence" or "affect" one another or that they "depend" on one another. But this convention proved to be too cumbersome, and I shall use "causes" to denote relations among both tokens and types.

Boldfaced lower case letters from near the beginning of the alphabet, **a**, **b**, **c**, etc., will denote *kinds* of events or kinds of tropes. Events of kind **a** all instantiate property *A* or consist in a variable having some value. *a*'s and *b*'s are tokens of the kinds of events **a** and **b**. When *a*'s cause *b*'s, I shall sometimes say that the property *B* depends on the property *A* or that the event-type **b** depends on event-type **a**.

Boldfaced capital letters refer to propositions and, since facts are true propositions, facts. Propositions stating that the event *a* occurs, that property *A* is instantiated, or that some event of kind **a** occurs will be represented by the boldfaced capital letter **A**. Figure 2.2 summarizes these conventions.

### 2.6.2 Probability

With the exception of chapter 9, I shall treat causation as a deterministic relation. (What I mean by this is explained in §3.2 below.) Nevertheless, I shall have a good deal to say about relations between causation and probabilities, and I need to say something about probabilities. Probabilities obey the axioms of the mathematical theory of probability. There are four main interpretations of probabilities, and many complicated problems arise in adjudicating among them. *Logical* and *subjective* probabilities attach to propositions. Logical (conditional) probabilities express the degree to which one proposition "partially entails" another. Subjective probabilities represent the degree of belief an agent has in a proposition. I shall *not* be concerned with logical or subjective probabilities.

On the interpretations of probability that are relevant to my concerns, what has a probability is a particular kind of event or trope in particular circumstances. One can also speak of the probability that a property will be instantiated in some population. To say that the probability that a coin lands heads is 0.5 is to state the probability that a particular kind of event occurs. In speaking of such a probability, one is implicitly referring to some set of circumstances or to some "chance set-up." Without such a reference, the probability is undefined. For example, if one tosses the coin onto a bed of rough gravel, it may have a substantial probability of landing with neither heads nor tails up. If *s* and *t* are the place and time of interest – which will pick out the exact chance set up – one might write Pr(**a**, *s, t*), Pr(*A*, *s, t*), or $\Pr((x = x^*), s, t)$. If one leaves *s* and *t* implicit, one might write Pr(**a**), Pr(*A*), or $\Pr(x = x^*)$. Unless there is some reason to distinguish between these or to make the place and time explicit, I shall simply write Pr(A). Pr(A) = *r* (where *r* is a real number on the closed interval [0,1]) should be read, "The probability is *r* that an event or trope of kind **a** occurs or that property *A* is instantiated."

Most relevant to this book are the *frequency* and *propensity* interpretations of probabilities. According to the frequency interpretation, the probability that property *B* will be instantiated by some set-up is a ratio of the number of times *B* is instantiated to the number of trials. This seems appealing – surely a fair coin should turn up heads about half the time – but this interpretation is also problematic, since the actual frequency with which a fair coin comes up heads in $k$ tosses may be any fraction $j/k$ where $j$ is an integer between 0 and $k$. A coin tossed only once and then melted down has a frequency of heads of 1 or 0. Any plausible version of a frequency interpretation is going to have to talk about frequencies in hypothetical infinite populations or series of tests, or about the limits of the frequencies in larger and larger samples or in greater and greater numbers of trials in samples of some fixed size. On the frequency interpretation, to talk about the probability that a coin lands heads is not to say anything about a particular throw. It is rather to say what will be the (limit of the) frequency of heads in some sequence of throws. Insofar as one wants knowledge of probabilities to guide judgments concerning single cases, frequency theorists need principles linking frequencies to judgments about single cases.

According to the propensity interpretation, on the other hand, a probability reflects a "chancy" disposition of an individual set-up. Claims about propensities are *tested* by observations of frequencies, but they are not themselves claims about frequencies. The assertion that when a die is thrown, it has a propensity of one-sixth of landing with a three showing is analogous to the claim that when a sugar cube is placed in water, it will dissolve. To say that this sugar cube will (or would) dissolve is to make a prediction (or a counterfactual claim) about a set-up involving placing this cube in water. The predictions or counterfactuals are made true by physical facts about sugar and water molecules and express the water solubility of sugar. According to the propensity interpretation of probability, to say that the probability is one-sixth that the die will land with a three showing is also to make a prediction or a counterfactual claim, this time concerning a die-throwing set-up. Such probabilistic predictions or counterfactuals also reflect physical facts and express the propensities of the set-up. But in the case of the die, the prediction and the consequent of the counterfactual are not deterministic. One cannot predict with certainty which face will turn up when the die is thrown. Knowledge of the propensity only justifies some degree of belief less than one in claims concerning how many threes will come up in any number of tosses. For reasons that arise in chapter 9, I hold a propensity interpretation. Except with respect to that chapter, what is crucial is simply that the probabilities I talk about are *objective* frequencies or propensities concerning event types, not subjective or logical probabilities of propositions.

Probabilistic causation will be discussed only in chapter 9. The rest of the book will be concerned with deterministic causation, and I shall hold that, except in cases of preemption and overdetermination, individual causes are in the circumstances necessary for their effects and the conjunction of all the causes of some kind of event **b** is sufficient for **b** in the circumstances *C*. So it would appear that the probabilities will all be 1's or 0's. When $\mathbf{a}_1, \ldots, \mathbf{a}_n$ are all the causes of **b** in circumstance *C*, then for all *i*, $Pr(B/{\sim}A_i\&C) = 0$ and $Pr(B/C\&A_1\&...\&A_n) = 1$. Yet one can also consider probabilities such as $Pr(B/{\sim}A_i)$, $Pr(B/(A_i\&C)$ or $Pr(B/{\sim}A_i\&C')$, where $C'$ is only a partial specification of the circumstances. Such "mixed" probabilities (Papineau 1989, 1990) need not be zero or one. Most of the probabilities I will be concerned with are in this sense "mixed."

I would also like to call attention to one potentially confusing fact about both propensities and hypothetical frequencies. To say "$(x)(Fx \rightarrow Gx)$," where "$\rightarrow$" is the truth-functional if-then connective, is to say that everything that is *F* is *G*. Whenever *F* is instantiated, so is *G*. But "$(x)(Fx \rightarrow Gx)$" says nothing about whether the connection between *F* and *G* is "accidental" and nothing about whether one should assert the counterfactual, "if something were *F* it would be *G*." Claims about actual frequencies are like this. To say that the frequency of *G*'s among some fixed group of *F*'s is .75 is to say that *G* is instantiated in three-quarters of the cases in which *F* is instantiated. To say, on the other hand, Pr(G/F) = .75 is, on both the hypothetical frequency and the propensity interpretations, already to commit oneself to counterfactuals (Spohn 1983b, pp. 74–5; Pollock 1984, p. 178). The conditional probability on such interpretations does not refer to any particular sample, nor even to the set of all the *F*'s there will ever be, and it implies that the relationship between instantiations of these properties is not accidental. In this sense there cannot be such a thing as an accidental probabilistic dependency. Claims about probabilities are stronger than are universal generalizations, and they presuppose that one can distinguish what is accidental from what is a matter of law.[8] The analogue to "$(x)(Fx \rightarrow Gx)$" is not a claim about conditional probabilities, but what one might call a "universal frequency claim": for example, that 3/4 of all *F*'s are in fact *G*'s. It is easier to relate causation to probabilities than to universal generalizations, because there is already a nomic element built into the interpretation of probability.

### 2.6.3 Causal Graphs

To facilitate the exposition I will make use of three kinds of causal graphs. A *directed acyclic graph* like the ones shown in figures 2.3a and 2.3b involves a set of vertices (points) with *directed edges* (arrows) connecting them. A graph is acyclic if following a series of arrows from tail to head

[8] They are in another regard weaker. On many formulations of the probability calculus, Pr(G/F) = 1 implies only that the set of *F*'s that are not *G*'s has measure zero, not that all *F*'s are *G*'s. To say that the set of *F*'s that are not *G*'s has measure zero means, roughly, that it is infinitesimally small.

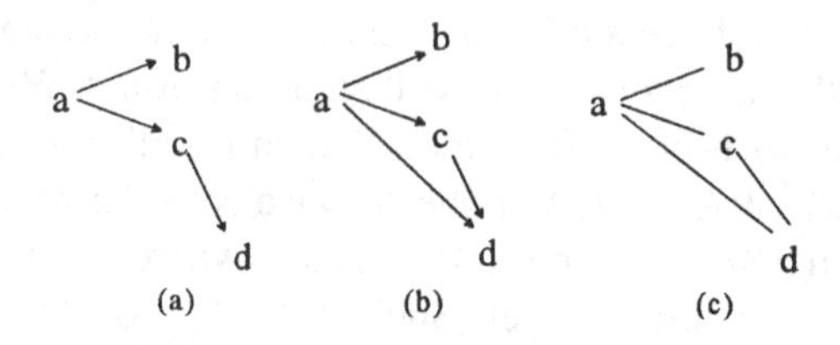


**Figure 2.3:** Causal graphs

will never bring one back to where one started. Directed graphs can represent causal and explanatory relations. One can interpret the vertices to represent tropes, events, types, or variables, and one can interpret a directed edge to be a causal or explanatory relation.

Two quite different kinds of directed acyclic graphs can be drawn. In the first, which I shall call "normal graphs," directed edges represent causal relations that are, relative to the variables or events considered, direct. If event *a*, Henry's call for help, causes Mary to turn the light switch (event *b*), which in turn causes the fire, *c*, then relative to the set of events $\{a, b, c\}$, *b* is a direct cause of *c* and *a* is an indirect cause, even though obviously there are unmentioned causal intermediaries as well. In the normal causal graph drawn in figure 2.3a, *a* is a direct cause of *b* and *c*, and *c* is a direct cause of *d*. If causation is transitive, then the *path* in figure 2.3a from "*a*" to "*d*" (that is, the series of adjacent edges always going from tail to head) represents the fact that *a* is an indirect cause of *d*. In the second kind of directed graph, edges represent causal relations that are direct *or indirect*. I shall call such graphs "path graphs," because they contain an edge if and only if there is a path in a corresponding normal graph. Figure 2.3b represents the path graph that corresponds to the normal graph in figure 2.3a.[9]

A third kind of graph I shall make use of is *undirected*. In these graphs the edges represent causal connections (intuitively the idea of a necessary or nomological connection). I shall call these graphs "connection graphs." Figure 2.3c is the connection graph corresponding to the normal and path graphs shown in figures 2.3a and 2.3b.

## 2.7 What Do Causal Relata Suggest About Causation?

This book presents a theory of (token) causation as a relation among events that obtains when there are explanatory relations among relevant aspects or

[9] Is figure 2.3b a path graph of the causal relations show in the normal graph of figure 2.3a, or is figure 2.3b a normal graph of a different causal set-up in which *a* is both a direct and an indirect cause of *d*? There is no way to tell from the figure itself. Both this normal graph and the graph in figure 2.3a have the same path graph. The relations between normal and path graphs is discussed in §12.4.

tropes. Is such a theory consonant with my initial picture? Does the thesis that causal relations among events obtain only if there are explanatory relations among relevant aspects undermine the objectivity of causation? Relevant aspects do not seem like objective, mind-independent inhabitants of the universe. They stand in the proper relations only if everything about them is "causally relevant," and one might suspect that causal relevance among properties cannot be fully explicated without bringing in human interests and conceptualizations. Explanation itself may be linked to human interests. Since the account of causation I shall defend derives from an account of causal explanation, it might appear that my initial picture must be scrapped.

This conclusion is hasty. Tropes may have more objectivity than might appear: The fact the people define variables does not imply that causal relations obtain only among the products of human thought. (Nor does it imply – as social scientists and economists are fond of saying – that causal claims are true only of a particular model.) If one picks out the wrong variables, one's causal claims will be mistaken. Furthermore, even if explanatory relations cannot be spelled out without reference to human interests, it may be that certain objectively specifiable structures of events turn out to be those that serve our explanatory interests. Much depends on the details.

Beneath these clouds, let us turn to the causal relation and its asymmetries.

# 3

# Causation, Regularities, and Time: Hume's Theory

It is fitting for both historical and philosophical reasons to begin the discussion of causation with David Hume's theory and its contemporary developments. Hume's theory is the starting point for most modern treatments of causation, and the problems his theory must surmount are problems for all theories of causation. Some philosophers believe that Hume was right, though they would add that his account of causation needs refinement. Others, including me, would reject Hume's theory. Nevertheless, my views owe a great deal to Hume.

## 3.1 Hume's Theory

Hume argues that causation involves a regular association between cause and effect with the cause contiguous with its immediate effects and preceding them. He also argues that the human psychological propensity to pass from an "impression" (perception) of a cause to an idea of its effects (or from an impression of an effect to an idea of its cause) leads people mistakenly to believe that there is a necessary connection between cause and effect. This account takes the mystery out of causation, while retaining the all-important links between causation and regularities and explaining the mistaken belief that there is more to causation than regularity, contiguity, and time order.

Hume claims to clarify the content of our "ideas" (mental contents) and of the relations among them, and he does not clearly demarcate philosophy from introspective psychology. His account of causation is constrained by his empiricism – that is, his view that all our ideas derive from preceding impressions of sense or reflection. He is saying that nonlogical terms have meaning only if one can tell by observation whether they apply to things and that the evidence for all claims concerning matters of fact derives from observation.

In Hume's view, the way to clarify ideas, including the idea of causation, is to determine what impressions they derive from.

> Complex ideas may, perhaps, be well known by definition, which is nothing but an enumeration of those parts of simple ideas, that compose them. But when we have pushed up definitions to the most simple ideas, and find still some ambiguity and obscurity; what resource are we then possessed of?. . . Produce the impressions or original sentiments, from which the ideas are copied. These impressions are all strong and sensible. They admit not of ambiguity. They are not only placed in a full light themselves, but may throw light on their correspondent ideas, which lie in obscurity. (1748, p. 41)

In the case of a complex idea such as causation, one must either find the complex impression of which it is a copy or determine its components and trace them to the simpler impressions from which they derive.

Hume maintains that within the complex idea of causation are the ideas of temporal priority, spatial and temporal contiguity, regularity, and necessary connection. In the *Treatise* Hume argues that all causes must come before their effects, or else none of them would (Book I, Part III, §2). I do not understand this argument, and Hume does not repeat it in the later account in the *Inquiry*.[1] In the later treatment, Hume takes it as obvious that causes precede their effects. It seems plausible, however, that causes may continue to exist after some of their effects have begun. For example, one may still be in the act of striking a match after it has begun to light. So one should take Hume's temporal priority condition to require only that a cause *begin* before its effects begin.

The spatial-temporal contiguity requirement creates difficulties for Hume, because he believes that mental things stand in causal relations even though they are not located in space (*Treatise*, Book I, Part IV, §5). Hume omits the contiguity requirement in his later account of causation in the *Inquiry*. Modern materialists do not have this objection to the spatial contiguity requirement, but many would question whether a theory of causation should rule out remote causation without more detailed inquiry (Spohn 1983b, p. 84). Developments in quantum theory also cast a cloud over contiguity requirements. As we shall see, however, it is hard to make a Humean account of causation plausible without relying on a contiguity condition. Even if contiguity is not built into the truth conditions for causal claims, it is an element of familiar causal processes.[2]

It is not clear from Hume's account whether the notion of a regularity is contained in people's actual idea of causation or whether instead it *should be* a part of their idea, since it is an aspect of causation itself. Hume speaks of a "constant conjunction" among "similar" objects. According to Hume,

[1] For an attempt to reconstruct this argument, see Beauchamp and Rosenberg (1981, pp. 192–5).

[2] I shall not explore the difficulties that are involved in the notion of contiguity. See Russell (1913), Taylor (1964), Brand (1980), and Beauchamp and Rosenberg (1981, ch. 5).

when *a* causes *b*, "objects similar to" *a* are invariably followed by objects similar to *b*. There are serious problems in this account. My waving my hand is similar to my striking a match, but my hand-waving is not usually followed by flames. Like most contemporary philosophers, I shall explicate the relevant notion of similarity in terms of the laws of nature in virtue of which *a* causes *b*. If there is a law of the form $(x)(Ax \rightarrow Bx)$, then objects similar to *a* are events of kind **a** – that is, other instantiations of *A* – and objects similar to *b* are events of kind **b**. Recall that I am assuming that one knows the relevant aspects of events and has thus already identified the "similarity."

The problems with Hume's notion of constant conjunction don't stop with the identification of the relevant tropes. Even if one has identified the tropes, the relation between them is not invariable. Turning a switch causes a bulb to light, but the bulb does not always light when the switch is turned. Hume maintains that when the relation is not invariable, then one has not fully specified the cause. Turning the switch is only a part of the cause of its lighting. The cause is turning the switch when the battery is charged, the bulb in good repair, and . . . .

This response to the fact that what we call causes are not invariably linked to their effects conflicts with ordinary usage and the scientific concern to identify "causal factors" that promote the occurrence of effects without by themselves requiring them. Furthermore, if the description of the cause is logically independent of the statement that the effect occurs, then it is virtually impossible to specify explicitly a set of conditions, the satisfaction of which will lead invariably to the effect's occurrence. Rather than calling the complete set of sufficient conditions – if there is one – "the cause," I shall call the separate causal factors "causes."

The fact that it is so difficult to specify a set of causes that is sufficient for an effect constitutes a reason to question whether the connection between cause and effect is deterministic. To say that smoking causes lung cancer seems to imply only that smoking increases the risk of lung cancer. Such considerations have led a number of philosophers to propose theories of probabilistic causation, which are discussed in chapter 9. Until then, I shall assume with Hume that causation is a deterministic relation. I defend this assumption in §9.5 and §9.7. To say that causes determine their effects is not to say that everything is caused or that nothing happens by chance, and it does not commit me to determinism. Taking causation to be a deterministic relation implies only that *if* something is caused, its causes are sufficient in the circumstances.

Passing over the problems concerning constant conjunction, to which I will return in the next section, let us turn to necessary connection, the final element in the idea of causation. Although necessary connection is, Hume maintains, the most important part of the *idea* of causation, he also maintains that it is impossible to have any impression of necessity. All we can ever observe are successions of separate occurrences.

> *What is our idea of necessity, when we say that two objects are necessarily connected together?* Upon this head I repeat what I have often had occasion to observe, that as we have no idea that is not derived from an impression, we must find some impression that gives rise to this idea of necessity, if we assert we have really such an idea. In order to [do] this, I consider in what objects necessity is commonly supposed to lie; and, finding that it is always ascribed to causes and effects, I turn my eye to two objects supposed to be placed in that relation, and examine them in all the situations of which they are susceptible. I immediately perceive that they are *contiguous* in time and place, and that the object we call cause *precedes* the other we call effect. In no one instance can I go any further, nor is it possible for me to discover any third relation betwixt these objects. I therefore enlarge my view to comprehend several instances, where I find like objects always existing in like relations of contiguity and succession. The reflection on several instances only repeats the same objects; and therefore can never give rise to a new idea. But upon further inquiry I find that the repetition is not in every particular the same, but produces a new impression, and by that means the idea which I at present examine. For, after a frequent repetition, I find that upon the appearance of one of the objects the mind is *determined* by custom to consider its usual attendant, and to consider it in a stronger light upon account of its relation to the first object. It is this impression, then, or *determination*, which affords me the idea of necessity. (1738, p. 154)

The idea of necessary connection is a mental (causal) consequence of regularities. The idea of necessity arises from the way in which human minds react to perceptions of regularities, not from any feature of the objects. All that causation consists of in the objects themselves (which is knowable to cognizers like human beings) is regularity, contiguity, and temporal priority of the cause to the effect. That is all there is to causation, although if one also wants to explain how people *think about* causation, it is important to add that the regular connection between cause and effect leads human minds to turn from the impression of one to the idea of the other and that this tendency of human minds makes people *feel* a further necessary connection. Thus Hume offers *two* definitions of causation in terms respectively of regularities and of the feeling of determination caused by the experience of regularities. Hume knows that his account will seem implausible, but having argued that power is unobservable and having given such a neat explanation for how regularities give rise to the idea of necessity, he thinks readers should find his arguments convincing.

### 3.2 Constant Conjunction

The notion of a constant conjunction between cause and effect – of a deterministic causal relation – requires clarification. Even if one has identified the relevant tropes, one must still characterize the regularities that obtain among them. J. L. Mackie, drawing on ideas of John Stuart Mill, argues that a cause, such as striking a match, need not be a necessary nor a sufficient condition for an effect, such as the match lighting. A cause is rather "at least an INUS condition" – that is, at least "an *insufficient* but *nonredundant* part of an *unnecessary* but *sufficient* condition" for the effect (1980, p. 62). Jonathan Bennett speaks more simply of "NS conditions" – necessary parts of sufficient conditions (1988, p. 44). For example, matches light in many conditions. They may be struck in the presence of oxygen in the right conditions of temperature, wind, and humidity. They may be kindled by the presence of a flame in other conditions. A match lights if and only if one or more of these sets of sufficient conditions obtains. Each of the sets of sufficient conditions is minimal in the sense that none of its conjuncts is redundant. The conjuncts of these minimal sufficient conditions are INUS conditions.

Mackie stresses that causal claims take for granted a "causal field," a background in which the INUS relation is situated.[3] For example, the facts that a mouse is alive, breathes, eats, and defecates matter to whether it contracts cancer, but one would not regard them as causes of its contracting cancer. Such facts are instead part of the causal field. "The causal field in this sense is not itself even part of a cause, but is rather a background against which the causing goes on" (Mackie 1980, p. 63). *I take constituents of the causal field to be themselves causes or causal conditions that are relegated to the background for pragmatic reasons.* I will usually not mention the causal field explicitly.

According to Mackie, the general form of an INUS condition is:

> For some $G$ and for some $H$ all $F(AG$ *or* $H)$ are $B$, and all $FB$ are $(AG$ or $H)$ (1980, p. 67 [my relettering]).

$F$ is the causal field, $A$ is a causal factor, $B$ is the effect, $G$ is a variable that ranges over conjunctions of causal factors, and $H$ is a variable that ranges over disjunctions of conjunctions of causal factors. When Mackie first introduces the notion of an INUS condition, he supposes that the other conjuncts, which together with $A$ make up a minimal sufficient condition for $B$, are known, as are the other minimal sufficient conditions. But he notes that this is an idealization and that the formulation with second-order existential quantification is the most we can say. One rarely if ever knows the full content of $F$ and what values of $G$ and $H$ make the biconditional true.

The logical form of INUS conditions requires clarification. Are "$A$" and

[3] Mackie credits John Anderson (1938) with introducing this notion.

"*B*" propositions, properties, or events? It does not make much sense to take them to be event *tokens*, since concrete particulars cannot be conjuncts in necessary or sufficient conditions. That leaves propositions or properties. (Event types can be treated as properties.) Suppose **A** and **B** are propositions stating truly that events $e_1$ and $e_2$ occur. Then by virtue of the definition of the truth-functional "if and only if" connection, it will trivially be the case that **A** if and only if **B**, and **A** will be necessary and sufficient for **B**. To require that a proposition asserting that a cause occurs be at least an INUS condition for a proposition that an effect occurs is much too weak. That leaves properties.

The idea to be explicated is of a constant conjunction or a regular association, and so it seems that INUS conditions must be (second-order) quantified statements: there exists $G$ and there exists $H$ such that $(x)[Fx \rightarrow (Bx \leftrightarrow Gx \,\&\, Ax \text{ or } Hx)]$. "In the causal field $F$, every match lights if and only if it is struck and some conditions $G$ are true of it or some other set of conditions is true of it."

If $a$ causes $b$, then it must also be the case that $B$ and $A$ and some $G$ are in fact true of the particular entity or set-up, and that no $H$ distinct from A&$G$ is, and that $A$ is not redundant in the necessary condition $A\&G$. Otherwise the regularity $(\exists G)(\exists H)(x)[Fx \rightarrow (Bx \leftrightarrow Gx \,\&\, Ax \text{ or } Hx)]$ will not be relevant. On this construal, Hume's constant conjunction condition requires that a property of the cause $a$ is in the actual circumstances (in which some $G$ is instantiated and no disjunct of $H$ is) necessary and sufficient for a property of the effect $b$. In cases of preemption or overdetermination some disjunct of $H$ will be instantiated.

One can restate this more formally as:

> $a$ is necessary and sufficient in the circumstances for $b$ in causal field $F$ if and only if in $F$ 1. $(\exists G)(\exists H)(x)(Gx \,\&\, Ax \text{ or } Hx) \leftrightarrow Bx)$, 2. $B(c, t')$, 3. $A(c, t)$, 4. $(\exists G)G(c, [t, t'])$, and 5. $(H)(x)[(Hx \rightarrow Bx \,\&\, H[c, (t, t')]) \rightarrow G \,\&\, A$ are conjuncts in $H]$.

$A$ and $B$ are properties instantiated by $a$ and $b$, $c$ is the particular set-up, $t$ the time when $a$ begins, $t'$ the time when $b$ begins, and $[t, t']$ the interval between $t$ and $t'$. (1) states that $A$ is an INUS condition for $B$. (2), (3), (4), and (5) state that $A$, $B$, and some $G$ are instantiated by the set-up at the proper time and that no $H$ distinct from $G \,\&\, A$ is. (5) also implies that $A$ is not redundant in the minimal sufficient condition for $B$ that is instantiated in the set up at the time. In order to keep the clarification of "constant conjunction" separate from Hume's insistence that causes always precede their effects, this account does not require that $t$ and $t'$ be distinct or that $t$ precede $t'$. One may want to make explicit the temporal and spatial elements in clause (1), but I shall ignore this complication here (see pp. 56–7). This account of necessity and sufficiency in the circumstances is not Mackie's.

He gives instead a counterfactual construal of the terms "necessary (or sufficient) in the circumstances" (1980, ch. 2).[4]

Although constant conjunction is an empirical proxy for necessary connection, the recognition that a property of one event is an INUS condition for a property of another does not always lead us to believe that the events have a necessary connection. For example, the law that light travels in straight lines establishes nomological connections between the height $h$ of a flagpole, the length of its shadow $s$, and the angle of elevation of the sun $a$. Propositions stating the value of any two of these (plus absence of interference, etc.) constitute a minimal sufficient condition for a proposition stating the value of the remaining variable. Thus a value of $a$, the angle of elevation of the sun is an INUS condition for a value of $s$, the length of the flagpole's shadow; and in this case we are inclined to "feel" a necessary connection. But a value of $a$ is also necessary in the circumstances for the value of $h$, the height of the flagpole, and obviously there is no necessary connection between the angle of elevation of the sun and the height of the flagpole. The most one can say is that **a** being at least an INUS condition for **b** is a necessary condition for the existence of a necessary connection in the circumstances between $a$ and $b$.

A serious problem with this account of constant conjunction remains. Suppose that a coin is flipped only twice before it is melted down and that it lands heads both times. Thus, whenever this coin is flipped, it lands heads. So tossing this coin is necessary and sufficient in the circumstances for the coin to land heads. Yet no one would say that tossing this coin and its landing heads are, in the relevant sense, constantly conjoined. The problem is that the regularity between the tossing and the coin's landing heads is "accidental." The INUS condition linking cause and effect must be "nomological" or "lawlike."

With this addition, one then can state the INUS condition construal of the constant conjunction requirement. For simplicity the following formulation leaves the reference to the causal field implicit and treats $G$ and $H$ as

[4] Statements of quantitative relations among variables resemble statements of INUS conditions (Cartwright 1989, p. 28; Hoover 1990, pp. 218–19). Suppose, for example, that the fact that $x$ has value $x^*$ (at $s$, $t$) causes $z$ to have value $z^*$ (at $s'$, $t'$). Then on this account of the regularities involved in causation it must be the case that the fact that $x$ has value $x^*$ is necessary and sufficient in the circumstances for the fact that $z$ has value $z^*$. What this means in a quantitative case (supposing for simplicity that the relations are linear) is (1) that a functional relationship obtains such as $(x)(y)(z)(Fxyz \rightarrow (z = j + kx + my))$ (2′) $Fx^*y^*z^*$ at $(s, [t, t'])$ (3′) $z = z^*$ at $(s, t')$, (4′) $x = x^*$ at $(s, t)$, and (5′) $k \neq 0$. $s$ refers to the "system" within which the quantitative relation holds. The quantitative relation has an antecedent condition ($Fxyz$) analogous to the causal field relativization mentioned above, because quantitative relations among variables typically hold only for particular set-ups and particular ranges of the values of the variables. The conditions are then analogous to those involving nonquantitative variables. In a particular causal field, the located value of $z$ is caused by the located value of $x$ in conjunction with other factors.

properties rather than second-order variables. One can then state the following necessary condition for a deterministic causal relation:

**DC** (*Deterministic causation – causes as INUS conditions*) If $a$ is a deterministic cause of $b$ in set up $c$ during the time interval $[t, t']$, then given laws of nature **L**, 1. $B(c, t')$ entails and is entailed by $\{A(c, t)$ & $G(c, [t, t'])$ or $H(c, [t, t'])\}$, but $B(c, t')$ is not entailed by $G(c, [t, t'])$, 2. $B(c, t')$ 3. $A(c, t)$, 4. $G(c, [t, t'])$, and 5. $\neg H(c, [t, t'])$.

$A$ and $B$ are the properties whose instantiations constitute aspects of the events $a$ and $b$, c is the particular set-up, $t$ is the time when $a$ begins, $t'$ is the time when $b$ begins, $[t, t']$ is the interval between $t$ and $t'$ and **L** some set of laws of nature. The entailment must not hold if **L** is empty. **DC** states that a deductive relationship among properties, given laws of nature, is a necessary condition for a causal relationship. It would be more faithful to Hume to formulate **DC** as a nomological generalization over token events, but the formulation here is preferable nevertheless, because it avoids the misleading suggestion that there are regular co-occurrences between events instantiating properties $A$ and $B$. The particular instance may be the only occasion in the whole of history that an event instantiating $A$ causes one instantiating $B$. To speak of a regularity theory of causation as many contemporary philosophers do or to speak as Hume does of constant conjunction is misleading. The tokens are linked by laws, and if it were possible to recreate the circumstances repeatedly, then a regularity would certainly appear, but there is nothing in the notion of causal connections as instantiating lawful connections that rules out unique sequences of cause and effect.

The claim that causes are nomic INUS conditions for their effects appears to capture what one intends by maintaining that causation is deterministic and that it involves regularities. Both here and in later chapters *I shall take **DC** as a formal explication of what it means to say that causation is deterministic*. Humeans (like empiricists generally) cannot be fully satisfied with this explication, because it is difficult to distinguish nomic from accidental regularities. How could one possibly observe whether a regularity was merely accidental? Moreover, even supposing that one could distinguish laws from accidental regularities, how is one to spell out the link between laws of nature and specific nomic INUS conditions? These difficult problems do not count against the Humean account when it is compared to alternative theories of causation, because these are equally problems for the alternative theories. So let us permit the Humean to rely on the distinction between lawlike and accidental regularities. Cause and effect are constantly conjoined if and only if a property instantiated by the first is nomically necessary and sufficient in the circumstances for a property instantiated by the second.

### 3.3 A Neo-Humean Theory of Causation

An updated Humean theory thus maintains that *a* is a direct cause of *b* if and only if *a* and *b* are distinct and contiguous, *A* is necessary and sufficient in the circumstances for *B*, and *a* precedes *b*. Although the distinctness condition may be implicit in the relation of *nomic* necessity and sufficiency, which precludes logical necessity and sufficiency, it should not be forgotten. If *a* and *b* are cause and effect, they must be different events, they can have no parts in common, and the relevant tropes cannot entail one another.

Causes need not be direct causes, and one can go on to say that *a* is a cause of *b* if and only if *a* is a direct cause of *b* or there is a unidirectional chain of direct causation from *a* to *b*. But should one say such a thing without good evidence that there are no "remote" causes that operate without the help of any connecting chain? To these definitions, a Humean can add a psychological explanation of why people mistakenly believe that causation also involves some sort of power, efficacy, or necessary connection.

Hume has a simple theory of causal priority: Causes precede their effects.[5] Many philosophers, myself included, have been dissatisfied with the stipulation that causes precede their effects. It may be true that causes always precede their effects, and it could be that the best theory of causal priority relies on time. But an acceptable temporal-priority theory of causal priority should rely on better evidence than the casual observation that causes typically precede their effects. Paul Horwich points toward the kind of argument that is needed. He maintains that temporal priority constitutes the asymmetry of causation, but that this claim is an *a posteriori* necessity, like the claim that water is $H_2O$ (1987, pp. 140–5). His account is not subject to the criticism that a theory of causation ought not to foreclose the possibility of simultaneous or backwards causation without attending to the evidence. Yet Horwich provides only a problematic sketch of the relations between time and other asymmetrical features of causation, and he makes no serious argument for the claim that the temporal priority constitutes the asymmetry of causation. Indeed his own defense of the possibility of simultaneous causation contradicts his claim that temporal precedence constitutes causal priority.

Many people believe that some causes are simultaneous with their effects (Taylor 1966). Suppose, for example, I take two boards and lean them against one another so that they support one another. After I walk away, one has the situation shown in figure 3.1: The left board holds up the right board

[5] Beauchamp and Rosenberg defend the view that Hume has no theory of causal priority at all! They interpret Hume to deny that causation itself is asymmetrical. The asymmetry people find in causal relations reflects the fact that events that are causally related are not simultaneous.

**Figure 3.1:** Simultaneous and reciprocal causation

and the right board holds up the left board. Doesn't the position of each causally depend on the position of the other?[6] Should a philosophical theory deny this possibility on the grounds that causation is always asymmetric or on the grounds that causes always precede their effects in time? For an example of asymmetrical simultaneous causation, consider the relationship between tastes and prices in general equilibrium theory (Hoover 1993). Tastes influence prices even though there are no temporal relations in the model.

One can dispute both these examples. The position of the right board at time *t* in fact depends on the position of the left board slightly *before t* (Frankel 1986, pp. 365–6; Mellor 1995, pp. 220–4). If one were to vaporize the left board at *t*, the right board would not yet have moved. So apparently simultaneous mutual causation reduces to asymmetric causation by temporally prior causes. In the second case, there is an implicit temporal story in the background.

These responses are not without their problems. Is there a time when the left board is vaporized and the right has not yet begun to move? It is not as if the right board takes its bearings before it begins falling. In the second case, the relations between the explicit model and the implicit theory remain unarticulated. In any case, the possibility of explaining away such cases settles nothing. Should reference to an implicit temporal story be necessary in order to make sense of the claim that tastes are supposed to determine prices, even though tastes do not precede prices? How much physics should be written into theories of causation? Speculative contemporary physical theories suggest that some causal influences go backward in time. Should a philosophical theory cast its fortunes with some physical theories and rule out others without considering the evidence? Finally, identifying causal and temporal priority makes it impossible to explain why the direction of causation and the direction of time match.

Defenders of a Humean theory might try to relax the requirement that causes precede their effects by embedding particular causal relations within

[6] This example is discussed by Pollock (1976, p. 173), Fair (1979, p. 230), and Frankel (1986, pp. 361, 365–6).

larger chains. One might, for example, say that *a* is causally prior to *b* if (1) *a* precedes *b* or (2) *a* is causally between *c* and *b* and *c* precedes *b* or (3) *b* is causally between *a* and *d* and *a* precedes *d*. This suggestion is, however, empty until one has provided some theory of the relation "is causally between." Reichenbach provided such a theory (1956, §5), but only for nonsimultaneous events.

Paul Horwich argues for a more promising version of this strategy. Consider figure 3.2a, which (apart from my lettering) reproduces a drawing of Horwich's (1987, p. 136). Suppose that there are basic laws connecting the pairs (*a*, *c*), (*a*, *b*), and (*b*, *d*). The temporal priority of *c* to *a* and *b* to *d* establishes that *c* directly causes *a* and *b* directly causes *d*. If we then "maximize causal continuity (so that causal priority in one part of the chain may be 'smoothly' extended to adjacent parts)" (Horwich 1987, p. 136), we should draw the remaining arrow from "*a*" to "*b*."

To see the problems with this argument, one should notice first that graphs of real situations do not look like Horwich's picture. The picture in figure 3.2b is more accurate. *a* and *b* will have many causes and effects. In particular, suppose that *e* is a direct cause of *b* and *f* a direct effect of *a*. If we draw the arrow between "*a*" and "*b*" the way Horwich does, there is a causal chain: $c \rightarrow a \rightarrow b \rightarrow d$ and *b* and *f* are related only as effects of their common cause *a*. If we draw the arrow in the other direction, there is a causal chain $e \rightarrow b \rightarrow a \rightarrow f$, and *a* and *d* are related only as effects of their common cause *b*. Which way should one draw the arrow between "*a*" and "*b*?" *If c* causes *d*, then there has to be a directed path from "*c*" to "*d*." If, in addition, there are no other causal relations among these variables, then the edge must go from "*a*" to "*b*." But the indirect causal relation between *c* and *d* is supposed to be constituted by the existence of a chain of direct causation. The direction of the edge between "*a*" and "*b*" is presupposed by the existence of the path between "*c*" and "*d*" and cannot be determined or constituted by it.

One might be able to infer that *a* is a direct cause of *b* if one had some way to know that *c* caused *d* that did not depend on knowing that there was a causal chain linking them. What motivates Horwich's argument is, I think, the sensible thought that there may be lawlike connections between events of kinds **c** and **d** which require a causal explanation, and it may be that one can rule out all the alternatives that do not depend on *c* being an indirect cause of *d*. Such a theory goes beyond Hume in asserting that lawlike connections require causal explanations, but appears in principle to permit one to relax the temporal priority requirement. The details of such a theory are not easy to fill in, and I shall not attempt to formulate explicitly the conditions for "causal betweenness." Although Hume's theory might thus in principle be revised to encompass some simultaneous causation, I shall

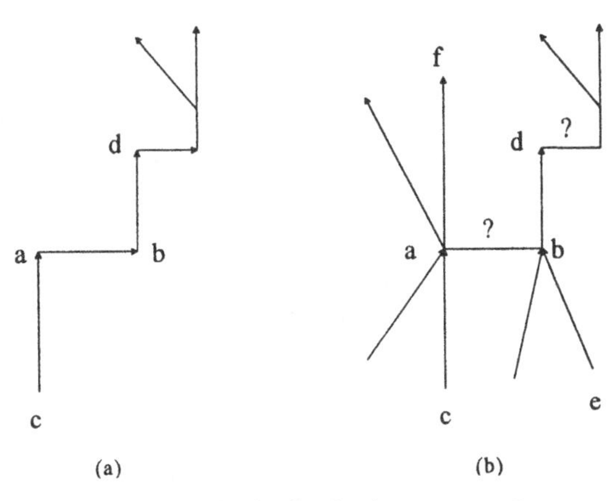


**Figure 3.2:** Allowing for simultaneous causation

ignore this complication in what follows and take Humean views to require that causes precede their effects.

For the record, here is an explicit formulation of the Humean theory of causation:

> **(H)** (*Humean theory*) *a* is a *cause* of *b* if and only if *a* is a direct cause of *b* or there is a unidirectional chain of direct causes running from *a* to *b*.
>
> *a* is a *direct cause* of *b* if and only if *a* and *b* are distinct, *a* is nomically necessary and sufficient in the circumstances for *b*, *a* and *b* are spatially and temporally contiguous, and *a* begins before *b* begins.

## 3.4 Spurious Causation

A thunder clap is not caused by a contiguous lightning flash, even though the two apparently satisfy all the conditions of Hume's account of causation. Hume's theory falsely implies that contiguous and temporally successive effects of a common cause are related as cause and effect. The complex of problems that arise here is called by David Lewis, "the problem of epiphenomena." One can also call this "the problem of spurious causation."

The Humean account of causation implicitly defines a symmetrical notion of "direct causal connection." Events are directly causally connected if they are distinct, contiguous, and necessary and sufficient for one another in the circumstances. To address the problem of epiphenomena, one must either find some way to deny that effects of a common cause are causally connected to one another, or one must amend Hume's temporal precedence account of causal asymmetry. Although I believe that the problem of spurious causation can be solved only by an improved account of causal priority, let us consider how one might attempt to rule out causal connections between effects of common causes. (The fact that their concomitance seems in some sense "causal" is not decisive, since one is free to take "causal connection" to be a technical term.)

On what grounds might one deny that effects of a common cause are causally connected? It is hopeless to argue that effects of a common cause are never truly contiguous. Consider, for example, the relations among images in a mirror (Edwards, cited in Faust and Johnson 1935, p. 335, Kim 1984, pp. 257–8). Perhaps, one can instead strengthen the sort of regularity required for a *direct* connection between cause and effects. Paul Horwich, for example, insists that the laws relating causes and their direct effects be "basic" and that the nomic connection between a cause and its direct effects not require reference to any third event or state of affairs (Horwich 1987, pp. 134–5; see also Clendinnen 1992, pp. 351–3). Such basic nomological connections will never obtain between effects of a common cause, and so one will never mistakenly take them to be related as cause and effect. I question whether any conditions on laws and on the relations between laws and causes can be strong enough to rule out causal connections between effects of a common cause without ruling out direct causal connections between genuine causes and effects. On Horwich's account, for example, it turns out that there is no direct causal relation between the length of a pendulum and its frequency of oscillation, because the law of the pendulum requires reference to the presence of the Earth and hence is not basic (1987, p. 166).

One might instead add some clause to Hume's account to the effect that if *a* and *b* are effects of a direct common cause, then they are not directly causally connected. An explicit stipulation to this effect will not serve, since *a* and *b* may be effects of a common cause and *also* be related as cause and effect. A more promising suggestion would be to say that *a* and *b* are directly causally connected if (1) they are distinct and contiguous, (2) *A* is nomically necessary and sufficient in the circumstances for *B*, and (3) when one "controls for" or "holds fixed" common causes of *a* and *b*, *A* is still necessary and sufficient in the circumstances for *B*. (Since causal connection is a symmetrical relation, the positions of *a* and *b* can of course be reversed.) The idea is that if the regular association between types **a** and **b** in the circumstances is due entirely to common causes, then once the occurrence or nonoccurrence of these common causes is specified, whether *a* occurs should be irrelevant to whether *b* occurs. If *A* appears to be necessary and sufficient in the circumstances for *B* because *a* and *b* are effects of a common cause *c*, then including that cause in the "circumstances" makes *a* redundant. If *A* is "truly" at least an INUS condition for *B*, then *a* and *b* cannot be effects of a common cause. Thus it seems that no revision to Hume's account is necessary after all. When *a* and *b* are effects of a common cause, *A* will not be necessary "in the circumstances" for *B*.

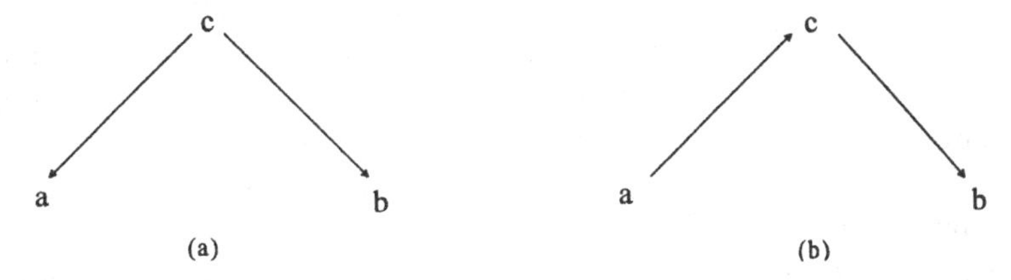


**Figure 3.3:** Common causes and causal chains

This suggestion fails. Suppose that causal relations are deterministic and that *c* is a common cause of *a* and *b*, which stand in no other causal relation to one another. The proposal is that once we include *C* in the circumstances, then *A* will not be necessary for *B*. But, given determinism, if one includes *A* among the circumstances, it may be that *C* is not necessary and sufficient in the circumstances for *B*. One needs some reason to deny that *a* is causally connected to *b* that is not equally a reason to deny that *c* is causally connected to *b*. To cope with this problem, one might make use of the theory of causal priority and argue that it is *a* rather than *c* that is irrelevant to the occurrence of *b*, because *c* precedes *a*. But suppose that $c \rightarrow a \rightarrow b$. One does not want the possibility that *C* is necessary and sufficient in the circumstances for *B* to imply that *a* is not a cause of *b*. Furthermore, there is a danger of circularity, since one is relying on the notion of causation to define causal connection.

It is questionable whether Humeans have any further response to the problem of spurious causation. There is more to causation in the objects than regularity, contiguity, and temporal priority *among those objects*. In figure 3.3, the only difference is the direction of the causal arrow between *a* and *c*. The theory of causal priority rather than the theory of causal connection ought to explain why effects of a common cause are not themselves cause and effect. (A key to causal asymmetry lies, I shall argue, in what is not shown in the figure – in the lawlike relations that aspects of cause and effect bear to other things.)

## 3.5 Redundant Causation

There are also problems concerning two varieties of *redundant causation*. The first involves *overdetermination*. The redundant causes are on a par: There are two or more events that are equally capable of bringing about the effect all by themselves. The shots of separate soldiers in a firing line might be an example. Since none of the shots is necessary in the circumstances for the effect, none would count as a cause on the Humean view of causation (**H**). Perhaps one can simply accept this implication of the analysis: None of the separate shots caused the death. The death was instead caused by a complex event constituted out of the separate shots.[7] Work remains to be done, because it is not clear what this complex event is. I believe that the problems concerning causal overdetermination do not point to any special difficulties with a Humean view, because they arise in similar ways for all the competing theories of causation.

In cases of *causal preemption*, the redundant causes play different roles. "One of them, the *preempting cause*, does the causing; while the other, the *preempted alternative*, waits in reserve" (Lewis 1986d, p. 199). Here is a memorable example:

> A man sets out on a trip across the desert. He has two enemies. One of them puts a deadly poison in his reserve can of drinking water. The other (not knowing this) makes a hole in the bottom of the can. The poisoned water all leaks out before the traveler needs to resort to this reserve can; the traveler dies of thirst.[8]

A correct analysis of causation ought to say that the preempted alternative – poisoning the canteen in this case – did not cause the traveler's death, but that the preempting cause – drilling the hole – did. The Humean analysis apparently implies that neither the poisoning nor the drilling caused the death, for in the circumstances, which include the presence of the other potential cause, neither is necessary for the death.

One way to cope with the difficulty is to note that it is only direct causes, not causes in general, that must be necessary and sufficient in the circumstances for their effects and that neither the poisoning nor the drilling is a direct cause of the traveler's death. The drilling (and not the poisoning) still counts as an indirect cause of death, because there is a unidirectional chain of direct causation between the drilling and the death, while there is no chain of direct causation connecting the poisoning and the death. This resembles the way David Lewis defends his counterfactual analysis of causation from the apparent counterexamples provided by cases of causal preemption (1973a). I shall argue in chapter 13 that attention to spatiotemporal chains of direct causation is the best way to respond to most cases of preemption, regardless of one's general theory of causation.

This solution to the problem of preemption depends on there being causal intermediaries between the preempting cause and the effect that cut the connection between the backup cause and the effect and then go on to bring about the effect. In the context of David Lewis's theory, in which links in a

[7] Causal overdeterminants need not be simultaneous. The bullets from two successive shots of soldiers who stand at different distances from the condemned prisoner can arrive at the same time and overdetermine the death.

[8] Quoted from Mackie (1980, p. 44), who based it on Hart and Honoré (1959, pp. 219–20), who in turn drew on an example of McLaughlin (1925–6, pp. 149, 155n).

causal chain need not be spatiotemporally contiguous, this assumption is not always satisfied. In cases of "late" preemption, only the occurrence of the effect itself may cut a chain of counterfactual dependence that would have led to that same effect, and hence Lewis's account mistakenly implies that the effect is causally dependent on the preempted alternative. It is plausible to maintain in contrast that one will never find a chain of direct spatiotemporally contiguous causation from a preempted alternative to the effect. Or so I shall argue in chapter 13.

Another way that has been suggested for dealing with the problems posed by cases of causal preemption might be tried here. Which factor is truly the cause may affect the character of the effect. In Mackie's canteen case, the traveler dies of thirst and not of poison, and drilling a hole in the canteen is necessary in the circumstances for dying of thirst, while poisoning the canteen is not. If one is precise enough in the description of the effect, then it may turn out that the preempting cause but not the preempted alternative will be necessary in the circumstances for the effect.[9]

This method of dealing with cases of late preemption seems to me less convincing than attempting to assimilate cases of late preemption to cases of early preemption. If one insists on describing effects in such detail, one will be committed to counting everything that affects any detail of an event or its time as a cause of the event. David Lewis argues that the results are absurd:

> Boddie eats a big dinner, and then the poisoned chocolates. Poison taken on a full stomach passes more slowly into the blood, which slightly affects the time and manner of the death. If the death is extremely fragile [that is, if it must have all of its detailed features], then one of its causes is the eating of the dinner. Not so. (1986d, p. 198)

Unlike Lewis, I think that one of the causes of (a feature of) Boddie's death is his eating a full dinner. Eating a full dinner does not explain why Boddie died or why Boddie died that evening, but it explains Boddie's slow death. Similarly, if one wants to know why an executed prisoner was struck by seven bullets and what caused his seven-bullet death, one can say that one of the causes was soldier #4's shot. One is free to consider what explains any particular aspect of the effect that one pleases.

This freedom does not, however, vindicate the fragile-events response to problems of overdetermination and preemption. When a coroner seeks the

[9] More radically, one might maintain that all the causes of an event are essential to it. If any of event *e*'s causes had been different in any way, *e* would not have occurred. So preempting or overdetermining causes are necessary to their effects, and some of the problems of overdetermination and preemption apparently disappear. The difficulties of distinguishing preempted from preempting causes and causes from causal overdeterminers have, however, only been swept under the rug. I am indebted to Gordon Barnes for clarifying my thinking here.

causes of Boddie's death, he is not seeking the causes of his slow death, and someone asking why the executed prisoner died is not asking why the prisoner died a seven-bullet death. Moreover, there are cases where a fragile construal of events apparently will not help. Suppose the second murderer in the story of the reserve canteen pours out the poisoned water in the mistaken belief that it was not poisoned and then refills the canteen with a poisoned solution of the same chemical constitution. The traveler would have died in exactly the same way at exactly the same time if the second enemy had not acted, yet the actions of the second enemy still caused the death.

Overdetermination and preemption constitute obvious complications for Hume's account of causation. In the case of overdetermination, the difficulty is not so much that Hume's theory implies something mistaken, as that we do not know what to say about such cases. In cases of preemption, there will often be a chain of direct causation only from the preempting cause to the effect and so Hume's account will work fine. I shall return to the problems of preemption and overdetermination in chapter 13.

## 3.6 The Problems of Causation

Hume's account appears to be unsatisfactory in the following regards:

1. It lacks a theory of laws and of the relations between laws and specific nomic INUS conditions.
2. It either implies falsely that successive and contiguous effects of a common cause are related as cause and effect or it implies falsely that many causal intermediaries are not causes.
3. It closes apparently empirical questions concerning the possibility of backwards causation and remote causation.

*If* the fundamental difficulties raised by successive and contiguous effects of common causes can be solved and *if* the distinction between lawlike and accidental generalizations can be drawn in a satisfactory way, and *if* we set aside the possibilities of remote and backwards causation, then the Humean theory of causation is attractive. Given these three ifs, it is a powerful theory. To appreciate its strengths, consider for a moment how it fares with respect to the criteria of adequacy set out in the first chapter.

1. *Intuitive fit*: A theory of causation should fit our intuitions. Obviously Hume's account is in serious trouble with respect to successive effects of common causes and in some trouble concerning simultaneous causation. Otherwise, it seems that Hume's theory fits most of our intuitions, except those concerning necessary connection, which it explains away. There are also the complications of overdetermination and preemption but the competitors have trouble with these, too.

2. *Empirical adequacy*: Like theories in physics or psychology a good theory of causation should fit the facts. Apart from problems with spurious causation, Hume's account apparently fits much of scientific usage and explains what role causal language plays. By taking some of the metaphysical mysteries out of causation, Hume's theory legitimates and explains the role of causal notions within science.

3. *Epistemic access*: A good theory of causation should explain how one finds out what causes what and, ideally, assist in learning more about causal relations. Hume reduces the problem of identifying causes to the problem of identifying nomic regularities in which the further conditions of contiguity and succession are satisfied. The only mysteries concerning causation are the mysteries concerning laws themselves, which are arguably unavoidable in science. Hume's account is explicitly motivated by empiricist scruples, and it responds to them as well as any alternative.

4. *Superseding competitors*: A good theory of causation should be better than competing theories and help explain why they succeed and fail. We have not yet examined any competitors, so the application of this criterion will have to wait.

5. *Metaphysical coherence and fecundity*: A good theory of causation should clarify the links between causation and other relations, such as temporal relations, and it should help one to answer other metaphysical questions, such as why people know so much more about the past than the future. The extent of coherence turns largely on what account can be given of the notion of a law of nature. The fecundity of Hume's definition is controversial, though any account that has been used by so many philosophers and scientists must be fecund.

It is thus obvious why Hume's theory still demands serious attention. In addition to presenting and refining a Humean view of causation and displaying its unresolved difficulties, this chapter has also introduced central problems for theories of causation that will recur throughout the book. The problems so far are:

*Regularity and necessity*

- *Singular and general*: What are the relations between token causal claims and regularities?
- *Laws*: What are the relations between causation and laws, and what are laws?
- *Necessity*: Can one make any sense of the necessity, efficacy, or power of causes?
- *Indeterminism*: Are there merely probabilistic causes?

*Contiguity*: Are direct causes always contiguous with their effects?

*Priority*: How do cause and effect differ? What is the connection between causation and time?

*Spurious causation*: Are effects of a common cause causally connected to one another? How do effects of a common cause differ from cause and effect?

*Redundancy*

*Overdetermination*: What should one say about cases of causal overdetermination?

*Preemption*: Can a theory of causation correctly identify preempting factors as causes and correctly rule preempted alternatives not to be causes?

This book is mainly concerned with causal priority, but I shall have to say something about the other problems, too. I shall set aside the difficulties involving preemption and overdetermination until chapter 13.

# 4

# Causation and Independence

This chapter presents what I believe to be the central asymmetry of causation. The ideas first occurred to me early in 1981 and derive from work by Herbert Simon (1953) and Douglas Ehring (1982). The central intuition is that causal priority consists in the causal connection among effects of a common cause and the causal independence among the causes of a given effect. For this reason I call this view the independence theory of causal priority.

## 4.1 Causal Connection and Probabilistic Dependency

J. L. Mackie suggests that one can factor the causal relation into a symmetrical relation of "causal connection" and an asymmetrical relation of "causal priority": For all events *a* and *b*, *a* causes *b* iff *a* is causally connected to *b* and *a* is causally prior to *b*. In Mackie's view, to say that *a* and *b* are causally connected is to say "*a* and *b* are connected by some fact of causation" (Mackie 1980, p. 85). Causal connection is the topic of this section and of chapter 12, while causal priority is the topic of the whole book. Humeans, for example, take *a* to be causally prior to *b* if *a* precedes *b*, and they take *a* to be directly causally connected to *b* iff *a* and *b* are distinct and contiguous and one is necessary and sufficient in the circumstances for the other.

I share Mackie's intuition that causation has both a symmetrical and an asymmetrical aspect. The symmetrical part is akin to Hume's notion of a necessary connection. But what is a necessary connection? Philosophers with empiricist leanings are uncomfortable with necessities in nature, and they have tried to analyze "necessary connection" in terms of nomological regularities. The hope is to say that *a* and *b* are causally connected only if their co-occurrence is not an accident, but is demanded by natural law. When things are causally connected, the presence of one (given the circumstances and laws of nature) demands the presence of the other.

The attempt to analyze causal or necessary connection in terms of nomological regularities has been a failure. As pointed out in chapter 3 (esp. p. 42), it is mistaken to claim that *a* and *b* are causally connected if and only if $A$ is an INUS condition for $B$ or $B$ is an INUS condition for $A$. The height of a flagpole is an INUS condition for the angle of elevation of the sun, but nobody believes that there is any sort of causal connection between them. The closest one can come to an analysis is, I believe, to link causal connections among tokens to unconditional probabilistic dependencies among types: There are no *unconditional* probabilistic dependencies between the flagpole's height and the sun's angle. On the other hand, there are probabilistic dependencies between a flagpole's height and the length of its shadow and between the angle of elevation of the sun and the length of the shadow. A nomological connection is that relation between tokens that typically manifests itself in probabilistic dependencies among their types. As we shall see, however, it is impossible to analyze causal or necessary connections in terms of unconditional probabilistic dependencies.

It might seem obviously mistaken to link a token-level relation of causal connection to *any* type-level relation, whether probabilistic or deterministic. Suppose $a_1$ causes $b_1$ and $a_2$ causes $b_2$, where $a_1$ and $a_2$ are tokens of type **a** and $b_1$ and $b_2$ are tokens of type **b**. **a** and **b** can be probabilistically dependent on one another without any causal connection between $a_1$ and $b_2$ or between $a_2$ and $b_1$.[1] The existence of a probabilistic dependency between **a** and **b** does not imply that there are causal connections between every token of the two types. For example, suppose that two identical electric circuits sit side by side on a table top. In each there is a switch, a battery, and a light. The existence of a probabilistic dependency between flipping a switch in a circuit on top of this table and a light going on in a circuit on top of this table does not establish that there is a causal connection between flipping a switch in circuit 1 and the bulb going on in circuit 2.

In response, consider the two event types: **a** (= switch flipped in a circuit on this table) and **b** (= bulb lights in a circuit on this table). The probabilistic dependency between **a** and **b** derives from the probabilistic dependencies between the flippings and lightings within each separate circuit – i.e., between $\mathbf{a}_1$ and $\mathbf{b}_1$ and $\mathbf{a}_2$ and $\mathbf{b}_2$. ($\mathbf{a}_1$ is closing a switch in circuit 1, etc.) $a_1$ is a token of both type $\mathbf{a}_1$ and type **a**, but it is not a token of type $\mathbf{a}_2$, while $b_2$ is a token of types $\mathbf{b}_2$ and **b**, but not of $\mathbf{b}_1$. The only genuine probabilistic dependencies are between types $\mathbf{a}_1$ and $\mathbf{b}_1$ and between $\mathbf{a}_2$ and $\mathbf{b}_2$. The dependency between **a** and **b** reflects a fluke of the sample – that is, the accident that the same kinds of circuits were placed on this table. There is no genuine probabilistic dependency between **a** and **b**. Recall that to speak about probabilities in the sense of propensities or limits of relative frequencies introduces a modal element. If **a** and **b** are genuinely probabilistically dependent, then this dependence is not accidental.

[1] I am indebted to Sungsu Kim for pressing this objection. See also Ehring's discussion of what he calls "the pairing problem" (1997, esp. pp. 18f).

The objection is not yet fully answered. Consider just $\mathbf{a}_1$ "a closing of the switch in circuit one" and $\mathbf{b}_1$ "the bulb in circuit one lighting." There is a probabilistic dependency between $\mathbf{a}_1$ and $\mathbf{b}_1$. Suppose that the switch is closed and opened several times and that the light goes on and off several times. If one picks a particular switch-flipping, only one of the bulb-lightings is causally connected to it, yet all of the lightings are of the same kind.

One can resolve this problem as follows. Divide the time interval during which the switch is turned off and on into a large number of subintervals. Let

> $x_i = 1$ if the switch is closed for at least part of the interval *i*,
> $x_i = 0$ if the switch is open during the whole of interval *i*,
> $y_i = 1$ if the bulb is lit for at least part of interval *i*, and
> $y_i = 0$ if the bulb is off for the whole of interval *i*.

Assume that the gap between the time when a switch is closed and a bulb begins to light is too small to measure. There will be an excellent correlation between $x_i$ and $y_i$. But there will be no correlation between $x_i$ and $y_j$, where the series $\{y_j\}$ is an arbitrary permutation of the series $\{y_i\}$.

One should deny that there is an unconditional probabilistic dependency between the types "a switch is closed in this circuit" and "the bulb lights in this circuit." There is, instead, a probabilistic dependency conditional on the proper temporal relations obtaining between tokens of the two types. When I speak of an unconditional probabilistic dependency from now on, I shall mean a probabilistic dependency conditional on the proper temporal relations obtaining between the tokens of the types and on nothing else.

Suppose one then endorses the following simple linkage between causal connections and probabilistic dependencies:

> **OA** (*Operationalizing assumption*) Events or tropes *a* and *b* are causally connected if and only if they are distinct and the kinds **a** and **b** or the properties *A* and *B* are in the background circumstances unconditionally probabilistically dependent.

I take the claim that *a* and *b* are distinct as implying that *a* and *b* exist. The distinctness and existence of *a* and *b* must be required, because probabilistic dependence between properties does not imply that the properties are distinct or that they are instantiated at the particular place and time that the tropes *a* and *b* occur. For more on distinctness, see page 59. **a** and **b** are unconditionally probabilistically dependent in background circumstances *C* if and only if $\Pr(A\&B/C) \neq \Pr(A/C).\Pr(B/C)$. The exact meaning of the phrase "in the background circumstances" and the reasons for including it will be discussed in §4.1* and in chapter 12. As a first approximation, it will do little harm to ignore this qualification.

I call **OA** an "operationalizing assumption" because it enables one to link the notion of a causal (or necessary) condition to something that is "closer"

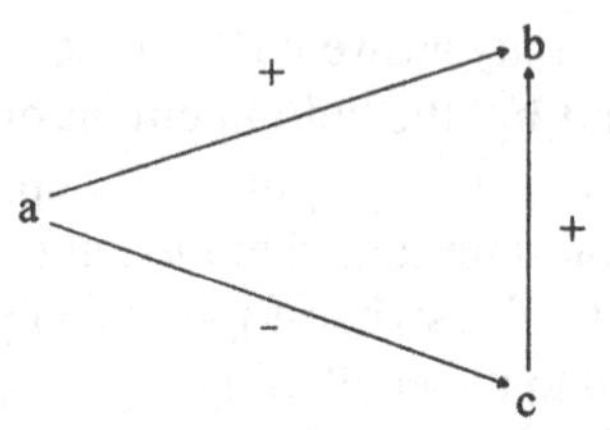

**Figure 4.1:** Causal connection without correlation

to observation. Genuine probabilities, as opposed to sample frequencies, are no more observable than are laws, but they are more empirically respectable than are causal or necessary connections. One might also regard **OA** as a sketch of a definition of a theoretical relation: Tokens are causally connected if they bear that relation to one another that typically issues in probabilistic dependencies between their types (Lewis 1972; Menzies 1996).

Unfortunately even with the understanding of "unconditional" dependencies as conditional on temporal conditions and with the understanding of types discussed above, the operationalizing assumption seems false. There are cases where *a* causes *b* even though **a** and **b** are probabilistically independent in the background circumstances, and there are cases in which **a** and **b** are probabilistically dependent even though *a* and *b* appear to be causally independent. For an apparent case of correlation without causal connection, consider the relation between bread prices in England and water levels in Venice or between any other causally unrelated increasing sequences.[2] Conversely, the causal structure illustrated in figure 4.1 demonstrates that cause and effect need not be correlated. Tokens of **a** and **c** cause tokens of **b** and tokens of **a** prevent tokens of **c**. If the strength of the influences is just right, **a** and **b** may be probabilistically independent even though *a*'s are direct causes of *b*'s. For a controversial example with roughly this structure (though the causal influences do not cancel out), let **a** be taking birth control pills, **c** be pregnancy, and **b** be blood clots (Hesslow 1976, 1981).

I assert, though I cannot prove this, that no better analysis of causal connection exists. In my view, for which further argument will be provided in chapter 12, there are important relations between causation and probabilistic dependencies, but these relations are not exact. A causal connection is a theoretical relation; the notion is rather like an *idealization* of the notion of a (lawful) probabilistic dependency. Causal connections are not reducible to probabilistic dependencies, but reasonable empiricists should not demand

[2]The example is borrowed from Sober (1987, p. 465; 1988, p. 215). I shall have more to say about it in §12.1. Yule (1926) pointed out that increasing sequences will be correlated.

a reduction. The imprecise link to probabilistic dependencies legitimates the notion of a causal connection.

Readers who are not satisfied with this account of causal connection can take the operationalizing assumption as *defining* causal connection. If causal connection between tokens is defined as probabilistic dependence among types, then the claims I shall shortly make linking causation and causal connection are false. But they are still useful approximations. Qualms about the notion of a causal connection should not discourage readers from proceeding.

## 4.2 The Connection Principle

The principal claim I shall make about causal connection is the following:

> **CC** (*Connection principle*) For all events *a* and *b*, *a* and *b* are causally connected if and only if they are distinct and either *a* causes *b*, *b* causes *a*, or *a* and *b* are effects of a common cause.

*a* and *b* may of course be effects of a common cause and also cause and effect.

Reichenbach takes the connection principle to define the notion of a causal connection (1956, p. 29). I prefer, in contrast, to take the notion of a causal connection as the undefined intuitive notion of a nomological linkage, that relation among tokens that typically manifests itself as a probabilistic dependency among types. In either case the connection principle (**CC**) implies controversial metaphysical claims.

1. **CC** asserts that if events are causally connected, then they must be *distinct*. (I shall take the claim that *a* and *b* are causally independent also as implying distinctness.) Events and tropes *a* and *b* are distinct if they have no parts in common and the proposition that *A* is instantiated in the circumstances does not entail that *B* is instantiated (or vice versa). Socrates' death is not causally connected to Xantippe's widowing, because in the circumstances the former entails the latter. Socrates' drinking half the hemlock is not causally connected to his drinking all of it, because it is a spatiotemporal part of his drinking all of it. These events or tropes are not distinct. In ordinary and scientific usage, causal connection implies distinctness, but the justification for requiring distinctness lies mainly in the theoretical development it makes possible. Whether events and tropes are distinct is not always obvious. Before the reduction of temperature and pressure to mechanical properties of gas molecules, instantiations of these properties appeared to be distinct. Now we know they are not, although in some contexts it is harmless to assume that they are.

2. The connection principle and the operationalizing assumption imply

that cause and effect are of types that are probabilistically dependent. Since a causal connection is supposed to be the symmetrical "part" of causation, this is as it should be – cause and effect ought to be causally connected. The claim that properties of cause and effect are probabilistically dependent is, as we have seen, only an approximation.

3. **CC** and **OA** imply that effects of a common cause are of kinds that are probabilistically dependent. This is more controversial.[3] If one maintains that effects of a common cause are causally connected, then, given the transitivity of causation, it seems that almost everything is causally connected to almost everything else. But for the continued influx of energy from the sun there would be no bread-price increases in England and no water-level increases in Venice. So these series have common causes. But Elliott Sober picked these two series specifically to illustrate the fact that *unconnected* increasing sequences will be correlated. Was he wrong to assume that bread prices in England and water levels in Venice are not causally connected?

No. One should understand the statement "*a* and *b* are causally connected" – like the statement "*a* causes *b*" – as carrying an implicit reference to a causal "field" (see p. 40). A connection that obtains merely in virtue of the causal field does not count. *a* and *b* are causally connected only when controlling for features of the causal field does not eliminate the connection. Causal claims concern relations *within* some sort of "system." Connections that hold in virtue of features outside the system demand enlargement of the system, or, holding fixed such external features, these connections disappear. For example, consider an adjustable flagpole standing in the sun. Suppose one finds a correlation between the flagpole's height **h** and the angle of elevation of the sun **a**, because someone wants to keep the length of the shadow constant. Either the efforts of the person need to be held fixed, in which case the correlation and the appearance of a causal connection between **h** and **a** disappears, or the system needs to be expanded so that the person's efforts are incorporated. The connection principle might accordingly be restated as: "For all events *a* and *b*, *a* and *b* are causally connected in system *W* if and only if *a* and *b* are distinct events in *W* and either *a* causes *b*, *b* causes *a*, or *a* and *b* are effects of a common cause that is in *W*." I shall usually leave this relativization implicit.

The relativization of the notion of a causal connection provides a further

[3]Though not unusual. Some examples of others who have held that effects of a common cause bear a causal connection: "Even though it is true that an earlier mirror image is not a cause of a later one, it is also true that there *is* a causal relation between the two – the two are successive effects of the same underlying causal process" (Kim 1984, p. 259). "I am causally connected to the condition I believe will obtain (patient's death [*a* is F]). . . [in that] I am caused to believe of *a* that it will be F by a condition, massive brain damage, that causes *a* to be F" (Dretske and Enc 1984, p. 520).

reason to regard causal relations as tied to explanatory concerns. Explanatory interests lead us to attend to particular causal systems, and only with respect to such systems are causal connections well defined. Relativizing causal connections and hence causal relations to systems does not undercut their objectivity or mind-independence, because explanatory interests do not call causal systems into existence.[4] Context and interests influence which systems to consider, not which systems exist.

Relativizing the understanding of causal connection resolves the obvious objections to the claim that effects of a common cause are causally connected, but it gives one no positive reason to make this claim. In defense of this proposition, first one can point to its conformity with everyday expectations. For example, one expects animals who resemble one another to have a common ancestor. Second, this expectation rests on fact: When *a* and *b* are effects of a common cause, **a** and **b** are in fact typically probabilistically dependent in the background circumstances. Third, if causes are probabilistically dependent on their effects, then it is provable in a wide range of circumstances that effects of a common cause will be probabilistically dependent on one another (see §4.1*). Given the association between probabilistic dependencies and causal connections, one has reason to believe that effects of a common cause are causally connected.

4. **CC** and **OA**, the connection principle and the operationalizing assumption, imply that types **a** and **b** are not probabilistically dependent if *a* and *b* are not related as cause and effect or as effects of a common cause. This proposition might seem unacceptable, because if one leaves a common cause of *a* and *b* out of a system, then **a** and **b** may be probabilistically dependent even though *a* and *b* are not causally connected in that system. But the relativization described above resolves this difficulty. If there is a correlation between the height of a flagpole and the angle of elevation of the sun, even though they are not cause and effect or effects of a common cause in the particular system, then one should either enlarge the system of interest to include the omitted common cause or one should control for omitted causes and examine only those probabilistic dependencies within the system that remain.

**CC** implies that if *a* and *b* are *only* related as two causes of some common effect, then they are not causally connected. **CC** does not rule out the possibility that all the causes of some event are *in fact* causally connected, because they are all related as cause and effect or as effects of a common cause. **CC** only denies that events are connected in virtue of being causes of a common effect. In defense of this implication of the connection principle, one can again point to its conformity with ordinary expectations. People do

[4] I am indebted here to Nancy Cartwright.

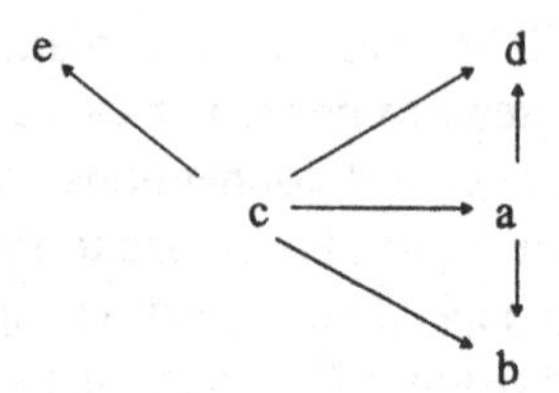


**Figure 4.2:** A necessary condition for causation

not take similarities in the traits of animals as evidence that they will have a common descendant. Second, as a matter of fact, causes of a common effect that are not also related as cause and effect or as effects of a common cause are almost always probabilistically independent of one another.

These considerations do not prove that the connection principle is correct, but they are good grounds to take **CC** seriously. If one accepts **CC** and takes **OA** to be a reasonable approximation, then one holds that distinct events are usually probabilistically dependent on one another if and only if they are related as cause and effect or as effects of a common cause.

### 4.3 A Necessary Condition for Causal Priority

If one assumes that cause and effect are distinct and that causation is transitive,[5] then a necessary condition for "*a* causes *b*" follows from the connection principle. That necessary condition says that *a* causes *b* only if *a* and *b* are causally connected and everything (in the system) connected to *a* is connected to *b*. The path graph shown in figure 4.2 shows why this is true. Since the directed edges in a path graph represent the relation of being either a direct or an indirect cause, **CC** implies that an event is causally connected to *a* if and only if it is related to *a* in the way that *c, d,* or *e* is. In a path graph like figure 4.2, there must be an edge between cause and effect whether the causal relation is direct or indirect. So given that *c* causes *a*, *a* causes *b* and *a* causes *d*, there must be arrows from *c* to *d* and from *c* to *b* whether or not *c* directly causes *b* or *d*.

This necessary condition immediately provides a way of testing whether *a* causes *b*. The absence of a probabilistic dependence between **a** and **b** or the existence of something correlated with **a** and independent of **b** is evidence that tokens of **a** do not cause tokens of **b**. Because **OA** is only approximate, this test is fallible. But rough principles of causal inference should not be despised. Chapters 10–12 will explore further connections between causal asymmetries and facts concerning probability distributions.

Suppose, as in figure 4.3a, that there is some event *c* causally connected

[5]As pointed out in chapter 2, transitivity is only defensible if events are described in terms of the relevant tropes. See p. 27.

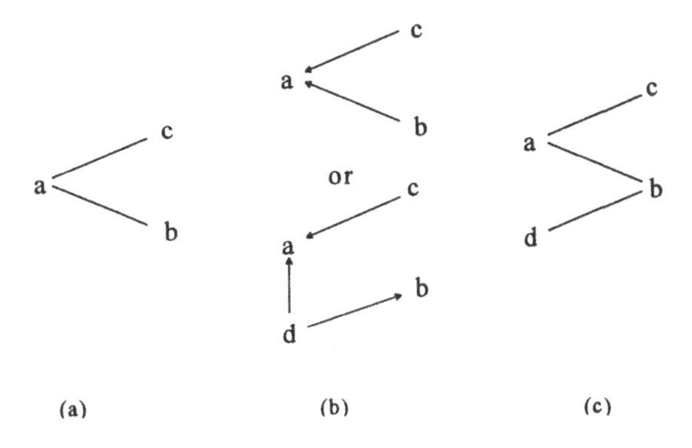


**Figure 4.3:** More implications of **CC**

to *a* but not to *b*. (The undirected edges here represent causal connections.) The necessary condition says that *a* does not cause *b*. If, in addition, *a* and *b* are causally connected, then it must be the case, as in figure 4.3b that *b* causes *a* or that *a* and *b* are effects of a common cause. Finally, if it is also the case, as in figure 4.3c that there is some *d* causally connected to *b* and not to *a*, then *b* also does not cause *a*. Figure 4.3c thus represents a sufficient condition for *a* and *b* to be related only as effects of a common cause: If *a* and *b* are causally connected and both the necessary conditions for *a* causes *b* and *b* causes *a* are unmet, then *a* and *b* must be related only as effects of a common cause. No necessary condition is derivable for "*a* and *b* are related only as effects of a common cause," nor is it possible to derive a sufficient condition for "*a* causes *b*."[6]

## 4.4 Independence

The connection principle says that if causes of a given effect are causally connected, their connection is not forged by their sharing in producing the effect. Causal connections among causes of a common effect are not *explained by* their causing a common effect. **CC** thus implies an explanatory asymmetry, but it does not rule out the possibility that all the causes of a given effect are causally connected as effects of a common cause or as cause and effect. The connection principle implies no sufficient condition for causation.

[6]If one had some other way of determining a relation of causal connection "in a direct line" – that is, a relation that obtains only among causes and their effects and not among effects of a common cause, then one could derive a sufficient condition for *a* causes *b*. But since I can see no way of defining such a relation except as "*a* causes *b* or *b* causes *a*," such a possibility is of little interest. Accounts such as Sanford's (1976) and Ehring's (1982) rely on the notion of a causal connection in a direct line. Other possible sufficient conditions will be explored in chapters 10–12.

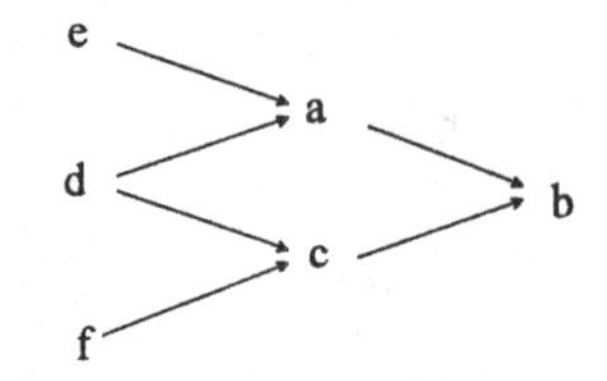


**Figure 4.4:** What independence implies

To get a sufficient condition, I shall defend this independence condition:

**I** (*Independence condition*) If *a* causes *b* or *a* and *b* are causally connected only as effects of a common cause, then *b* has a cause that is distinct from *a* and not causally connected to *a*.

I shall say that events are *causally independent* if and only if they are both distinct and not causally connected. So one can express **I** more compactly as the claim that if *a* and *b* are causally connected and *b* does not cause *a*, then *b* has a cause that is causally independent of *a*. Like the connection principle, the independence condition needs to be relativized so as not to be falsified by the existence of remote causal connections. To make the relativization explicit, I should say that for all events *a* and *b*, there is some causal field or system *F* such that if *a* causes *b* or *a* and *b* are causally connected only in virtue of possessing a common cause in *F*, then *b* has a cause in *F* that is distinct from *a* and not causally connected to *a* in *F*. I shall usually leave this relativization implicit.

Unlike the connection principle, which (if true) either expresses a definition or captures a "deep" truth about causation, the independence condition *seems* to have a de facto character. If it is true, its truth seems to be happenstance. The *possibility* that **I** is true reflects the claim that causes are not causally connected merely in virtue of cooperating to produce a common effect. The *truth* of **I** on the other hand (if it is true), seems to reflect nothing but the coincidence that it never happens that all the causes of a given effect are related as cause and effect or as effects of a common cause.

As a metaphysical claim about patterns of lawlike connections found in nature, **I** seems incredible, and its truth miraculous. But there is another reading:[7] *The independence condition states a necessary condition for the possibility of causal attributions and causal explanations*. When **I** fails, there are no specifically *causal* relations. There may be nomic relations, but there is no causal asymmetry, and causal explanations cannot be given.

[7]Although I toyed with this reading in earlier drafts, I did not appreciate its significance until Leslie Graves pointed it out to me. The discussion that follows and indeed the central thesis of this book owe a great deal to her.

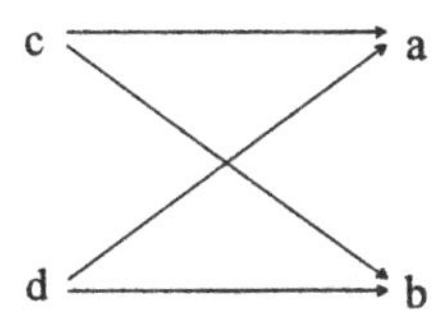


**Figure 4.5:** Independence and effects of a common cause

Defending this ambitious claim takes lots of work, only a little bit of which is done in this chapter. The next four chapters demonstrate that alternative theories of causation presuppose **I** or even stronger independence conditions and that human explanatory practices take **I** for granted. The argument that independence is essential to causation is mainly in chapters 5, 7, 8, 12, and 13.

Before offering a preliminary appraisal of **I** in §4.5, let me clarify further what **I** says. First, although **I** does not rule out the possibility of events that have no causes at all, it implies what I call

> $\mathbf{I}_w$ (*Weak independence*) If *a* causes *b*, and *b* does not cause *a*, then *b* has a cause that is causally independent of *a*.

$\mathbf{I}_w$ implies that every event that has any causes has at least two causes that are causally independent of one another. Notice that transitivity implies that if *b* has at least two causes that are causally independent of one another, then so does every consequence of *b,* although the independent causes may, as it were, dissolve into the causal field. So (again ignoring the relativization to causal fields or systems) if weak independence fails and *b* does not have two causes that are causally independent of one another, then no cause of *b* can have two causes that are causally independent of one another. Notice that neither the independence nor the weak independence condition requires independence among the *proximate* causes of events. The causal arrangement shown in figure 4.4 satisfies **I** even though *b*'s only proximate causes, *a* and *c*, are causally connected.

**I** is stronger than $\mathbf{I}_w$ because it implies that if *a* and *b* are related only as effects of a common cause, then *a* has a cause that is independent of *b* and *b* has a cause that is independent of *a*. Consider figure 4.5. If *c* and *d* are causally independent, then *a* and *b* both have two causes that are causally independent of one another. So weak independence is satisfied, but **I** is not, because *b* does not have a cause that is independent of *a*.

**I** does not imply that every event *b* has some cause that is independent of *all* its other causes. For example, in figure 4.4 the independence condition is satisfied even though none of the five causes of *e* is independent of all the other four. In chapter 4*, p. 82, I define a stronger independence condition that will be relevant to formal results in later chapters.

## 4.5 Why Believe I?

The independence condition is central to this book and far from self-evident. **I** appears to be satisfied in typical causal relations in everyday life and the sciences, but it seems an unlikely candidate for a metaphysical truth. Consider, for example, some dynamical system. Can't one say that a state variable at a particular time, such as the total energy at *t* ($E_{S,t}$) (causally) depends on the value of that variable at an earlier time ($E_{S,t'}$) and on nothing else? Is this not a counterexample to the independence condition?

In reply, one might say that there is no violation of independence, because the values of such state variables at particular times are not "natural" tropes. It would then be up to the theory linking causal relations among arbitrary tropes to causal relations among natural tropes to vindicate the claim that total energy at one time causally depends on total energy at an earlier time. This response places a heavy burden on an unelucidated theory of "natural" tropes. (One can make the same response – more plausibly – to a critic who argues that if *f* is the "fusion" of all of the proximate causes of event *e*, then *e* has no causes that are causally independent of *f*. The possibility of event fusions does not refute **I**.)

Alternatively, one might argue that physical systems are not in fact closed, and that a value of a state variable such as $E_{S,t}$ depends not only on $E_{S,t'}$ but also on various intervention or disturbance variables having "off" values. The independence condition is implied by the view that there are no perfectly closed systems – that disturbing causes are always possible. If a system is open with respect to some endogenous variable *x*, then there must be some causal factor outside the system and thus not causally connected to the other causes of *x*. Such a claim cannot, of course, be made with respect to the universe as a whole. But it might be plausible to deny that the value at time *t* of some variable pertaining to the state of the whole universe, such as its total energy, counts as an event or as a "natural" trope, which can stand in a causal relationships. Moreover, explanatory questions typically arise with respect to particular systems, which are not themselves closed. So **I** may be defensible.

More worrying is the fact that the independence condition implies that there are no closed deterministic *causal* systems. A closed deterministic system with time-symmetric laws of nature could "run" just as well either forward or backward. So there can be no nontemporal difference between causes and effects, and it cannot be the case that independence obtains among the causes of *e* but not among its effects. Even if, contrary to what current physics suggests, the laws of nature are not symmetric in time, a set of *n* causes that have no causal connections to one another can have at most $n(n-1)/2$ effects without a violation of independence (see footnote 4, p. 83).

If there are $n$ state variables, after $(n-1)/2$ periods there will be too many effects for the independence condition to be satisfied.

Of course there may be no closed systems, but it seems objectionable for a theory of causation to decide this matter a priori. Just as I criticized Humean theories on the grounds that theories of causation should not without evidence preclude the possibility of backward causation, so one might reasonably object that theories of causation should leave open the possibility that there are closed systems. But fortunately **I** does not imply that there are no closed systems. What it implies is that there are no closed systems *with causal relations among the state variables*. It implies that one cannot make causal attributions concerning closed systems or offer causal explanations of features of closed systems, not that closed systems do not exist. In chapters 8, 12, and 13, I argue that this surprising implication is correct. Note that **I** also does not rule out causal relations in open subsystems of closed systems.

**I** (or $\mathbf{I}_w$) and **CC** also imply that causation must be asymmetric (see theorem 4.3, p. 82), and this might also be thought objectionable. Shouldn't the question of whether causation is sometimes symmetric be left to empirical inquiry (Lewis 1973a, p. 170)? In §12.7 I shall argue that this implication is welcome.

At this point in the discussion, I will ask the readers to bracket their qualms and to entertain **I** as a hypothesis. The demonstrations in later chapters of the ways in which **I** is involved in alternative theories of causal asymmetry are intriguing in themselves; and they suggest the conclusion that I will argue for later – that cases in which the independence condition apparently breaks down are cases in which our causal concepts break down, too.

Let me say a few more things to make the suspension of qualms about **I** more palatable. First, something like **I** is implicit in inferential practices. For example, people take common causes to screen off their effects even when those effects have other causes. But when there are multiple causes, individual common causes will not in general screen off their effects. One sufficient condition for such screening off is that the strong independence condition (p. 82) is satisfied (see theorem 2, pp. 11–12 of Sober and Barrett 1992). This condition is not necessary, but there is no other simple necessary condition; and it is arguable that independence is taken for granted.

Second, one cannot refute a theory of causal asymmetry with the mere phrase, "Consider a possible world in which the theory is false." Hume's theory, for example, is not refuted by postulating a possible world in which *a* causes *b* and *b* precedes *a*. Similarly, one cannot refute the independence condition merely by gesturing toward a possible world in which an event *e* has only a single proximate cause *c* and in which *e* does not cause *c*. If no more is said, such a purported refutation begs the question. The existence of such possible worlds must be established, not assumed. A critic of **I** must make the case that there are (causally) possible worlds in which the independence condition is violated.

Third, one can consider in more detail how the independence condition might be criticized. Possible counterexamples to **I** fall into the following two categories.

1. An event *b* has only a single proximate cause.
2. An event *b* has more than one proximate cause, and at least one of its causes is causally connected to all of its other causes.

Consider the following three cases that purportedly fall into the first category of events that have only a single proximate cause:

1. An amoeba divides in two. The two amoebae that result have only a single proximate cause. So the independence condition fails. The premise of this criticism, that there is only a single proximate cause, is false. In fact there are many causally relevant factors. But the critic might maintain that the causally relevant factors are all causally connected to one another, and so one still has a counterexample to **I**. Again the objection relies on a false premise. The relevant tropes – for example, instantiations of properties of chemical compounds – are not all causally connected to one another. Furthermore, the emergence of the two amoebae depends on features of the environment that are causally independent of properties of the dividing amoeba.[8]

2. The Big Bang. It raises two separate problems. First, if it is the cause of everything, then everything is causally connected to everything else and no event ever has independent causes. I am not sure whether the Big Bang should be regarded as single event that causes everything, but in any case this problem is resolved by the relativization discussed on p. 60. Except with respect to the first seconds of the universe, if anything is a feature of the causal field, surely the big bang is. I do not know what to say about the first few seconds of the universe.

3. Suppose a single radium nucleus decays into a radon nucleus and an alpha particle. The sole cause of the emission of the alpha particle is the radioactive decay. No other causal factor has any role to play. So **I** fails. In response, one may reasonably question whether the emission and the decay are distinct. Consider just the decay. There is genuine scientific controversy here. On some interpretations of the phenomena, **I** is vindicated. Decay

[8]Richard Scheines has argued that at the token level a nuclear detonation is the only cause of the disintegration of a paper cup four feet away. He suggests that the independence condition is plausible only if one considers counterfactuals or relations at the level of types or variables, where many factors are relevant to the disintegration. I would still maintain that there are independent token-causal factors such as the molecular constitution of the cup.

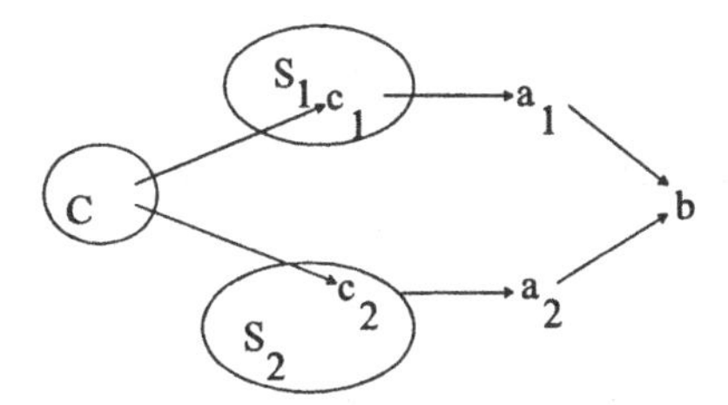


**Figure 4.6:** Failures of independence with multiple proximate causes

depends not only on the state of the nucleus but on independent fluctuations. Suppose one holds, however, that decay depends on nothing except the state of the nucleus (which arguably contains no relevant aspects that are causally independent of one another). In particular there appears to be no way to promote or to retard radioactive decay. It may thus be plausible to deny that the decay has multiple independent causes. If one argues that radioactive decay happens by chance and has no causes, the objection to **I** can be reformulated. The decay propensity of an atom does not have multiple causally independent causes.

Suppose that radioactive decay is a counterexample to independence and consider what the example reveals about a situation in which an event apparently has only a single proximate cause. If $c$ were the single proximate cause of $e$, all causes of $c$ would be causes of $e$ and all causes of $e$ except for $c$ itself would be causes of $c$. Against the background of these common causes, one trope $c$ must all by itself *cause* another. No other property of the particular region of space and time is relevant. Is such a situation possible? To what extent is an event with only one proximate cause *explicable*? The phenomena we identify as causes and effects are not at all like this.

Turn now to the second category in which an event $b$ is supposed to have more than one proximate cause and some cause of $b$ is causally connected to all of $b$'s other causes. Figure 4.6 depicts such a case. Assume that $a_1$ is causally connected to every cause of $b$. Let $C$ be the set of all the causes of both $a_1$ and some other proximate cause of $b$, $a_2$. I have not shown in the diagram other proximate causes that $b$ might have. Let $S_1$ and $S_2$ be the sets of all the causes of $a_1$ and $a_2$ respectively that are not in $C$. By assumption, every cause $c_1$ in $S_1$ that is distinct from $a_2$ is causally connected to $a_2$. Given **CC** and the assumption that $c_1$ does not cause $a_2$ (because it is not in $C$), there are three possibilities: (1) $c_1$ is not distinct from $a_2$, (2) $a_2$ causes $c_1$, and (3) $a_2$ and $c_1$ have a common cause. In each case $c_1$ is caused by some member of $C$. So every event in $S_1$ or $S_2$ must be caused by events in $C$. If one traces back the causes of $a_1$ and $a_2$, eventually one reaches exclusively common causes. Although there may be intermediaries between members of $C$ and $a_1$ that are not causes of $a_2$ and intermediaries between members of $C$

and $a_2$ that are not causes of $a_1$, $a_1$ and $a_2$ ultimately trace back to exactly the same causes.

Note what an odd state of affairs this is. Every member of $C$ causes both $a_1$ and $a_2$, and nothing that is not a member of $C$ or a consequence of a member of $C$ plays any role. How then could $a_1$ and $a_2$ be distinct? What could causally explain the different properties of $a_1$ and $a_2$ or the different times or places of their instantiation of the same properties?

These arguments do not prove that the independence condition is true. They merely suggest that **I** is presupposed by causal beliefs and inferential procedures, that apparent counterexamples are peculiar, and that one can explain away some of them. I hope this is enough to persuade readers to entertain the independence condition as not obviously absurd and to see what follows.

### 4.6 A Theory of Causal Priority

The connection principle and the independence condition imply a sufficient condition: If *a* and *b* are causally connected and nothing causally connected to *a* is independent of *b*, then *a* causes *b*. In other words, if *a* and *b* are causally connected, and everything that is causally connected to *a* and distinct from *b* is causally connected to *b*, then *a* causes *b*. **I** can be restated as: If *a* and *b* are causally connected, then either *a* causes *b* or *a* has a cause that is causally independent of *b*. Thus, if *a* and *b* are causally connected and nothing causally connected to *a* is causally independent of *b*, *a* causes *b*.

If one puts together this sufficient condition and the necessary condition of §4.3, one has:

> **CP** (*Independence theory of causal priority*) *a* causes *b* if and only if *a* is causally connected to *b* and everything causally connected to *a* and distinct from *b* is causally connected to *b*.

To make clear exactly what the independence theory says, one needs to recall many details:

1. "Causes" include all causally relevant factors, regardless of whether one would call them "causes," "causal conditions," "preventatives," or whatever, and there is no concern with whether a particular causal factor is "the" (salient) cause. There is no distinction between direct and indirect causes (see §12.4).
2. The "things" that are causally connected are distinct located values of natural relevant variables (§2.4 and §2.5) or events that are causally connected in virtue of the connection between their aspects.
3. Arbitrary tropes and events are at this point in the discussion ruled out of consideration. Their causal relations, which will be discussed in chapter 13, depend on causal relations among natural tropes.
4. Tropes are *aspects* of events, but the examination of causal or explanatory relations

among aspects is part of a treatment of causal relations among events rather than a competing view.
5. Events are distinct when no part of one is a part of another. The restriction to natural tropes does not render the distinctness clause unnecessary, because when *a* causes *b*, *b* itself will be causally connected to *a* but not to itself.
6. Causal connection is a primitive – what people have in mind when they think of necessary connections.
7. Causal connections are indicated by probabilistic dependencies and relativized to the causal field (pp. 40 and 60).
8. **CP** (unlike **CC** and **I**) does not preclude the possibility that all the same distinct events are causally connected to both *a* and *b*. **CP** leaves open the possibility that *a* may cause *b* and *b* may cause *a*. In §12.7 we shall see why this possibility should be ruled out, but for now I shall leave it open.

Given transitivity and the connection principle, the independence condition not only implies **CP** but is implied by **CP**. The independence theory of causal priority depends on the connection principle, the assumptions that causation is transitive and that cause and effect are distinct, and on the independence condition. The most controversial part is the independence condition **I**.

## 4.7 Should One Accept the Independence Theory?

This chapter constitutes an argument for the independence theory of causal priority (**CP**), since it defends the connection principle and the independence condition, and **CP** is deducible from these axioms plus transitivity. The argument is not conclusive, and the independence theory may be false. Perhaps, for example, radioactive decay *causes* the resulting entities, even though there is nothing causally connected to any of the effects that is not causally connected to their causes. I am not sure whether **CP** is true, but I think that the arguments above establish that exceptions are rare. Such a claim should raise metaphysical eyebrows. Can a respectable metaphysical theory claim as a virtue that it is not often wrong? I shall have more to say about this methodological point later.

I maintain that the independence theory of causal priority is superior to the neo-Humean view presented in chapter 3. Hume's theory faced three main difficulties: (1) distinguishing lawlike from accidental regularities, (2) avoiding the mistaken implication that contiguous and successive effects of a common cause are cause and effect, and (3) prematurely foreclosing all inquiry about whether there could be noncontiguous or backwards causation and about why causation and time run in tandem. **CP** fares no better or worse with respect to the first of these problems. Its notion of a causal connection should offend empiricist sensibilities no more than the notion (which is unavoidable in a Humean theory) of a nomic INUS condition.

**CP** fares decidedly better with respect to the other two problems, which Humean theories could not solve. Consider the moving spot of illumination of a light along a wall. Hume's theory mistakenly implies that the successive points of illumination are related as cause and effect. The independence theory does not. Let *a* and *b* be two contiguous illumination events with *a* preceding *b*. Since *a* and *b* are effects of a common cause, the connection principle implies that they are causally connected. But it is not the case that everything causally connected to *a* is causally connected to *b*. For example, *a* depends on that point of the wall reflecting rather than absorbing light of a particular wavelength, while *b* does not. So *a* does not cause *b*. Similar evidence shows that *b* does not cause *a*, and so one can conclude (correctly) that *a* and *b* are causally connected only as effects of a common cause.

The independence theory of causal priority does not preclude the possibility of simultaneous, backwards, and noncontiguous causation a priori, in advance of evidence. In the case of the two boards leaning against one another, it is arguable that everything causally connected to the position of either board is causally connected to the position of the other and that their positions simultaneously depend on one other.

**CP** is consistent both with the possibility of backwards causation and with the arguments that show that backwards causation is rare. The so-called bilking argument maintains that if it is possible to intervene to prevent events from occurring and an effect occurred before its cause, then one could intervene, prevent the cause, and "bilk" the purported causal relation (Dummett 1964; Brown 1992). When it is possible to intervene to prevent things from happening, effects cannot precede their causes. Interventions are not always possible, and so backwards causation is not ruled out. But it will not be common, because one can typically intervene. The connections between the independence condition and the possibility of intervention will be explored in chapters 5 and 7. As shown there, the independence condition does not imply that interventions are always possible, and consequently it permits backwards causation. Since it is only in exceptional circumstances that the independence condition is satisfied but interventions (abstractly conceived) are impossible, **CP** helps explain why causes generally precede their effects.

So with respect to the three major problems facing the neo-Humean view, **CP** fares no worse with respect to one and much better with respect to the other two. Consider how the two theories compare with respect to the criteria of adequacy set out in chapter 1.

1. *Intuitive fit*: **CP** is less intuitive: What does independence have to do with causation? Yet the independence theory matches causal judgments much better than does Hume's, and I have argued that its claims concerning independence and causal connection are in accord with fundamental causal

intuitions. In showing how independence is linked to notions of intervention and counterfactual dependence, chapters 5 and 6 will help make **CP** more intuitively appealing. There remain difficulties of overdetermination and preemption, which I shall not discuss until chapter 13. These problems are no more (or less) serious for the independence theory than for Humean theories.

2. *Empirical adequacy*: **CP** is more empirically adequate than Hume's theory. **CP** fits much of scientific usage and, as will be more apparent after chapter 8, it helps explain the role causal language plays in science. One crucial advantage over Humean theories is that (as shown in chapter 12) it readily explains some of the methods scientists and statisticians employ to infer causal relations from statistical data. This advantage would be matched by a damaging deficiency if **CP** failed to underwrite a connection between causal and temporal order, but as sketched above, **CP** coupled with the bilking argument implies that effects scarcely ever precede their causes. By taking the metaphysical mysteries out of causation, **CP**, like Humean views, is able to legitimate and to explain the role of causal notions within science.

3. *Epistemic access*: Hume's theory reduces the problem of identifying causes to the problem of identifying nomic regularities in which the further conditions of contiguity and succession are satisfied. This is easier than determining whether in some system everything causally connected to one thing is causally connected to another. But the cost of this advantage is of course the mistaken implication that contiguous and successive events linked by a nomic regularity are automatically cause and effect. **CP** avoids this mistake, and it is tightly linked to actual methods of making causal inferences.

4. *Superseding competitors*: So far we have examined only one competitor – Hume's theory – and we are in the midst of an argument that **CP** is superior to it. As just argued, **CP** implies that causes generally precede their effects. But can't a Humean also explain why **CP** works? Given the connection principle, which a Humean may consistently accept, it is obvious that everything causally connected to a cause will be causally connected to an effect. Given (1) the temporal priority of cause to effect and (2) the possibility of exerting a causal influence *between* a cause and its effect, it will be the effect and not the cause that is modified. Consequently, there will be some cause (at the very least the absence of an intervention) that will be causally connected to an effect and not to its cause. So causes, unlike effects, will be independent. This argument requires, however, not just the Humean theory of causality, but also an additional thesis concerning "the possibility of exerting a causal influence between a cause and its effect." I argue in chapter 7 that this additional thesis entails **I**. So the fact that Hume's theory plus this thesis may "explain" why **CP** works is not much of

an argument for Hume's theory – all the elements of **CP** have been smuggled into the "explanation."

5. *Metaphysical coherence and fecundity*: The coherence and fecundity of **CP** remain to be shown. The remaining chapters will, I hope, demonstrate that **CP** has these virtues.

The independence theory of causal priority represents, I believe, a big step toward understanding the asymmetry of the causal relation and the sense in which causes are "prior" to their effects. But there are other significant things to be said about the asymmetry of causation, and chapter 5 turns to the most important of these.

# 4*

# Causation, Independence, and Causal Connection

This chapter will clarify details, especially concerning logical relations among **CC**, **I**, and **CP**.

## 4.1* The Operationalizing Assumption and the Connection Principle

Recall that the connection principle says:

> **CC** (*Connection principle*) For all events *a* and *b*, *a* and *b* are causally connected if and only if they are distinct and either *a* causes *b*, *b* causes *a*, or *a* and *b* are effects of a common cause.

"Causal connection" is a primitive – but I defended the following approximation:

> **OA** (*Operationalizing assumption*) Events or tropes *a* and *b* are causally connected if and only if *a* and *b* are distinct and the kinds **a** and **b** or the properties *A* and *B* are in the background circumstances probabilistically dependent.

As I shall explain at greater length in chapter 5*, generalizations about relations among types or properties, whether these are deterministic like **DC** or probabilistic like **OA,** must carry a reference to the "circumstances." For example, in most circumstances there is a correlation between consuming highly acidic foods and stomach pains. In circumstances in which one has already ingested an alkali, this correlation can be reversed. The overall association depends on the de facto frequency of different circumstances. Although the unconditional probabilistic dependence between *A* and *B* is a reasonable guide to whether *a* and *b* are causally connected, onc's operationalization of the notion of causal connection ought not to depend on de facto frequencies. So one should link causal connection to probabilistic dependence "in the background circumstances." On the other hand, if the "circumstances" are *fully* specified (as in **DC**) and causes are, as I have assumed, deterministic – that is, necessary and sufficient in the circumstances – then it seems **OA** reduces (disastrously – see p. 42) to the view that *a* and *b* will be causally connected if and only if *a* is an INUS condition for *b*.

What one needs to avoid relying on the happenstance of actual frequencies of different circumstances and to avoid inducing probabilistic dependencies by conditionalizing on too much is to conditionalize on only a partial characterization of the circumstances. If one has an ordinary physiology and has not ingested neutralizing substances, then there will be a correlation between ingesting acids and feeling sick to one's stomach. Just as causal connections are relativized to a causal field (see p. 60), so the probabilistic dependencies to which they correspond must be relativized to some set of *background* circumstances.

One can also define causal independence:

> Tropes or events are causally *independent* if and only if they are distinct and not causally connected.

I asserted above that if one accepts **OA**, then effects of a common cause will be causally connected if cause and effect are causally connected. Here is the argument. Suppose that the only causal connection between $b$ and $c$ is their possession of a common cause $a$. One can prove that

$\Pr(B\&C) \neq \Pr(B)\cdot\Pr(C)$ whenever
1. $\Pr(B/A) \neq \Pr(B/{\sim}A)$,
2. $\Pr(C/A) \neq \Pr(C/{\sim}A)$,
3. $\Pr(B/A\&C) = \Pr(B/A)$,
4. $\Pr(B/C\&{\sim}A) = \Pr(B/{\sim}A)$, and
5. $0 < \Pr(A) < 1$.

The first two premises follow from the assumption that causes make some difference to the probability that their effects will occur. The third and fourth premises restate the assumption that the probabilistic dependence between **b** and **c** disappears when one controls for the common cause. The last premise says that the common cause is neither always nor never instantiated.

Actually one proves slightly stronger results. Suppose first that instead of the first two premises, we have $\Pr(B/A) > \Pr(B/{\sim}A)$ and $\Pr(C/A) > \Pr(C/{\sim}A)$. Then one can prove that $\Pr(BC) > \Pr(B)\cdot\Pr(C)$. Let

$\Pr(B/A) = r, \quad \Pr(B/{\sim}A) = r',$
$\Pr(C/A) = s, \quad \Pr(C/{\sim}A) = s', \quad \Pr(A) = p.$

Then:

1. $r > r'$ (premise)
2. $s > s'$ (premise)
3. $0 < p < 1$ (premise)
4. $sp(1-p)(r-r') > s'p(1-p)(r-r')$ (from 1, 2, 3)
5. $srp(1-p) - sr'p(1-p) > s'rp(1-p) - s'r'p(1-p)$ (from 4)
6. $srp - srp^2 - sr'p(1-p) > s'rp(1-p) - s'r'[(1-p)-(1-p)^2]$ (from 5)
7. $srp + s'r'(1-p) > srp^2 + s'rp(1-p) + sr'(1-p) + s'r'(1-p)^2$ (from 6)

8. $srp + s'r'(1-p) > [rp + r'(1-p)]\cdot[sp + s'(1-p)]$ (from 7)
9. Pr(B/A&C) = $r$ (premise)
10. Pr(B/C&~A) = $s$ (premise)
11. Pr(B&C) = Pr(B&C&A) + Pr(B&C&~A) (probability theorem)
12. Pr(B&C&A)/Pr(C&A) = $r$ (from 9)
13. Pr(B&C&A) = $srp$ (from 12)
14. Pr(B&C&~A) = $s'r'(1-p)$ (from 10)
15. Pr(B) = $rp + r'(1-p)$ (probability theorem)
16. Pr(C) = $sp + s'(1-p)$ (probability theorem)
17. Pr(B&C) > Pr(B)·Pr(C) (from 13, 14, 15, 16)

When Pr(C/A) < Pr(C/~A) and Pr(B/A) < Pr(B/~A), all the steps in the proof are just the same and Pr(BC) > Pr(B)·Pr(C). When Pr(C/A) > Pr(C/~A) and Pr(B/A) < Pr(B/~A) or Pr(C/A) < Pr(C/~A) and Pr(B/A) > Pr(B/~A), then the direction of the inequality in step 4 and all later steps is reversed and Pr(B&C) < Pr(B)·Pr(C). The proof is due to Reichenbach (1956, pp. 160–1). As Richard Scheines pointed out to me, the proof is trivial in the case of linear relations among continuous variables. Suppose $y$ and $z$ depend on $x$. Let $\rho_{xy}$, $\rho_{xz}$ and $\rho_{yz}$ be the three correlations and $\rho_{yz.x}$ be the partial correlation between $y$ and $z$ given $x$. If $\rho_{yz.x} = 0$, then $\rho_{yz} = \rho_{xy}.\rho_{xz}$, and so if $\rho_{xy}$ and $\rho_{xz}$ are not zero, neither is $\rho_{yz}$.

This proof supports the intuitive view that probabilistic dependencies among (properties of) effects of a common cause "are no accident." But the argument is not conclusive. The proofs apply only if there are linear relations among continuous variables or if the variables are dichotomous. If the common cause variable $x$ can have three values, while its two effects $y$ and $z$ are binary, then it is possible for there to be a probabilistic dependency between $x$ and $y$ and between $x$ and $z$ and for all three values of $x$ to screen off $y$ and $z$ but for $y$ and $z$ nevertheless to be probabilistically independent.[1]

In thinking about token causation among events, it may be reasonable to limit oneself to binary variables. But even in the case of binary variables, one might question the force of this proof, because one of the premises is that the causes screen off their effects, and cases in which causes fail to screen off their effects are common. In particular, Elliott Sober (1988) has proven the following two things:

1. If proximate common causes screen off their causes from their effects and their effects from one another (and all the probabilities are intermediate), then indirect common causes do not). One can use the same algebra. Let *r, r', s, and s'* be defined as before so steps 1 – 8 are unaffected. But now let $p$ be Pr(A/D), where $d$ is a cause of $a$ that (indirectly) causes $b$ and $c$ only through $a$ and is screened off from **b** and **c** by **a**. Then

[1] Richard Scheines pointed this out to me, and Andrew Banas has produced an example.

$Pr(BC/D) = srp + s'r'(1-p)$,
$Pr(B/D) = rp + r'(1-p)$, and
$Pr(C/D) = sp + s'(1-p)$.

So if **a** screens off **b** and **c** and the signs of Pr(B/A) - Pr(B/~A) and Pr(C/A) - Pr(C/~A) are the same, then **b** and **c** are positively probabilistically dependent conditional on **d**. When the signs differ, then **b** and **c** are negatively dependent on one another.

2. Suppose that *b* and *c* have two causes, *a* and *e*, that the probability of each kind of effect is an increasing function of the number of its causes present, and that **a&e**, **a&~e**, **e&~a**, and **~a&~e** all screen off **b** and **c**. Then it cannot be the case that **a** screens off **b** and **c**. One can use the same algebra again. Let

$Pr(B/A\&E) = r$, $Pr(B/A\&{\sim}E) = r'$,
$Pr(C/A\&E) = s$, $Pr(C/A\&{\sim}E) = s'$, and $Pr(E) = p$,

The assumption that the probabilities of **b** and **c** are increasing functions of the number of the causes of each present enables us to follow the same proof through step 8. A slightly different formulation of the screening-off assumption gives us

$Pr(B\&C/A\&E) = Pr(B/A\&E) \cdot Pr(C/A\&E) = rs$ and
$Pr(B\&C/A\&{\sim}E) = r's'$.
Thus $Pr(B\&C/A) = rsp + r's'(1-p)$.

Similarly

$Pr(B/A) = rp + r'(1-p)$ and $Pr(C/A) = sp + s'(1-p)$.

Substituting into step 8, one has the result that $Pr(B\&C/A) > Pr(B/A) \cdot Pr(C/A)$. The sign of the inequality is reversed if the two causes *a* and *e* promote one of their common effects and inhibit the other. Similar arguments show that **~a**, **e**, and **~e** also fail to screen off **b** and **c**. One can also invert the argument and show that if any of **a**, **~a**, **e**, or **~e** screens off **b** and **c**, then it cannot be the case that **a&e**, **a&~e**, **e&~a**, and **~a&~e** do.

Once one thinks about them, Sober's demonstrations are not surprising. Given the distal cause *d* (or its absence), the occurrence of *b* increases the probability that the proximate cause *a* will have occurred, and thus a probabilistic dependence between **b** and **c** remains. Given only one of two causes of *b* and *c*, the occurrence of *b* makes it more likely that its other cause will have occurred and hence that *c* will have. Although these arguments show that common causes often do not screen off their effects, they pose no problem for the claim that if (binary) effects of a (binary) common cause are probabilistically dependent on their common cause, then they are probabilistically dependent on one another. In all the cases in which Sober has demonstrated that some causes do not screen off their effects from one another, their effects are still probabilistically dependent on one another. Effects $b$ and $c$ of a common cause will fail to be probabilistically dependent on one another only when there is no common cause $a$ that satisfies the conditions of the original proof, and in Sober's cases, there is always some screening-off common cause.

Furthermore, suppose that $a$ causes $b$ and $c$, $b$ causes $f$, the implicit variables are binary, and the following probabilistic relations hold: (1) effect types are probabilistically dependent on their cause types, (2) **b** and **c** are probabilistically dependent, and (3) **b** and ~**b** screen off **f** and **c**. Then Pr(F/C) ≠ Pr(F/~C). Let

| | |
|---|---|
| Pr(F/B) = $u$, | Pr(F/~B) = $u'$, |
| Pr(B/C) = $p$ and | Pr(B/~C) = $p'$. |

Then

| | |
|---|---|
| 1. $p > p'$, $u > u'$ and all probabilities are intermediate | (premise) |
| 2. $u(p-p') > u'(p-p')$ | (from 1) |
| 3. $up + u'(1-p) > up' + u'(1-p')$ | (from 2) |
| 4. Pr(F/C) = $up + u'(1-p)$ | (premise) |
| 5. Pr(F/~C) = $up' + u'(1-p')$ | (premise) |
| 6. Pr(F/C) > Pr(F/~C) | (from 3, 4, 5) |

Premises 4 and 5 follow from the assumptions that **b** and ~**b** screen off **f** and **c**. If $p < p'$ and $u < u'$, the proof goes through exactly the same way. If $p > p'$ and $u < u'$ or vice versa, then the inequalities are reversed and **f** and **c** are negatively dependent. But in any case Pr(F/C) ≠ Pr(F/~C). The assumption that **b** and ~**b** screen off **f** from **c** is extremely plausible. Given the presence or absence of $b$, the occurrence or nonoccurrence of $c$ cannot possibly make it more or less likely that $f$ occurs. This result can of course be iterated on either the "$b$ branch" or "$c$ branch," and it establishes that remote effects of a common cause should be probabilistically dependent. The magnitude of the probabilistic dependence between **f** and **c** ($u(p-p')-u'(p-p')$) will always be less than the magnitude of the probabilistic dependence between **b** and **c** ($p-p'$), and the probabilistic dependence between remote effects of a common cause will typically be undetectable.

In many of the cases in which effects of a common cause are not probabilistically dependent, their independence will itself have a causal explanation. The simplest case in which **b** and **c** are probabilistically independent even though $b$ and $c$ are effects of a common cause, $a$, is one in which there is also a direct causal influence of $b$ on $c$ or of $c$ on $b$, which counteracts the dependence that would otherwise obtain between **b** and **c**. In such a case the probabilistic independence between **b** and **c** results from the existence of a specific causal structure with links of just the right strength. It thus seems appropriate to say in such a case that $b$ and $c$ are connected by some fact of causation. Since *b* and *c* are in such cases also related as cause and effect, one has indeed no other choice. Another simple case in which effects of a common cause will not be probabilistically dependent arises when **b** (or **c**) is not probabilistically dependent on **a**, even though *a* causes *b* and *c*. In such a case it seems that one should be as willing to regard *b* and *c* as causally connected to one another as one is willing to regard *b* and *c* as causally connected to *a*. Exactly the same explanation that accounts for the probabilistic independence between **a** and **b** or between **a** and **c** accounts for the probabilistic independence between **b** and **c**. There is good reason to maintain that if *a* and *b* are effects of a common cause, then they are causally connected.

## 4.2* A Necessary Condition

Suppose we accept two further principles concerning causation:

**D** (*Distinctness*) If *a* causes *b*, then *a* and *b* are distinct.

**T** (*Transitivity*) If *a* causes *b*, *b* causes *c*, and *a* and *c* are distinct, then *a* causes *c*.

As explained in chapter 2, transitivity obtains only if the causal relations obtain in virtue of the properties used to refer to *a*, *b*, and *c* (p. 27). The clause in **T** requiring *a* and *c* to be distinct guarantees that there will be no conflict between transitivity and distinctness.[2] The meaning of distinctness was discussed on p. 59. The *irreflexivity* of causation follows from distinctness. I shall treat distinctness as implicit in causation and shall not mention it again as a separate premise. If *a* causes *b*, then *a* and *b* are distinct, and so *a* cannot cause itself.

One can also formulate:

**A** (*Asymmetry*) If *a* causes *b*, then *b* does not cause *a*.

Although **I** and **CC** imply **A**, **CP** is formulated so as to leave open the possibility that causation is sometimes symmetrical. Like David Lewis's account (1973a, p. 170), it treats the question of whether causation is ever symmetrical as an empirical question. In chapter 12, I shall argue that asymmetry should be built into the theory of causation.

The first condition to be proved is:

[2] Frankel criticizes this provision as an ad hoc or question-begging maneuver introduced in order to admit mutual causation without having also to admit self-causation (1986, p. 362). His reasoning seems to be that if one countenances mutual causation, then one must give some reason why one should not also countenance self-causation. The reasons lie in the first instance in the intuitions supporting the distinctness requirement and in the second instance in the theoretical fruitfulness of the requirement.

**N** (*Necessary condition*) If *a* causes *b*, then *a* and *b* are causally connected, and everything causally connected to *a* and distinct from *b* is causally connected to *b*.

In what follows let "*a* cc *b*" abbreviate "*a* and *b* are causally connected."

**Theorem 4.1: T** and **CC** entail that if *a* causes *b*, then everything causally connected to *a* and distinct from *b* is causally connected to *b*.

Proof: **CC** establishes that *a* cc *b*. Suppose *c* cc *a* and *c* is distinct from *b*. Given **CC** there are three ways that *c* cc *a*:

1. *c* is a cause of *a*. Then by transitivity *c* is a cause of *b* and by **CC**, *c* cc *b*.
2. *a* is a cause of *c*. Then *c* and *b* are distinct effects of a common cause and so by **C**, *c* cc *b*.
3. Some *d* is a cause of *a* and *c*. If *d* is distinct from *b*, then by transitivity *d* is also a cause of *b*. Then *b* and *c* are effects of a common cause and by **CC**, *b* cc *c*.

One loose end remains. Suppose in case 3 that the common cause *d* of *a* and *c* is not distinct from *b*. Must it be the case that *b* cc *c*? This question cannot be broached without a general consideration of event fusions, and at this point I will rest on the plausibility of taking the fact that *c* is caused by a cause of *b* or by something that is not distinct from *b* as implying that *b* cc *c*.

**Theorem 4.2:** Given **T** and **CC**, if *a* is distinct from *b* and not causally connected to *b*, then it is not causally connected to any cause of *b*.

Proof: Let *c* be any cause of *b*. Given **N**, if *a* cc *c* and *a* is distinct from *b*, then *a* cc *b*. So if *a* is distinct from *b* and ~(*a* cc *b*) then ~(*a* cc *c*).

## 4.3* Independence Conditions

The independence condition can be restated:

**I** (*Independence condition*) If *a* causes *b* and *b* does not cause *a*, then there is some *c* such that *c* causes *b* and *a* and *c* are causally independent.

Given **CC**, this more compact formulation is equivalent to the version given on p. 64. This independence condition is only one of many claims about causal independence one can make. **CC** itself implies:

$\mathbf{I}_0$ If *a* and *b* are distinct and *a* does not cause *b*, *b* does not cause *a*, and *a* and *b* are not effects of a common cause, then *a* and *b* are causally independent. Furthermore, if *a* and *b* are causally independent, every cause of *a* is causally independent of every cause of *b*.

If some cause of *a* is causally connected to some cause of *b*, *a* and *b* must have a common cause. So if *a* and *b* are not causally connected, none of their causes can be causally connected either.

**I** implies the following more restrictive independence claims:

$\mathbf{I}_w$ (*Weak independence condition*) If *a* causes *b* and *b* does not cause *a*, then *b* has a cause that is causally independent of *a*.

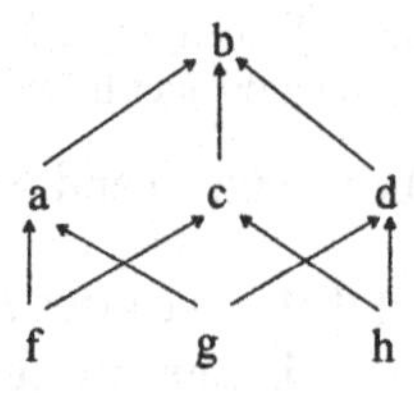


**Figure 4.7:** Independence and open back paths

> $\mathbf{I}_{cc}$ (*Independence and common causes*) If *a* and *b* are effects of a common cause and neither causes the other, then *a* has a cause that is causally independent of *b* and *b* has a cause that is causally independent of *a*.

As demonstrated by figure 4.5, $\mathbf{I}_w$ does not imply $\mathbf{I}_{cc}$. If every cause has multiple effects, then $\mathbf{I}_{cc}$ entails $\mathbf{I}_w$, and one could replace **I** with $\mathbf{I}_{cc}$.

It is also worth calling attention to the following consequence of **CC** and **I**.

**Theorem 4.3: CC, I, and T entail A.**

Proof: If *a* causes *b*, then, by theorem 4.1, nothing causally connected to *a* is independent of *b*. So *a* has no cause that is causally independent of *b*. **I** then implies that *b* does not cause *a*.

There are also stronger independence conditions than **I**. Consider in particular:

> $\mathbf{I}_s$ (*Strong independence condition*) Every event *b* that has at least one cause has another cause, *f*, such that for all events *a*
> 1. If *b* does not cause *a* and there is a causal path that does not go through *f* from *a* to *b* or from some cause of *a* to *b*, then *a* and *f* are causally independent.
> 2. If *b* causes *a*, then there is no path that does not pass through *b* from *f* to *a* or from any cause of *f* to *a*.

Consider the set *S* of all those events that would be causally connected to *b* if its independent cause *f* never occurred. Strong independence says that when *f* occurs, it must not be causally connected to any events in *S* except those that are effects of *b*. Given transitivity, *f* will of course be causally connected to the effects of *b*, but $\mathbf{I}_s$ also says that *f*'s only connection to the effects of *b* must be via causing *b*. As illustrated in figure 4.4 (p. 64), $\mathbf{I}_s$ is stronger than **I**. *b* has only two causes *a* and *c*, which are causally connected as effects of the common cause *d*, even though they both have independent causes. **I** is satisfied but $\mathbf{I}_s$ is not.

$\mathbf{I}_s$ obviously entails **I**. It is worth formulating $\mathbf{I}_s$ explicitly, because it helps clarify the content of **I** and because versions of the strong independence condition are assumed in alternative accounts of causation. $\mathbf{I}_s$ entails a variant of Cartwright's "open back path" condition.

**OBP** (*Open back path condition*) Every cause *a* of *b* that has any causes has at least one cause *d* such that the only path from *d* to *b* is via *a*.[3]

Suppose that *a* causes *b*. Given $\mathbf{I}_s$, if *a* has any causes, then it has a cause *c* that is independent of all other causes of *a* that are not causes of *c*, no cause of *c* will cause *a* except via causing *c*, and neither *c* nor its causes will cause *b* except via causing *a*. Thus all causes of *b* will have open back paths. **OBP** is itself an independence assumption. It is weaker than $\mathbf{I}_s$ but neither stronger nor weaker than **I**. As figure 4.7 illustrates, **I** does not imply **OBP**. None of *a*, *c*, or *d* has an open back path; yet **I** holds. Nor does **OBP** imply **I** (or hence $\mathbf{I}_s$), since **OBP** does not imply that events must have multiple causes and since **OBP**, unlike **I**, is satisfied by a structure such as that shown in figure 4.5, where *a* and *b* are each caused by *c* and *d*, which are independent of one another.

Both **I** and $\mathbf{I}_s$ imply that causal explanations of finite-state closed deterministic systems will not be possible. Although *n* variables that are independent of one another can, without violating the independence condition have $n(n-1)/2$ effects, eventually a finite-state closed system will have more effects than can be traced back to any finite number of initial causes.[4]

I have little to say in defense of **I** here except to point out that there are no arguments analogous to those in §4.1* showing that if causes are probabilistically dependent on their effects, then causes of a common effect must be probabilistically dependent. The arguments there showing that effects of a common cause are probabilistically dependent relied on the premise that effects of a common cause have a screening-off common cause. There is no justification for a parallel premise concerning causes of a common effect. The consequence of conditioning on a common effect is commonly to *induce* a probabilistic dependence among causes. For example, the light in my upstairs hall is controlled by both a switch upstairs and a switch downstairs.[5] If exactly one of the switches is up, the light is on; otherwise it is off. Whether the upstairs switch is on is causally and probabilistically independent of whether the downstairs switch is on. Conditional on the upstairs light being on or on effects of its being on, the positions of the switches are correlated (see §10.2 and §12.5). The induced probabilistic dependence of

[3] This is not exactly Cartwright's condition (1989, p. 33); her formulation requires that causes be earlier in time than their effect.

[4] The maximum number of effects that **I** permits from any finite set of causes that are not causally connected to one another occurs when each effect has two causes. Since there are $n(n-1)/2$ distinct pairs of causes, there can be no more than $n(n-1)/2$ effects without violating independence. I am indebted to Jean-François Laslier for calling my attention to these combinatorial matters.

[5] I am indebted to Malcolm Forster for this example.

causes of a common effect, given that effect, blocks any demonstration that causes of a common effect will be unconditionally dependent.

### 4.4* Derivation of CP and Other Results

**Theorem 4.4: CC** and **I** imply

> **S** (*Sufficiency*) If *a* cc *b* and everything causally connected to *a* and distinct from *b* is causally connected to *b*, then *a* causes *b*.

Proof: Suppose *b* causes *a* or that *a* and *b* are effects of a common cause only. Then by **I**, some cause of *a* is independent of *b*, and by **CC**, *a* cc *b* and something distinct from *b* is causally connected to *a* and not to *b*. So if everything causally connected to *a* and distinct from *b* is causally connected to *b*, either ~(*a* cc *b*) or *a* causes *b*. So if *a* cc *b* and everything causally connected to *a* and distinct from *b* is causally connected to *b*, then *a* causes *b*.[6]

The conjunction of **N** and **S** provides the following simple truth condition:

> **CP** (*Independence theory of causal priority*) *a* causes *b* if and only if *a* and *b* are causally connected and nothing causally independent of *b* is causally connected to *a*.[7]

Since by theorem 4.4 **CC** and **I** entail **S** and by theorem 4.1 **T** and **CC** entail **N**, the following theorem follows trivially:

**Theorem 4.5: T, CC,** and **I** entail **CP**.

Actually one can say something more:

**Theorem 4.6: T, CC,** and **I** entail **CP′**.

[6] This provides a simple alternative proof to the one provided by Nancy Cartwright (1989, pp. 37–8) that $\mathbf{I_s}$ entails that *a* causes *b*.

[7] This theory was suggested to me in long and helpful conversations with Douglas Ehring, who has developed the same basic intuition in a different direction. Ehring maintains, "*c* is *causally prior* to *e* if and only if (A) (1) *c* and *e* are causally connected, and (2) there is some condition of the causal connection between *c* and *e* that is not connected to *c* and is causally connected to *e*, and there is no condition causally connected to *c* but not to *e* or (B) *c* is causally connected to some event *f*, and *c* is a direct condition of a causal connection between *f* and *e*, and *f* is causally prior to *e* (1997, pp. 148–9)."

This is not an exact quotation, because Ehring states the sufficient conditions separately, before claiming that their disjunction is necessary (1997, p. 150). Two events are "causally connected" in Ehring's usage, if they are direct cause and effect. Ehring analyzes causal connection in terms of trope persistence, but his account of causal asymmetry is independent of that analysis. Intuitively some factor *f* is a "direct condition of a causal connection" between *c* and *e* if and only if *c* and *e* would not have been causally connected if *f* had not obtained. Ehring is articulating the same basic intuition developed here: The priority of causes to effects rests on the independence of causes.

**CP′** (*Revised independence theory of causal priority*) *a* causes *b* if and only if *a* is causally connected to *b*, everything causally connected to *a* and distinct from *b* is causally connected to *b*, and something causally connected to *b* is independent of *a*.

Given **CC** and **I**, **CP** and **CP′**are obviously equivalent. **CP′**, unlike **CP**, entails all by itself that causation is asymmetric, and it is consequently harder to relate to alternative accounts of causation, such as David Lewis's, that leave open the possibility of symmetric causation. For this reason I shall focus during most of this book on **CP**. There is a sense in which **CP** and **I** are, given **CC**, equivalent:

**Theorem 4.7: T, CC, CP,** and **A** (asymmetry) imply **I**.

Proof: Assume *a* causes *b* or that *a* and *b* are effects of a common cause only. Then by **CC** and **A**, *a* cc *b* and ~(*b* causes *a*). By **CP** there is some *c* distinct from *a* such that *c* cc *b* and ~(*c* cc *a*). Given *c* cc *b* and **CC**, either (i) *b* causes *c*, (ii) *c* causes *b*, or (iii) some *d* causes *b* and *c*. (i) is ruled out by ~(*c* cc *a*). In (ii) *b* has a cause that is causally independent of *a*. In case (iii) by theorem 4.2 ~(*a* cc *d*). So if *a* causes *b* or *a* and *b* are only causally connected as effects of a common cause, then *b* has a cause that is causally independent of *a*.

In the same way one can prove that **T**, **CC**, and **CP′** entail **I**. The account of causation given in this chapter is equivalent to **T**, **CC**, and **I**, and the most controversial part is **I**.

# 5

# Agency Theory

One crucial asymmetry of causation concerns *action*. Knowing the causes of an event helps one to bring it about. Knowing its effects does not. Knowing what things are related to an event as effects of a common cause does not help one to bring it about either. Causes are *means* and *tools*. People can use them to bring about their effects. This observation suggests a theory of causal asymmetry – something like: *a* causes *b* if and only if *a* can be used as a means to bring about *b*. This view has appealed to scientists and to statisticians. For example, the economist Guy Orcutt writes, ". . . we see that the statement that . . . $z_1$ is a cause of $z_2$, is just a convenient way of saying that if you pick an action which controls $z_1$, you will also have an action which controls $z_2$" (1952, p. 307).

Such a theory links the asymmetry of causation to a method of determining which way the causal arrow points: If there is a correlation between **a** and **b** and one wants to know what the causal connection is, *intervene* to find out. *a*'s cause *b*'s just in case human interventions that bring about tokens of kind **a** lead tokens of kind **b** to occur as well, while human interventions that bring about *b*'s do not lead to tokens of **a** occurring. This chapter will be concerned with this theory of causal asymmetry. Following Price and Menzies, I shall call it "the agency theory." It might also be called "the manipulability theory."

The claim that *a* causes *b* if and only if *a* can be used as a means to bring about *b* is crude. Taken literally it implies that no past events have causes, for nothing can be used as a means to bring about what is already past. The "can" in the formulation suggests some sort of modal claim. The last paragraph spoke of one *kind* of event causing another *kind* of event. Refinement is called for.

An agency theory can be formulated either as a counterfactual theory concerning causal relations between tokens or as a noncounterfactual theory concerning causal relations between types. Each version has its own advantages and drawbacks. Because the counterfactual formulation states a theory of causal relations among tokens, it is immediately comparable to the Humean and independence theories. But the token formulation has the disadvantage that it relies on counterfactuals, which bring many problems with them. The type formulation avoids counterfactuals, but in explaining how one *kind* of event can be a cause of another, it brings in counterfactuals through the back door.

This chapter is devoted to a type version of agency theory. Counterfactual token-level formulations will be considered in chapter 7. Variants of the arguments in this chapter apply to a counterfactual formulation. In this chapter, I shall take agency theories to say something like: "Events of kind **a** cause events of kind **b** in circumstances *K*, if and only if *a*'s can be used in *K* to bring about *b*'s" or, in the case of quantitative relations, "*y* depends on *x* in circumstances *K* if and only if in *K* one can affect the value of *y* by intervening to influence the value of *x*."

Claims about causal relations among types are, I maintain, generalizations of counterfactual claims about causal relations among tokens: Events of kind **a** cause events of kind **b** in circumstances *K* if and only if in *K* every event with property *A* that might occur would cause an event with property *B*. Claims about causal relations among variables are generalizations of such generalizations: *y* depends on *x* in circumstances *K* if and only if a whole set of such type-causal relations obtains: *y* depends on *x* in circumstances *K* if and only if, for all values of *x* within some range, possible instantiations of the property of having some value $x^*$ in circumstances *K* would cause instantiations of the property of having the value $y^*$.

Causation is not a relation among properties. Without tokens arrayed in particular ways, there is no time ordering, no manipulability, no asymmetrical counterfactual dependence. There is no asymmetry of independence either, because the notion of causal connection contains an ineliminable reference to a causal "field" or to background circumstances. As pointed out in §2.4, knowledge of a law, such as the ideal gas law, does not tell one whether the pressure of gases ever *causally* depends on their temperature. Knowledge of explicitly causal "laws" on the other hand, such as laws governing embryonic development in biology, refer to or quantify over particular systems. Without reference to some "system" or set of circumstances, there are no causal asymmetries and hence no causation. Thus causal relations among types are generalizations of causal relations among tokens. The argument sketched in this paragraph is developed more carefully in §5.1* and §5.2*. On this construal of type-causal claims, a type-level version of an agency theory offers truth conditions for generalizations of the form "In conditions *K*, all *a*'s would cause *b*'s."

## 5.1 Formulating an Agency Theory of Causation

Let us now see how one might formulate an agency view more precisely. Hermann Wold states the "facts" of manipulability as follows: "[I]f the experiment reveals that an observed variable varies systematically as the controlled variables are allowed to vary, this relationship is a type case of a causal relation" (1954, p. 165). G. H. von Wright writes, "What makes $p$ a cause-factor relative to the effect factor $q$ is, I shall maintain, the fact that by *manipulating p*, i.e. by producing changes in it 'at will' as we say, we could bring about changes in $q$."[1] As a first approximation, agency theorists maintain that the value of $y$ is dependent on the value of $x$ in circumstances $K$ if and only if, given $K$, if human interventions change the value of $x$, then the value of $y$ changes but not vice versa.

A little reflection reveals that this formulation is inadequate. It could be the case that an intervention that changes the value of $x$ from $x^*$ to $x^{*\prime}$ leaves the value of $y$ unchanged, because $y = f(x, z)$ and $f(x^*, z^*) = f(x^{*\prime}, z^*)$. For example, the range of a cannon obviously depends on the angle at which the cannon is fired, but it is not the case that interventions that change the angle always change the range. The possible relationship between birth control pills and thrombosis provides a different kind of example. An intervention consisting in giving a woman birth control pills may leave the probability that she experiences blood clots unchanged because its causal influence along different paths cancel out. Rather than saying that every intervention that changes the value of $x$ is accompanied by a change in the value of $y$, one must say something like, "The value of $y$ causally depends on actions that change the value of $x$."

Most of the argument of this chapter will go through if one takes agency theory as stating that $a$'s affect $b$'s (given $K$) if all actions that affect $a$'s affect $b$'s, while some actions that affect $b$'s leave $a$'s unaffected. In a previous treatment of agency theory (Hausman 1986), I employed such a formulation, but it fails to make explicit the most important concept in agency views: the concept of a *human intervention*.

By employing the notion of a human intervention, one can provide a somewhat more intuitive formulation of an agency theory:

> **AT$_g$** (*Type-level agency theory*) $y$ depends on $x$ in circumstances $K$ if and only if $x$ and $y$ are distinct and in $K$ every intervention with respect to $x$ affects $y$.[2]

[1] 1975, p. 107. In addition, see Collingwood (1940, pp. 296–7); Gasking (1955, p. 483–7), and Freudlich (1977, p. 476).

[2] The circularity is more blatant than it needs to be. See p. 106. Another version that is discussed in chapter 11 holds (roughly) that $x$ causes $y$ if and only if $y$ is a nonconstant function of $x$ that remains invariant as the value of $x$ is altered by intervention. **AT$_g$** states a truth condition for "$y$ depends on $x$ directly or indirectly." As Malcolm Forster showed me, agency theorists can use the notion of an intervention to determine whether causal relations are direct (relative to the variables under consideration). Suppose one knows that $y$ depends on $x$ and $z$ and that $z$ depends on $x$. One wants to find out whether $y$ depends on $x$ both directly and indirectly or only indirectly via $z$. One can find out by fixing $z$ with one intervention and changing the value of $x$ with another. See §12.4.

Agency theories say that *x* causes *y* if and only if *x* can be used to manipulate *y*. *x* can be used to manipulate *y* if and only if the value of *y* depends on human interventions that do nothing except set the value of *x*. So *x* causes *y* if and only if interventions that set the value of *x* affect the value of *y*. If one wants to build asymmetry into the agency account, one can add the clause "and some interventions with respect to *y* do not affect *x*." But an agency theorist can leave it to nature to say whether causation is asymmetric. I have stated $\mathbf{AT}_g$ in terms of variables, but one could just as easily restate it in terms of properties or kinds of events. I added a "g" subscript as a reminder that this is a formulation in terms of types or generalizations.

Despite the evident appeal of agency views, they face three serious difficulties. First, the claims of agency theorists have an unreasonably limited scope. It is not true that *x* causes *y* if and only if human interventions that affect *x* always affect *y*. Suppose that among two effects of a common cause, only one of them is subject to manipulation. For example, let *x* be a dichotomous variable whose values are the presence or absence of a storm and let *y* be the earlier reading of a barometer. Since there are (or were until recently), no human interventions that affect *x*, all interventions that affect *x* also affect *y*. $\mathbf{AT}_g$ mistakenly implies that the storm onset causes the earlier barometer reading. Fusion reactions in the sun provide light and heat on Earth, even though there are no human interventions that can directly manipulate the fusion reactions within the sun. In response to these difficulties, agency theorists have suggested that there are analogies between cases where manipulation is possible and cases where it is not and that people use those analogies to extend their causal judgments to cases where manipulations are not possible.

Second, agency theories appear to be circular. This is true of all versions of agency theory, including $\mathbf{AT}_g$, which relies on the notion of a human intervention and on the obviously causal notion of "affecting." I do not think that any noncircular analysis of causation is possible, and circularities are not necessarily uninformative or unimportant, but in this case the circle seems distressingly small.

Third, the anthropomorphism of agency theories is disturbing. Does the asymmetry of causation depend only on human perspectives? Would there be no causal asymmetries if there were no agents? Could agents call anything they liked causes and effects? What is it "in the world" that our perspective as agents "grabs onto"?

## 5.2 Menzies and Price's Reformulation and Defense

Peter Menzies and Huw Price have recently defended a sophisticated reformulation of an agency view of causation (Menzies and Price 1993; see also Price 1991; 1993). They hold that causation is a "secondary quality" – an extrinsic property of relations among events that is explained by its connections to human experience. In the same way that an object is red just in case it would look red to a normal observer under standard conditions, so "an event *A* is a cause of a distinct event *B* just in case bringing about the occurrence of *A* would be an effective means by which a free agent could bring about the occurrence of *B*" (Menzies and Price 1993, p. 187). Menzies and Price often formulate their theory counterfactually, but since they spell out the notion of "effective means" in terms of "agent probabilities" – the probability of **b** given that **a** comes about via an agent's intervention (1993, p. 190) – it seems that they are committed to a type theory; and in a type formulation, one must specify the circumstances. Most of their arguments carry over naturally into the deterministic and noncounterfactual context in this chapter, and I believe that $\mathbf{AT}_g$ is a reasonable type-level restatement of what their views imply in deterministic circumstances. I shall return briefly to their position in §7.1 and §7.2.

In response to the objection that agency theories are circular, Menzies and Price argue that if the notion of a human intervention or of an action *ab initio* is taken as a primitive, then an agency theory need involve no circularity: Agent probabilities require no causal notions in their definition (other than the notion of realizing some event via a human intervention), and causal relations are then defined in terms of agent probabilities (1993, p. 194). I question whether it is reasonable to take the notion of an intervention as a primitive and whether agent probabilities avoid relying on causal notions.

Menzies and Price cope with the remaining difficulties by deploying the analogy with secondary qualities. They point out that the objection that causal relations may obtain among things that are not subject to human manipulation has the same structure as the objection that things may be red even if there are no observers to whom they look red. Menzies and Price maintain that the answer to the objection is the same in both cases. Objects that are red but never appear red may have the same physical properties as objects that appear red. They possess the objective "basis" for the disposition to appear red to human observers in normal lighting. Similarly, things that cannot be manipulated may have the same objective relations to their effects as things that can be. Menzies and Price thus write,

> [W]hen an agent can bring about one event as a means to bring about another, this is true in virtue of certain basic features of the situation involved, these features being essentially noncausal though not necessarily physical in character. . . . In its weakened form, the agency account states that a pair of events are causally related just in case the situation involving them possesses intrinsic features that *either* support a means-end relation between the events as is, *or* are identical with (or

closely similar to) those of another situation involving an analogous pair of means-end related events. (1993, p. 197)

This response resembles the familiar line that causal claims concerning unmanipulable things involve analogical extensions of causal judgments concerning manipulable things, but at the beginning of the quotation Menzies and Price make a different point. They argue that the asymmetry of manipulability is true "in virtue of certain basic features of the situation involved." Causation derives from agency but extends beyond circumstances in which agency is possible because the asymmetry of agency is linked to "basic features" that may be true of circumstances even when human interventions are not possible.

This line of thought answers the other objections too. Even though every human intervention that affects the onset of a storm affects a barometer's reading (because no interventions affect the onset of a storm), objective features establish that barometer readings and storm onsets are related as effects of a common influence. Similarly, in response to the anthropomorphism objection, Menzies and Price can insist that making references to human agents is consistent with recognizing that there are extra-human factors in virtue of which things have colors or events stand in causal relations.

According to this defense, **b** may depend on **a** even when realizing *a*'s is not an effective way, nor indeed any way at all, of bringing about *b*'s. Consequently, in the passage quoted above, Menzies and Price revise the agency view. A compact noncounterfactual formulation of the view that results would read as follows:

**a** influences **b** in circumstances *K* if and only if the intrinsic features of **a** and **b** in *K* are in relevant respects the same as the intrinsic features of event types **d** and **e** in circumstances *G* in which *d*'s are effective means of bringing about *e*'s.

When *a*'s are manipulable in circumstances *K*, then of course *a*'s themselves are effective means for bringing about *b*'s.

Why should one take this revised account as an agency theory? Agency seems to play its part in identifying the intrinsic conditions in virtue of which causal relations obtain. Whether causal relations obtain depends on whether those conditions are met: **a** affects **b** in circumstances *K* if and only if certain intrinsic conditions obtain. Agency plays an important role in explaining how one acquires causal notions and how one finds out what causes what. But does agency have any role in a theory of what causation is?

Agency would be irrelevant to the analysis of causation, if the intrinsic features in virtue of which *a*'s affect *b*'s were simple and uniform. For example, someone who combined an agency view such as Menzies and

Price's with a Humean theory of the intrinsic conditions in virtue of which causes can be used as means to bring about their effects would be committed to a Humean theory of causation, not to an agency theory. But suppose that the existence of causal relations depended on different features in different cases. Leaving out reference to agency, one could then say that $a$'s influence $b$'s in circumstances $K$ if and only if $O_1$ or $O_2$ or . . . or $O_n$, where each $O_i$ is a set of intrinsic conditions in virtue of which properties, types, or variables depend on one another. And it might also be that every such list was incomplete. (In just the same way reference to the disposition to appear red may not be dispensable in the theory of colors if there is no simple physical property of surfaces in virtue of which they appear red.) If the only feature linking the various $O_i$'s to one another is that each obtains in circumstances where one event can be used as a means to bring about another, then it would seem that the theory remains an agency theory. Similarly, one would have an agency theory if all such lists were incomplete and additions were always justified in terms of agency. The only intelligible general characterization of the circumstances in which causal relations obtain would be in terms of agency. But I maintain that there is in fact a general characterization of the circumstances in which causal relations obtain in terms of independence relations.

Whether agency is a stepping stone toward discovering the nature of causation or whether it is a constituent of the notion of causation itself thus seems to depend on whether there is any intelligible, systematic, and reasonably unified characterization of the intrinsic features in virtue of which causal relations obtain. Agency theory seems like a default position; one is left with it if all other theories of causal priority fail. Thus Price argues that an asymmetry of the sort that **CP**, the independence theory, exploits "is not a sufficiently basic and widespread feature of the structure of the world to constitute the difference between cause and effect" (1992, p. 502). In particular, he argues that this asymmetry breaks down in the microscopic realm, while the asymmetry of agency (referring as it does to the doings of macroscopic beings) does not (1992, p. 516; see also §7.3).

### 5.3 Manipulability and Independence

This elaboration of Menzies and Price's defense of agency theory fails, because agency theories break down whenever the independence theory breaks down. Given further presuppositions *to which agency theorists subscribe*, $\mathbf{AT_g}$ entails a type-level version of **CP**. I shall also show that $\mathbf{AT_g}$ is deducible from a definition of an intervention coupled with the assumption that interventions are possible.

The extent to which $\mathbf{AT_g}$ captures the intentions of agency theorists

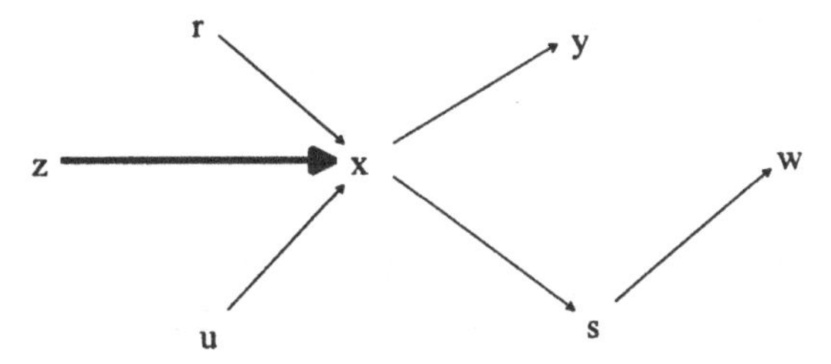


**Figure 5.1:** Human interventions

depends on how the notion of a human intervention is spelled out. The basic idea, illustrated in the normal graph of figure 5.1, is that there is a human intervention that affects only the value of *x* if and only if the value of the intervention variable *z* is a kind of human action that directly affects the value of *x* and has no causal connections to any of the other variables apart from directly influencing the value of *x*. In particular, one can idealize and suppose that there are no arrows into *z* and only one arrow out of *z*.

Interventions may do more than manipulate variables. They may change the causal structure and even alter the direction of causal relations. Most interventions can nevertheless be regarded as variable-setting interventions, and agency theorists are typically thinking in terms of such interventions. I shall not address the problem of whether all interventions can be represented as setting the value of variables.

What people *take* to be variable-setting interventions do not always satisfy the definition of an intervention offered above. When they do not, experimenters may mistakenly believe that their actions are causally influencing the variables they are studying when in fact their actions are determined by those variables or by other determinants of them. One sorts experimental subjects randomly into experimental and control groups to help ensure that what one takes to be an intervention is truly an intervention. If one's sorting of rats, for example, were affected by something correlated with their propensity to develop bladder cancer, one's administration of saccharin would depend on this preexisting propensity. One would not be truly intervening in feeding the rats saccharine, and one would not be justified in concluding that saccharin increases the incidence of bladder cancer in rats from the increase in cancer subsequent to saccharine consumption. In performing experiments to determine causal order, one takes precautions to make sure that one is truly intervening.

In order to draw conclusions concerning the results of interventions, one must also assume that there are intervention variables, values of which can affect the values of each variable separately. One might say that one must assume that interventions are "possible." This terminology may be puzzling, since there is nothing explicitly modal in the type-level formulations. Yet I would argue that it is possible to intervene if and only if there is *in fact* a causal structure containing an intervention or policy variable satisfying the definition of an intervention. Such a variable exists if and only if an (abstract) intervention is possible. Interventions change the value of intervention variables (from some "off" value to some "on" value). If there were not *in fact* an "opening" for an intervention – which is all that an intervention variable abstractly is – then no interventions would be possible. This line of thought is pursued further in §7.5 and 7.2*.

Given the transitivity of causation and a trivial technical condition, one can deduce this particular formulation of agency theory – $\mathbf{AT_g}$ – from the definition of an intervention and the assumption that interventions are possible. This is theorem 5.1, and it shows that in the circumstances in which interventions are possible, there is nothing in agency theory to which defenders of Humean or independence views should object. Agency theory adds little, and in any case human interventions are not always possible. The only reason to regard $\mathbf{AT_g}$ as a competing theory of causal asymmetry rather than as merely an implication of the possibility of intervention is its purported "default" status.

This reason to regard agency theory as a competing theory of causal asymmetry is undermined by a second theorem. Theorem 5.2 says that $\mathbf{AT_g}$, the definition of an intervention, and type-level reformations of transitivity and the connection principle entail a type-level reformulation of the independence theory (which I call $\mathbf{CP_g}$). This entailment of $\mathbf{CP_g}$ by $\mathbf{AT_g}$ means that if the independence theory is false, then either the definition of an intervention, the connection principle, transitivity, or agency theory itself must be false. It is hard to see how one could defend agency theory by pointing an accusing finger at the definition of a human intervention. Since it is difficult to reject transitivity (see pp. 27, 107, and 196–7) or the connection principle, it is hard to accept agency theory without also accepting the independence theory. The problems of an objective theory such as $\mathbf{CP_g}$ thus do not constitute an argument for agency theory. Agency theory faces all the same problems as well as further difficulties arising from the constraints on interventions.

These demonstrations constitute arguments for the independence theory. When interventions are possible, $\mathbf{AT_g}$ adds nothing to $\mathbf{CP_g}$, and if $\mathbf{CP_g}$ were not generally correct, at least in the domain in which human action and intervention are possible, then it would not be true that one could use causes to manipulate their effects and not vice versa. $\mathbf{CP_g}$ inherits the plausibility of agency theories, while overcoming their parochial and anthropomorphic character. It opens the circle of agency theory without taking the notion of a free action as a primitive. $\mathbf{CP_g}$ is more general than $\mathbf{AT_g}$, and it fills in the "intrinsic features" that agency theorists invoke to explain how there could be causal relations among things that are not subject to human manipulation.

### 5.4 Abstract Intervention Theory

Agency theorists take interventions to be human interventions. But one might explore the consequences of the more abstract notion of an intervention that results when one drops the requirement that interventions be human actions. An intervention with respect to some variable $x$ is simply a variable upon which $x$ directly depends that has no causal relations to any of the other variables except those that follow from its effect on $x$. The structure is the same as in figure 5.1. The only difference is that the intervention need not be a human action. This is no small difference. On the contrary, in substituting the abstract notion for the notion of a specifically human intervention, one is taking a big step away from the views of agency theorists and rendering the theory much less anthropomorphic.

If one drops the reference to human actions, then the wording of the assumption that interventions are possible and of agency theory do not need to be changed, but their content changes. Given the abstract notion of an intervention, it is no longer obviously true that interventions are not always possible and that the necessary condition stated by $\mathbf{AT}_g$ is false. The two theorems discussed above also continue to hold. Agency theory still follows from the assumption that interventions are possible. Stripped of specifically anthropomorphic claims about human action, the central metaphysical principle to which defenders of agency theories are committed is that abstract interventions are possible.

The two theorems mentioned above – (1) that agency theory, transitivity and the connection principle entail a type formulation of the independence theory and (2) that the assumption that interventions are possible entails the agency view – imply that one can derive $\mathbf{CP}_g$ from the connection principle, transitivity, and the assumption that interventions are possible. But to derive **CP** in chapter 4, I needed the independence condition (**I**), and theorem 4.7 states that the connection principle, transitivity, and **CP** imply **I**. So, it appears that the existence of intervention variables implies a type-level version of the independence condition. In fact, theorem 5.6 states that given the connection principle, asymmetry, and one further technical condition, the existence of intervention variables implies a type-level version of the *strong* independence condition.

Intervention variables are independent variables that vindicate the independence condition. Defenders of agency theories would thus be ill-advised to attack the independence condition. For they are committed to a stronger condition. If the independence condition fails, then intervention variables do not always exist, and a type-level formulation of an agency account also fails.

It might seem that the close relation between interventions and strong independence is a two-edged sword. Does it not show that **CP**$_g$, the type-level version of the independence theory of causal priority, will break down when agency theory fails? No. Even though strong independence will typically fail whenever interventions are not possible,[3] strong independence is closely tied to the existence of *abstract* interventions, not to the existence of specifically human interventions. The possibility of human interventions implies strong independence, but not vice versa. Furthermore, **CP** depends only on **I**, not on the strong independence condition.

## 5.5 Intervention, Independence, and Causation

This argument for the independence theory of causal asymmetry can be strengthened. One criticism of theories of causal asymmetry is that even if they enable one to distinguish causes and effects correctly, they fail to capture what is at the heart of the distinction. Even if, as Hume maintained, causes always precede their effects, this difference in time does not seem to be of the right "character." As Hume recognizes, he still owes us some account of the effectiveness of causes or the sense in which causes *make* their effects happen but not vice versa. Hume pays this debt with a psychological explanation of our mistaken belief in the existence of necessary connections. Similarly, one might object that even if the effects of an event are causally connected to one another, while not all of its causes are, this difference cannot lie at the heart of the difference between causes and effects. Ted Honderich makes this criticism of David Sanford's analysis of causal asymmetry, which bears a close resemblance to **CP** (Sanford 1976; Honderich 1982, p. 314; see also §8.1*).

Agency theory seems to strike close to the heart of the distinction between causes and effects. The fact that causes can be used to manipulate their effects, but not vice versa, seems to be linked to the idea that causes are "effective" and that they necessitate their effects. Indeed, I suggest that the thought that causes *necessitate* their effects is a metaphysical pun on the fact that agents "make" effects happen by means of their causes. What do people need causal explanations or a notion of causation for? Why isn't it enough to know the lawlike relations among types? Because human beings are actors, not just spectators. Knowledge of laws will guide the expectations of spectators, but it does not tell actors what will result from their interventions. The possibility of abstract intervention is essential to causa-

[3] Which does not follow from the result cited but is nevertheless true. See pp. 109-10.

tion, and its anthropomorphic exaggeration gives rise to the notion of causal necessitation.

If these speculations are correct, then the entailment of $\mathbf{CP}_g$ by $\mathbf{AT}_g$ and the other conditions shows something surprising and very important. At first sight it appears mere happenstance that the independence principle is true of so many common causal set-ups. But the conditional entailment of $\mathbf{CP}_g$ and $\mathbf{AT}_g$ shows that the independence of causes is – so to speak – "the objective content of causal necessitation." It would be nice if **CP** or $\mathbf{CP}_g$ were *obviously* of the right character – if it were obvious that independence goes to the core of causation. But if effectiveness or necessitation is, as I have suggested, an anthropomorphic notion, then no objective account of causation could be obviously of the right character. The asymmetry of independence is, I maintain, the intrinsic difference between causes and effects that manifests itself to agents like us as the effectiveness of causes and the necessitation of effects. The independence of causes *constitutes* specifically causal relations. Without independence, there can still be lawlike relations, but there is no causation, because there is no possibility of (abstract) intervention. The conditional entailment of $\mathbf{CP}_g$ by $\mathbf{AT}_g$ coupled with the hypotheses that necessitation is at root an anthropomorphic notion and that the possibility of abstract intervention is essential to causation give one strong reason to believe that **CP** captures the central features of the priority of causes to their effects.

## 5.6 What Role for Agency?

Agency theories satisfy the criterion of epistemic access well, and at first glance they appear to satisfy the criterion of intuitive fit, too. But there are anomalies, and agency theories do not appear to be adequate empirically. They are in need of qualifications, and they appear to be unduly anthropomorphic. Rather than superseding competitors, they are superseded by them. The theorems above show that those who are committed to agency theory are also committed to $\mathbf{CP}_g$. Given their commitment, there is no good reason to develop agency theory as an *alternative* account of causal asymmetry, and many good reasons not to do so. $\mathbf{CP}_g$ captures the basic asymmetry upon which the facts of manipulability depend in a much more general way. Humeans can, of course, also offer explanations for the limited validity of the claims of agency theorists. But the connections are not as precise and systematic as are those expressed in the theorems mentioned here and proved in chapter 5*. Rather than providing a fecund and coherent metaphysics, agency theories give rise to difficult questions concerning the nature of agency, which are not germane to a theory of causation.

Agency theory has few virtues as an alternative account of causal asymmetry, but that does not mean that agency considerations are irrelevant to understanding causation. On the contrary, the relations between agency and causation are of the utmost importance. They are crucial in testing causal claims and in understanding the acquisition of causal notions and their grip on our thinking. And only via an exploration of the links between causation and agency does it become evident just how central independence is to causation.

# 5*

# Causal Generalizations and Agency

## 5.1* Token and Type Causation

How should one interpret claims that a property *A*, a variable *x*, or kind of event **a** is a cause of another? In chapter 2, I argued that cause and effect are events. Since properties, variables, and kinds are not events, they cannot be causes or effects. Either they stand in some different kind of causal relation, or the truth of the claim that **a** causes **b** derives from the causal relations obtaining among tokens. My view is that claims concerning causal relations among types, properties, and variables are generalizations concerning causal relations among tokens.

Many authors have held, in contrast, that there are two varieties of causation, one relating tokens and the other relating types. Peter Menzies, for example, writes, "What kinds of entities do causal sentences relate? This question is readily answered in the case of *general causation*, since a distinguishing mark of a general causal sentence is that its causal relata are properties. The question is much harder to answer in the case of *singular causation*" (1989b, p. 59). In *Probabilistic Causality* (1991), Ellery Eells develops different theories of "type-level" and "token-level" causation. Eells argues extensively for countenancing two distinct kinds of causation, and I will accordingly focus on his case (see also Sober 1985, 1986).

Eells presents his argument for the existence of a distinct type-level causal relation in the context of a theory of probabilistic causation, and some of his concerns cannot be addressed until §9.2. Eells motivates the distinction between type and token causation by pointing out that the surgeon general's type-level claim that smoking is a positive causal factor for lung cancer leaves the facts about the token-level effects of smoking and the token-level causes of lung cancer almost completely open. Indeed he points out that smoking can be a cause of lung cancer without ever causing any individual to get lung cancer!

> Consistent with human physiology being just as it actually is (so that the surgeon general's claim is still true), is the possibility that everybody's causal field happens (improbably enough) to be such that, if they were to become smokers, they would, just before the time lung disease had a chance to develop, die from some other

cause that, given the causal field, is deterministically token causally related to smoking. (1991, p. 11)

Eells offers the following arguments against the claim that type-causal claims are generalizations concerning token causation.

> First, it is consistent with a type-level probabilistic causal claim, as I understand such claims, that the cause and effect factors involved never in fact happen to be exemplified. Thus, as I understand type-level probabilistic causal claims, they are not generalizations over instances of token causation. Second, in motivating the kind of interpretation of type-level probabilistic causal claims that I am endorsing, I have described examples, possible situations, in which the surgeon general's type-level claim is intuitively true, yet in which there are no cases in which a token of the cause type ever causes, or even would cause, a token of the effect type. In these cases, there are no instances of the relevant kind of token causation to generalize over. Finally, there is a problem for the suggestion that type-level causal claims be understood as generalizations over instances of token causation: What are the formal properties of the generalizations? In order for a type **c** to be a type-level cause of type **e**, must a token of **e** be "token caused" in *all* cases in which a token of type **c** occurs – or only in most such cases, or only many, or only some? Also, could **c** be a positive causal factor for **e** if instances of **c** sometimes cause the *absence* (that is, sometimes *prevent*) **e**? On the one hand, until all this is made precise, we have only a very minimal theory; and on the other hand, it seems that any way of making it precise makes the theory either too weak, too strong, or arbitrary.[1]

One can summarize and extend Eells's case for the existence of a distinct type-causal relation as follows:

1. If *a* is a (token) cause of *b*, then
   a. *a* occurs.
   b. *b* occurs.
   c. There is in fact a causal connection between *a*'s occurring and *b*'s occurring.
2. A type **a** may be a cause of a type **b** even though:
   a. No token *a* of kind **a** occurs.
   b. No token *b* of kind **b** occurs.
   c. Tokens of kind **a** that occur are never causes of occurrences of tokens of kind **b**.
3. The fact that *a* is a token cause of *b* does not imply that **a** is a type-level cause of **b**.

Eells does not cite point 3 explicitly, but he would readily endorse it. In addition Eells challenges those who hold that type causation derives from token causation to specify exactly what the relationship is. Eells appears to have a strong case. The first quotation from Eells already gives one reason to accept points 1 and 2. Obviously the surgeon general believes that people smoke and get lung cancer and that instances of smoking token-cause lung cancer in lots of people, and if he or she did not believe these things, it would be more natural to say something like "If anyone were to smoke he

[1] 1991, pp. 15–16. I substituted lowercase bold letters for the capital letters Eells uses.

or she would be more likely to get lung cancer." But it is questionable whether the surgeon general's claim strictly *implies* that there are instantiations of smoking that cause instantiations of lung cancer. Similar considerations support point 3. George's smoking might have led him to meet Jim, who found George a job in an asbestos factor, which caused George to get lung cancer. But the fact that his smoking was here an indirect (token) cause of his contracting lung cancer does not imply that smoking is a (type-level) cause of lung cancer.

I shall argue that the case against any reduction of type causation to token causation is much weaker than it appears. Points 1–3 do not provide good reason to countenance a distinct type-level causal relation.

## 5.2* Type Causation and Causal Generalizations

One serious problem with the notion of type-level causation is that, according to many construals of causal asymmetry, causal relata must be located in space and time.[2] But once one locates properties or types in space and time (via their instantiations or instances), then the relata of type causal relations turn out to be the same as the relata of token causal relations. This problem is especially clear in the case of Eells's theory, which requires that causes precede their effects. How can one *property* or *type* precede another?

Eells's solution to this difficulty involves defining properties that refer to times (1991, ch. 5). Let $s_t$ be a temporal "slice" of an individual substance or set-up $s$ at time $t$, and suppose that it is true that $s_t$ has the property $A$. Eells defines the time-dependent property $A_t$: $(x)[A_t(x) \leftrightarrow A(x_t)]$. George has the property of being-a-smoker-at-$t$ if the $t$ time slice of George has the property of being a smoker. The asymmetry of causation is secured by stipulating that $A_t$ causes $B_{t'}$ only if $t'$ is later than $t$. $A_t$ remains a property, not a particular. As such, it is not located in space and time, and cannot literally precede the property $B_{t'}$. The asymmetry instead rests on the time reference of $A_t$ preceding the time reference of $B_{t'}$.

The new time-dependent properties that Eells postulates are unenticing, and his account of causal asymmetry in terms of the temporal relations among time references in properties is idiosyncratic and undeveloped. It is questionable whether the considerations supporting a temporal priority theory of causal asymmetry will support this new theory. Moreover, suppose that something has the property $A_t$. This implies its $t$-slice has the property $A$. So if anything has the property $A_t$, then the property $A$ must be instantiated, and there must be a trope of kind **a**. Unless the properties that stand in

[2] Another problem, emphasized by Carroll 1988, is that the relations between properties depend on *which* individuals have the properties.

type-causal relations to one another have empty extensions, type-level causal relations always imply some sort of token-level relations among tropes.

A further reason for disquiet concerning time-indexed properties is (as Eells notes on p. 150) that type-level causal claims are not well construed as $t$-Smoking causes $t'$-Cancer. Such a claim relates smoking in one particular time period to lung cancer in a later time period. The surgeon general's claim in contrast applies to any time periods (provided that the circumstances are unchanged). Type-causal claims must involve quantification over the time-indices in properties. In a deterministic context (which is *not* Eells's), type-level causal claims construed as relating time-dependent properties will presumably have something like the following form:

**Type** For all $t$ and $t'$, $A_t$ is a cause of $B_{t'}$ in circumstances $K_t$ if and only if given laws of nature **L** and a specification of the circumstances, $K_t$ $B_{t'}$ entails and is entailed by $A_t$, when $t'$ stands in the right temporal relations to t.

Notice that **Type** does not give truth conditions for $A_t$ is a cause of $B_{t'}$ regardless of the circumstances. It only gives truth conditions for $A_t$ is a cause of $B_{t'}$ *in circumstances* $K_t$. This is no accident. One property $A$ is rarely nomically sufficient for another property $B$, regardless of the circumstances. As discussed in §3.2, reference to the circumstances is unavoidable.

Compare this construal of Eells's position in deterministic cases to the necessary condition on deterministic causation of tokens I called **(DC)** (p. 43):

**DC** If $a$ is a deterministic cause of $b$ in set-up $c$ during the time interval $[t, t']$, then given laws of nature **L**, 1. $B(c, t')$ entails and is entailed by $\{A(c, t)$ & $G(c, [t, t'])$ or $H(c, [t, t'])\}$, but $B(c, t')$ is not entailed by $G(c, [t, t'])$, 2. $B(c, t')$ 3. $A(c, t)$, 4. $G(c, [t, t'])$, and 5. $\neg H(c, [t, t'])$.

Clauses 4 and 5 of **DC** specify the circumstances within which $A$ is nomically necessary and sufficient for $B$. According to **DC**, the token causal claim, "$a$ causes $b$" implies that, given laws of nature $A_t$&$G_t$ entail $B_{t'}$, $B_{t'}$&~$H_t$ entail $A_t$&$G_t$, and $t$ and $t'$ bear the right relations to one another. **DC** thus implies that token causation is sufficient for type causation. One might then simply assert that **a** is a cause of **b** in circumstances $K$ if and only if in circumstances $K$ every event of kind **a** causes some event of kind **b** that bears the right temporal relations to it. But what if there are no events of kind **a** in circumstances $K$? Some reliance on counterfactuals is unavoidable. What one needs to say is something like:

**G** (*Counterfactual generalization view*) **a** is a cause of **b** in circumstances $K$ if and only if in $K$ all events of kind **a** that might occur would cause some event of kind **b** that would bear the right temporal relations to it.

Like **Type**, **G** offers a necessary condition for "**a** is a cause of **b**" in circumstances *K*, not for "**a** is a cause of **b**" full-stop. **G** avoids postulating a new category of time-relevant properties and analyzes type-causal claims (relativized to circumstances) in terms of token causation.[3] **G** is my response to Eells's challenge to specify how type causal claims derive from token causal claims.

One might object that **G** must be wrong because token-level causal claims, unlike type-level claims, are extensional. If *a* causes *b*, and "*c*" and "*d*" refer respectively to the same events as "*a*" and "*b*," then *c* causes *d*. On the other hand, from "smoking causes lung cancer" and "people have yellowed teeth if and only if they smoke," one cannot infer that "having yellowed teeth causes lung cancer." But the problem is merely apparent, because the truth of *a* causes *b*, where *a* is an instantiation of property *A* and *b* of *B*, does not imply that there is a law of nature relating *A* and *B*. It implies only that there is a property of people in the state of having yellowed teeth that bears a nomological relationship to lung cancer.

Even if **G** survives this objection, don't Eells's arguments refute **G**? No. What "saves" **G** is the relativization to the circumstances *K*. Given this relativization, **G** fits all the facts Eells adduces concerning claims such as "smoking causes lung cancer." **G** does not imply that there are any tokens of kinds **a** or **b**. If there are instances of **a** and **b**, **G** does not imply that they must be related as cause and effect, because it may be that tokens of **a** never occur in circumstances *K*. In this way it can be the case that given *K*, **a** is a cause of **b**, even though there are *a*'s and *b*'s, and none of the *a*'s cause *b*'s. The only sticking point concerns whether it is possible that *a* cause *b* without type-causal relations among the relevant properties or kinds.

Given determinism and a sufficiently detailed description of the circumstances *W* on some occasion when *a* is a cause of *b*, **G** implies that, given *W*, tokens of kind **a** always cause tokens of kind **b**. Yet George's smoking could cause his lung cancer via his meeting with Jim and his employment in an asbestos factory without smoking being a type-level cause of lung cancer. Does it not follow that **G** is mistaken?

Let *a* be the event of George's smoking, *b* be the event of his contracting lung cancer, and *W* be the properties (in all relevant detail) of the circumstances in which *a* led to *b* via the unfortunate meeting with Jim. Given a

[3] In this account, unlike Carroll's (1988, p. 314), the relationship between type and token causal relations is not probabilistic. This is not only because I am here abstracting from the possibility of probabilistic causation. See the arguments in § 9.2 and 9.5. Unlike Carroll's later attempt to eliminate property-level causation (1991), **G** claims that type-level claims are universal generalizations or claims about conditional property entailment rather than some sort of "generic" claim. Furthermore, the possibility of analyzing property-level claims should not, I would maintain, lead one to deny the reality or importance of type-level claims.

deterministic view of causation, it follows that, given *W*, *a*'s always cause *b*'s. **G** implies only that event type **a** is a cause of event type **b** in conditions *W*. But this generalization is not what the surgeon general maintains. The surgeon general claims that smoking causes lung cancer in a different set of circumstances. So one can agree that George's smoking led to his contracting lung cancer and deny the surgeon general's claim.

Furthermore, the surgeon general's claim should be understood as implying that there are exclusively physiological and chemical links in the causal chain between inhaling smoke and changes in the lungs, and in George's case there are different kinds of links. **G** does not imply that the truth of a singular causal claim entails any causal generalization of interest. Although these considerations suffice to refute Eells's arguments in the deterministic context in which I am working, I have not yet done justice to Eells's position, which is particularly concerned with probabilistic causation. I shall return to these issues in chapter 9.

One might object that the reduction of type- to token-causal claims puts things exactly backwards. Causal relations obtain in virtue of lawful relations among *properties*. How then can one turn around and claim, as I have, that causal relations among the properties are generalizations of relations among the tokens?

This objection rests on an equivocation. Causal relations among tokens presupposes *nomic* relations among properties. They do not presuppose asymmetrical *causal* relations among properties. It is the latter, not the former, that are generalizations of relations among tokens. The token-causal dependence of the pressure of a particular gas cylinder on its temperature rests upon the laws relating temperature to pressure. The claim that (in specified circumstances) pressure *causally depends* on temperature is a generalization of token causal relations and concerns a particular kind of system.

Even if readers conceded that the above remarks undermine Eells's critique of the reduction of type to token causation, they might question whether they demonstrate that the reduction should be accepted. What positive reason have I offered in support of **G**? Apart from considerations of parsimony, I have made three arguments. Two are directed specifically against Eells's view and maintain (1) that the relata of type-causal relations do not differ from the relata of token-causal relations and (2) that it is implausible to postulate time-dependent properties and to quantify over their time indices. What forces Eells to conjure up time-dependent properties is his view that the asymmetry of causation is a matter of temporal priority. If one were to adopt a view of causal asymmetry that does not make reference to time, then one might be able to avoid this undesirable feature of type causation. The third argument has nothing to do with the details of Eells's theory. The main point is that *without concrete systems or sets of circumstances, there is no such thing as causation.* The asymmetry and directionality of the causal relation exists only in connection with arrays of tokens.

Causal generalizations may nevertheless be more important and more interesting than are singular causal claims. For example, as I have argued elsewhere (Hausman 1989b, 1990), the law of demand, unlike the ideal gas law or Newton's law of gravitation, is a causal generalization: an increase in $p_x$, the price of $x$, causes less of $x$ to be demanded. The law of demand cannot be captured by the statement that the quantity demanded of $x$ is a decreasing function of $p_x$, because this statement is falsified whenever $p_x$ increases because of some independent increase in quantity demanded. In this instance, economists are concerned with a causal generalization and are largely uninterested in its particular instances. In this example – as in many others – people are more interested in patterns of dependence among variables than in the causal relations among particular values. Consequently, scientists often talk about causal relations among variables rather than causal relations among values of variables.

When scientists say that variables are causally related, I shall take them to be expressing generalizations concerning the relations among the located values of the variables. The values of variables are properties, and located values are tropes. Taking type relations to be generalizations of token relations does not, however, commit one to my construal of token relations (for an alternative, see Mellor 1995). I shall sometimes use the same terms for the relations among types as for the relations among tokens, but I shall also use terms such as "influence," "affect," "determine," and "depend" to talk about type-level relations.[4] Properties, types, or variables are nomically connected when, in the circumstances, instantiations, tokens, or values are nomically connected.

The fact that **a** depends on **b** in circumstances $K$ is perfectly consistent with **b** depending on **a** in other circumstances $K'$. In one set of circumstances, an increase in the pressure of a gas can depend on an increase in its temperature. In a second set of circumstances, an increase in the temperature of a gas can depend on an increase in its pressure. In token causal claims the circumstances are given, although the fineness of our description of them is not (Spohn 1990, p. 126). In causal generalizations or type-causal claims, they have to be specified. It is precisely because causal relations involve tokens that *causal generalizations must specify the relevant circumstances.* Furthermore, the universal conditionals implicit in INUS conditions

[4] Mellor 1995, ch. 12 points out that in ordinary language there is a contrast between causing (to exist) and affecting (an inessential property of something already existing). This distinction is not important to my argument, and I shall accordingly appropriate the term "affect" to talk about the causal relations among variables, properties or types.

and causal generalizations are supposed to be lawlike claims. Although generalizations of the form $(x)(Fx \rightarrow Gx)$ are true if there are no $F$'s, the fact that $F$ happens not to be instantiated will not make them laws. I shall not address the serious epistemological questions that concern how one distinguishes laws from true generalizations.

### 5.3* Definitions, Conditions, and Theorems

I formulate agency theory as:

> **$AT_g$** (*Type-level agency theory*) $y$ depends on $x$ in circumstances $K$ if and only if $x$ and $y$ are distinct and there is some set of variables $V$ including $x$ and $y$ such that in $K$ every intervention with respect to $x$ given $V$ affects $y$.

Since the notion of an intervention is relativized to a set of variables, the characterization of agency theory must also be relativized, or else one needs to quantify over sets of variables. I omitted this complication in the formulation in §5.1. This formulation is more blatantly circular than it needs to be. **$AT_g$** can easily be restated as: $y$ depends on $x$ in circumstances $K$ if and only if $x$ and $y$ are distinct and there is some set of variables $V$ including $x$ and $y$ such that in $K$ every intervention with respect to $x$ given $V$ is lawfully connected to $y$. It makes little difference to the formal results which version one uses.

The following definition of a human intervention is implicit in the views of agency theorists:

> **$DHI_g$** (*Type-level definition of a human intervention*) In circumstances $K$, $z$ is a human intervention variable with respect to $x$ only (given a set of variables $V$) if and only if in $K$ (1) $z$ is distinct from all variables in $V$, (2) $z$ directly affects $x$, (3) all other connections between $z$ and members of $V$ follow from $z$ being a direct influence on $x$, and (4) values of $z$ are kinds of human actions.

**$DI_g$**, the type-level definition of an abstract intervention, omits the last clause but is otherwise identical. In a causal graph of the variables that includes an intervention variable $z$ (such as figure 5.1), there is a directed edge from the vertex "$z$" representing the intervention variable to the vertices representing directly manipulated variable(s), and there are no other edges to or from "$z$." The "g" in the names **$DHI_g$** and **$DI_g$** is a reminder that the principle is stated in terms of causal generalizations or relations among variables.

The following marks the boundaries of where agency theories apply:

> **$PI_g$** (*Possibility of intervention*) Given the set $V$ of variables and the circumstances $K$ of concern, there is, for each variable $x$ in $V$, some $z$ that is an intervention variable with respect to $x$ only.

As explained on p. 94, the terminology may be misleading, since there is nothing explicitly modal in $\mathbf{PI_g}$. Yet I believe that intervention variables exist if and only if interventions are possible.

In addition I need to define a type-level version of transitivity:

$\mathbf{T_g}$ (*Type-level transitivity*) If in circumstances $K$, $z$ depends on $y$, $y$ depends on $x$, and $z$ and $x$ are distinct, then $z$ depends on $x$.

As already noted (p. 27), transitivity even at the token level requires characterization of cause and effect in terms of relevant properties. At the type level, it might appear that transitivity breaks down altogether. For example, suppose that flipping a switch (**a**) simultaneously closes electric circuit 1 and opens a gap in circuit 2. When circuit 1 closes it activates a solenoid switch (**b**), which (in circumstances in which there is no gap in circuit 2) causes a light bulb to go on (**c**). **a** is a type-level cause of **b** and **b** is a type-level cause of **c** (in circumstances in which there is no gap in circuit 2), but **a** is not a type-level cause of **c**, because **a** eliminates the circumstances in which **b** causes **c**. Transitivity at the type level holds only if **a** causes **b** in the same circumstances in which **b** causes **c**.

Even when the circumstances are held fixed, it might appear that type-level causation is not transitive. For example, employing birth control may be a type-level cause of sexual activity, which in turn causes pregnancy, but employing birth control does not cause pregnancy (Hitchcock 1995a, p. 275). Examples such as this one show that "positive causation" or "causal promotion" is not transitive. They do not show that "is causally relevant to" fails to be transitive, and I am concerned with the latter relation. Taking birth control pills is causally relevant to pregnancy. Provided that the circumstances along the chain are held constant, type causation, in the sense of causation with which I am concerned, is transitive. For further argument for the transitivity of type causation, see pp. 196–7.

The first theorem requires one further technical condition:

$\mathbf{PP_g}$ (*Type-level path principle*) If $z$ causes $y$, then in a directed graph correctly representing the causal relations, there is a path between the vertex denoting $z$ and the vertex denoting $y$.

I called this condition the "path principle" because it comes close to restating the definition of a path. $\mathbf{PP_g}$ is entailed by a restatement of **H** with respect to causal generalizations.

**Theorem 5.1: $\mathbf{DHI_g}$, $\mathbf{PI_g}$, $\mathbf{T_g}$, and $\mathbf{PP_g}$ imply $\mathbf{AT_g}$.**

Proof: Suppose that $y$ depends on $x$ in circumstances $K$. By $\mathbf{PI_g}$ there will be some intervention with respect to $x$. By $\mathbf{DHI_g}$ that intervention affects $x$ and is distinct from $y$. By $\mathbf{T_g}$ that intervention will thus influence $y$. To prove the converse, suppose that all interventions that directly affect $x$ affect $y$. By $\mathbf{PP_g}$, there is a path from $z$ to $y$. By

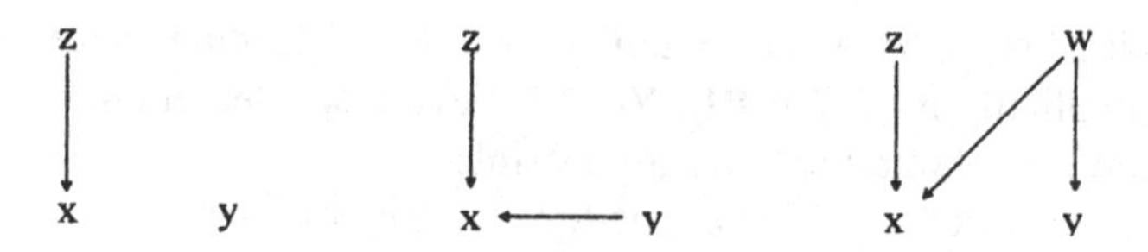


**Figure 5.2:** Deriving $\mathbf{CP_g}$ from the possibility of intervention

$\mathbf{DHI_g}$, *z* does not directly cause anything except *x*, and so any causal path from *z* to *y* must go through *x*.

To consider the relations between $\mathbf{AT_g}$, and the propositions of chapter 4, one needs to reformulate those propositions at the type level:

$\mathbf{CP_g}$ (*Type-level independence theory*) y depends on *x* in circumstances *K* if and only if in *K*, *x* and *y* are causally connected and everything connected to *x* and distinct from *y* is connected to *y*.

$\mathbf{CC_g}$ (*Type-level connection principle*) In circumstances *K*, *x* and *y* are causally connected if and only if they are distinct and in *K* either (i) *y* depends on *x*, (ii) *x* depends on *y*, or (iii) there is some variable *z* that *x* and *y* both depend on.

**Theorem 5.2: $\mathbf{DHI_g}$, $\mathbf{CC_g}$, $\mathbf{T_g}$, and $\mathbf{AT_g}$ entail $\mathbf{CP_g}$.**

Proof: Suppose that in *K*, *x* and *y* are connected and everything distinct from *y* and connected to *x* is connected to *y*. If there is in *K* an intervention *z* with respect to *x* only, then (by $\mathbf{DHI_g}$ and $\mathbf{CC_g}$) it is distinct from *y* and connected to *x*. So it is causally connected to *y*. Since (again by $\mathbf{DHI_g}$) there can be no arrows into *z*, $\mathbf{CC_g}$ implies that *z* affects *y*, and by $\mathbf{AT_g}$ it follows that *y* depends on *x*. The converse (if *y* depends on *x*, then everything distinct from *y* and connected to *x* is causally connected to *x*) follows from $\mathbf{T_g}$ and $\mathbf{CC_g}$ (see theorem 4.1).

The proofs of theorems 5.1 and 5.2 go through in exactly the same way if $\mathbf{DI_g}$ is substituted for $\mathbf{DHI_g}$. So one can state:

**Theorem 5.3: $\mathbf{DI_g}$, $\mathbf{PI_g}$, $\mathbf{T_g}$, and $\mathbf{PP_g}$ imply $\mathbf{AT_g}$.**

**Theorem 5.4: $\mathbf{DI_g}$, $\mathbf{CC_g}$, $\mathbf{T_g}$ and $\mathbf{AT_g}$ entails $\mathbf{CP_g}$.**

Given Theorems 5.3 and 5.4, it is unsurprising that one can prove:

**Theorem 5.5 $\mathbf{DI_g}$, $\mathbf{PI_g}$, $\mathbf{CC_g}$, and $\mathbf{T_g}$ entail $\mathbf{CP_g}$.**

Proof: A version of theorem 4.1 shows that $\mathbf{CC_g}$ and $\mathbf{T_g}$ by themselves entail the necessary condition for causality in $\mathbf{CP_g}$. The task is to prove the analogue to the sufficient condition, that if in circumstances *K*, *y* does not depend on *x* then in *K* there is something causally connected to *x* and distinct from *y* that is not connected to *y*. Consider figure 5.2, which presents path graphs of the three possible structures in which *x* does not cause *y* (given $\mathbf{CC_g}$). By $\mathbf{PI_g}$ there is an intervention *z* respect to *x*. By $\mathbf{DI_g}$ there are no arrows into *z* and only the single edge out to *x*. By $\mathbf{CC_g}$, *z* is not connected to *y*. $\mathbf{PP_g}$ is not needed in the proof.

In theorem 4.4, **T**, **CC**, and **I** together entail **CP**. In theorem 5.5, in contrast, $\mathbf{CP_g}$ follows from **T**, **CC**, $\mathbf{PI_g}$ and $\mathbf{DI_g}$ So $\mathbf{PI_g}$ and $\mathbf{DI_g}$ are doing the work in the proof of $\mathbf{CP_g}$ that **I** did in the proof of **CP**. This fact suggests that there are close relations between independence and interventions. To explore these, let me reformulate **I** and $\mathbf{I_s}$ in terms of generalizations:

$\mathbf{I_g}$ (*Type-level independence condition*) If in circumstances *K*, *x* causes *y* or *x* and *y* are causally connected only as effects of a common cause, then in *K y* depends on something that is not causally connected to *x*.

$\mathbf{I_{gs}}$ (*Type-level strong independence condition*) In circumstances *K*, every nonexogenous variable *y* has some additional cause *z* such that for all variables *x*,

1. If *y* does not cause *x* and there is a causal path that does not go through *z* from *x* to *y* or from a cause of *x* to *y*, then *x* and *z* are causally independent.
2. If *y* causes *x*, then there is no path that does not pass through *y* from *z* to *x* or from any cause of *z* to *x*.

A variable is exogenous if and only if it depends on no other variables. $\mathbf{I_{gs}}$ says every event *y* has a cause *z* that is independent of everything causally connected to *y* and not an effect of *y* whose causal connection to *y* is not "via" *z*, and that *z* is connected to the effects of *y* only via *y*.

The following theorems require one further assumption:

$\mathbf{NIC_g}$ (*Nonintervention causes*): If there is an intervention variable *z* with respect to some variable *x*, then *x* depends on some variable other than *z*.

One can then state and prove:

**Theorem 5.6:** Given $\mathbf{CC_g}$, and $\mathbf{NIC_g}$, $\mathbf{I_{gs}}$ is entailed by $\mathbf{DI_g}$ and $\mathbf{PI_g}$.

Proof: Given $\mathbf{DI_g}$ and $\mathbf{PI_g}$, for every *y*, there is some intervention variable *z*. Given $\mathbf{NIC_g}$, there must be some other variable that *y* depends on. Given $\mathbf{CC_g}$, *z* is not causally connected to anything connected to *y* that is not an effect of *y*, and *z* is causally connected to effects of *y* only via *y*. Thus $\mathbf{I_{gs}}$ follows.

The converse does not follow. In figure 5.3, *z* is a strongly independent cause of *x*, but not an intervention variable. $\mathbf{DI_g}$, which is relativized to the set of variables of interest, denies that an intervention with respect to *x* has any causal consequences except via causing *x*, while strong independence, which is not relativized only to the set of variables that are of interest, quite reasonably permits an independent cause of *x* to have other separate causal consequences. $\mathbf{DI_g}$ also requires that the intervention not be caused by any of the variables of interest, while $\mathbf{I_{gs}}$ does not require that an independent cause of a variable be itself uncaused. Only in circumstances of these kinds will one satisfy the strong independence condition without also having an (abstract) intervention. Even though $\mathbf{DI_g}$ and $\mathbf{PI_g}$ are not entailed by $\mathbf{I_{gs}}$ (given $\mathbf{CC_g}$ and asymmetry), one will seldom satisfy strong independence without satisfying $\mathbf{PI_g}$. Indeed it is arguable that interventions should be

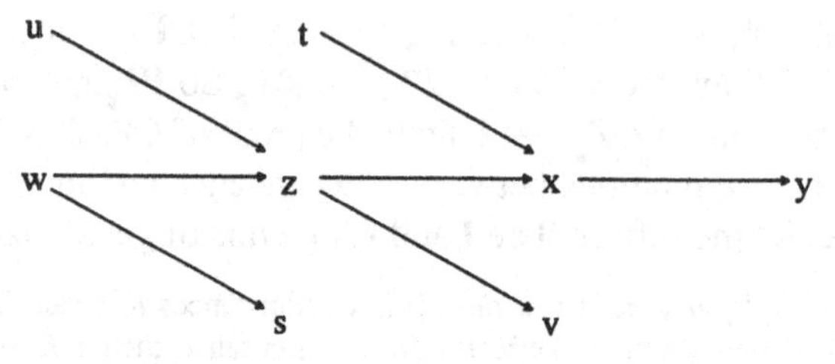

Figure 5.3: Strong Independence and interventions

defined so as to permit *z* in figure 5.3 to count as an intervention (provided that the causes of *z* and the other variable *z* is shown as causing (*v*) are not "of interest." The strong independence condition and that existence of abstract interventions almost coincide. As Richard Scheines pointed out to me, when an error term in a structural equation model satisfies standard independence assumptions, then it is an abstract intervention according to $\mathbf{DI_g}$.

Since strong independence entails independence, the following corollary is obvious:

**Theorem 5.7:** $\mathbf{CC_g}$, $\mathbf{PI_g}$, $\mathbf{DI_g}$, and $\mathbf{NIC_g}$ entail $\mathbf{I_g}$.

# 6

# The Counterfactual Theory

Chapter 3 explored the possibilities of explaining the asymmetry of causation in terms of the relations between causation and time. Chapter 5 explored the possibilities of generating a theory of causal asymmetry from the connections between causation and agency. This chapter explores another attractive set of connections – those between causation and counterfactuals. When one asserts that one event *a* causes another distinct event *b*, then it seems that one is committed to the counterfactual: "If *a* had not occurred, then *b* would not have occurred either." Hume himself wrote ". . . we may define a cause to be *an object, followed by another, and where all the objects similar to the first, are followed by objects similar to the second.* Or in other words, *where if the first object had not been, the second had never existed*" (*Inquiry* p. 51). The "other words" here are indeed "other words." Hume seems mid-definition to leap from a regularity to a counterfactual theory of causation. This chapter explores the possibility of constructing a counterfactual theory of causation.

### 6.1 Lewis's Theory

According to David Lewis, if *a* and *b* are distinct events that actually occur, then *b* causally depends on *a* if and only if, if *a* were not to occur, then *b* would not occur either (1973a). If the match had not been struck, then it would not have lit. In Lewis's usage (in contrast to mine), "causal dependence" is a relation among token events, which is sufficient for causation. It is not the same thing as causation, because in cases of preemption it is not transitive. Lewis takes causation to be the ancestral of causal dependence: *a* causes *b* if and only if *b* causally depends on *a* or there is a chain of causal dependence linking *a* and *b*. One should not take causal dependence to be "direct causation." The collapse of the Soviet Union may have causally depended on Lenin's death seventy years earlier.

Since I am postponing considering problems concerning preemption until chapter 13, I shall simplify Lewis's theory and take his account of causal dependence to be an account of causation. This begs no questions and eases the comparisons between his theory and the accounts discussed in previous chapters.[1] I shall formulate this simplification of Lewis's theory as:

**L** (*Lewis's theory*) *a* causes *b* if and only if *a* and *b* are distinct events and if *a* were not to occur, then *b* would not occur either.

I am taking the occurrence of *a* and *b* as implicit in the claim that they are distinct events. Let us say that *b* is *counterfactually dependent* on *a* if and only if both the following counterfactuals are true: "If *a* were to occur, then *b* would occur" and "If *a* were not to occur, then *b* would not occur" (Lewis 1973a, p. 166). If *a* and *b* both occur, then the first counterfactual is automatically true, since the closest possible world in which *a* occurs is the actual world and in that world *b* occurs too. **L** can be restated as the claim that counterfactual dependence among distinct events is necessary and sufficient for causation. Like Lewis I shall take counterfactuals to be statements that may be true or false.

To evaluate the counterfactual, "if *a* were not to occur, then *b* would not occur," one considers "possible worlds" in which *a* does not occur. Some of these possible worlds will be more similar to the actual world than are others. The counterfactual is true if some possible world without *a* (some "non-*a* possible world" ) in which *b* does not occur is more similar to the real world than any non-*a* possible world in which *b* occurs. In Lewis's theory, understanding of counterfactuals in this way grounds understanding of causality.

The "comparative overall similarity" of possible worlds is a vague relation, and Lewis argues that we should fill it in so as to fit our intuitions concerning which counterfactuals are true. When Lewis does this, he finds that the comparative overall similarity of possible worlds depends on several factors (Lewis 1979, 1986c). Widespread and diverse differences between the laws of nature count against similarity, while "perfect match" over a time interval counts heavily for similarity. Small differences between laws of nature are relatively unimportant. These appear from the perspective of the actual world as localized and minor miracles. Possible worlds that differ from the actual world only with respect to small numbers of such miracles are very similar to the actual world. In Lewis's view, the closest non-*a* possible world should have exactly the same history as the actual world until shortly before the time at which *a* occurs in the actual world. If *a* is causally determined, there will then be some relatively isolated violation of the laws of the actual world – so that *a* fails to occur. If the most similar possible worlds had the same laws of nature and determinism were true, the most similar possible worlds would have to have completely different histories. Permitting "miracles" avoids this absurdity. When one considers

[1] Taking causation to be the ancestral of causal dependence also allows *c* to be a cause of *e* when the influence of *c* on *e* along one path cancels out its influence along another path. See §12.2.

what would be true if *b* had not occurred, one holds fixed the past (including the occurrence of its cause, *a*) and "gets rid" of *b* by means of a "small miracle" just before *b* occurs.

Laws of nature are not sacrosanct in evaluating counterfactuals. Since perfect match counts so heavily in determining overall similarity among possible worlds, the closest possible world without the effect, *b*, will not diverge from the actual world until after the cause, *a*, occurs. So causes are not counterfactually dependent on their effects. Similarly, effects *e* and *f* of a common cause *c* are not counterfactually dependent on one another, because possible worlds without *e* that diverge from the actual world only after *c* has occurred will be closer to the actual world than are possible worlds without *c*. Thus Lewis purports to solve what he calls "the problem of effects" and "the problem of epiphenomena." Lewis argues that one must resist the temptation to say that causes are sometimes counterfactually dependent on their effects and that effects of a common cause are sometimes counterfactually dependent on one another.

This account of the asymmetry of causation might appear arbitrary. If perfect match between possible worlds is so important in considering their similarity, why isn't the most similar world without *a* one in which *a* fails to happen owing to one miracle and in which the future then goes on just the same as in the actual world (i.e., includes *b*), owing to another miracle? One might also ask why a possible world with just the same past and then, owing to a miracle, a different future is more similar to the actual world than a possible world with a different past and then, owing to a miracle, the same future.

Lewis's answer is an empirical one: A single small miracle will not erase the consequences of *a*'s failure to occur, nor will a single miracle lead a possible world with a different past to have just the same future. There is an asymmetry of miracles, which results from an asymmetry of overdetermination (1979, pp. 48–51). Causes leave many traces, which require many miracles to erase. The fact that the most similar possible worlds are just like ours until just before an event fails to occur is not an arbitrary stipulation, and it does not presuppose that causes must precede their effects. On the contrary, facts about overdetermination coupled with a correct understanding of the semantics for counterfactuals imply that the most similar possible worlds involve miracles just before the supposed occurrence or nonoccurrence. §6.5* shows that the asymmetry of overdetermination is a consequence of the independence condition (**I**) when causation is deterministic.

It is undeniable that people sometimes say that if an event had failed to occur, then one or another of its causes must have failed to occur. Lewis responds that these counterfactuals involve a nonstandard "backtracking" interpretation. Given the standard understanding of counterfactuals, such claims are, Lewis maintains, false. This response still seems ad hoc, and I shall argue that it is unnecessary.

## 6.2 Asymmetry Without Miracles

There is another way in which a defender of **L** might respond to the problem of effects and to the problem of epiphenomena. Consider exactly what Lewis writes:

> [1] The proper solution to both problems [of effects and of epiphenomena], I think, is flatly to deny the counterfactuals that cause the trouble. [2] If *e* had been absent, it is not that *c* would have been absent. . . . [3] Rather, *c* would have occurred just as it did but would have failed to cause *e*. [4] It is less of a departure from actuality to get rid of *e* by holding *c* fixed and giving up some or other of the laws and circumstances in virtue of which *c* could not have failed to cause *e*, rather than to hold those laws and circumstances fixed and get rid of *e* by going back and abolishing its cause *c*. (1973a, p. 170)

Focus particularly on the second and third sentences in this quotation. The second denies the counterfactual, $\sim O(e) > \sim O(c)$, while the third asserts the counterfactual, $\sim O(e) > O(c)$ (where "$O(e)$" is the proposition that *e* occurs, and ">" represents the counterfactual conditional). If the law of the conditional excluded middle holds, then the second sentence entails the third. (The law of the conditional excluded middle states: **P** > **Q** or **P** > ~**Q**. See Lewis (1973b), pp. 79–82.) But Lewis rightly denies the law of the conditional excluded middle and may deny $\sim O(e) > \sim O(c)$ without asserting, "If *e* had been absent, . . . *c* would have occurred just as it did." In order to account for the asymmetry of causation, Lewis need only deny the counterfactual dependence of specific causes on their effects. He does not need to assert that specific causes would still occur if their effects failed to occur.[2]

This suggests a way to defend **L** without invoking last-minute miracles. Indeed Lewis's wording in the last sentence in the quotation suggests that miracles between the cause and effect one is concerned with may not be necessary. He writes, "It is less of a departure from actuality to get rid of *e* by holding *c* fixed and giving up some or other of the laws *and circumstances* in virtue of which *c* could not have failed to cause *e* . . . [my italics]." If *e* had not occurred, then possible worlds without *c* are no more (or less) similar to the actual world than are possible worlds without some other cause or causal condition of *e*.

[2] One must thus reject Michael McDermott's revision of Lewis's theory (1995, p. 137), which says that if *c* and *e* occur, then *e* causally depends on *c* if and only if if *c* had not occurred, then *e* *might* not have. For the negation of such a might counterfactual (with the nonoccurrence of an effect as its antecedent) is identical to the assertion that if the effect had not occurred, the cause would have occurred just the same.

Suppose that *e* has two causes, *a* and *c*, and consider the three counterfactuals:

1. If *e* had not occurred, either *a* or *c* would not have occurred.
2. If *e* had not occurred *a* would not have occurred.
3. If *e* had not occurred *c* would not have occurred.

(1) does not imply (2) or (3). If non-*e* possible worlds without *a* are no more or less similar to the actual world than are non-*e* possible worlds without *c*, then both (2) and (3) are false. There is no need to deny (1). These claims are not epistemic. The problem is not that we are unable to *find out* whether (2) or (3) is true: If non-*e* possible worlds without *a* are no more or less similar to the actual world than are non-*e* possible worlds without *c*, then both (2) and (3) are false.

If there are such ties in similarity, Lewis could agree that if an effect determined by its causes were not to occur, then some of its causes wouldn't have occurred without having to assert that any one of its causes failed to occur.[3] He could still deny claims such as "if *e* had not occurred, then *a* would not have occurred." And this denial is all that a counterfactual theory needs in order to account for the asymmetry of causation. The solution to the problem of epiphenomena is similar. Lewis need not maintain that if one effect of a common cause failed to occur, then the others would still occur. He need only deny all specific counterfactual dependence of one effect of a common cause on another.

## 6.3 Why Causes Are Not Counterfactually Dependent on Their Effects

A specific cause *a* of an event *b* will not be counterfactually dependent on *b*, if *b* has other independent causes, and there are non-*b* possible worlds without one of *b*'s other causes that are at least as similar to the actual world as are non-*b* possible worlds without *a*.[4] If the independence condition of chapter 4 holds, then there will be a multiplicity of independent causes. It is not particularly plausible to maintain that there will always be ties among the closest possible worlds without one or another of these causes, but the implausibility of this claim is no worse than the implausibility of Lewis's views on similarity among possible worlds.

For example, suppose George jumps off the Brooklyn Bridge and plunges

[3] As Christopher Hitchcock pointed out to me, to accept the revised version of his theory, Lewis would have to give up his account of causal preemption. Since that account fails in any event in cases of so-called late preemption (1986d, pp. 193–212) and since a simpler and more powerful solution to the problem of preemption is available (see §13.2), this seems no great loss.

[4] Artificial compound events create problems here, which are discussed on pp. 135–6.

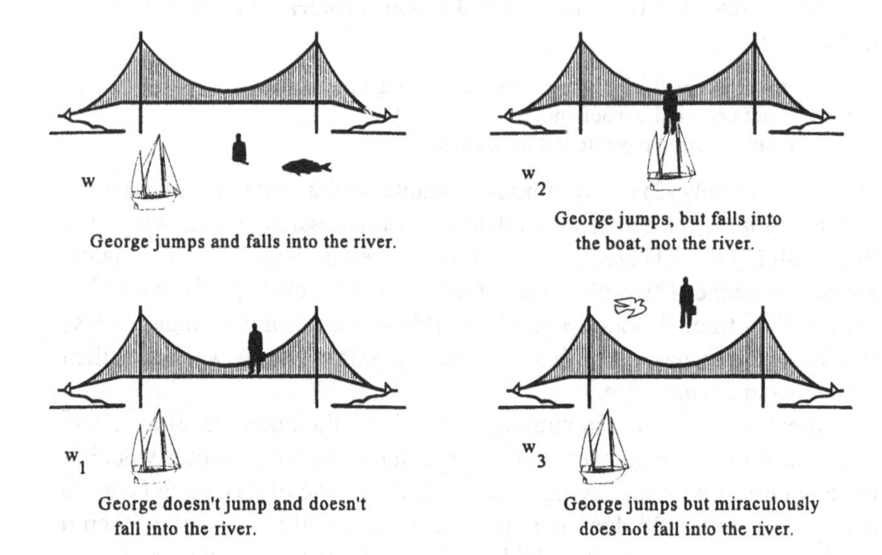


**Figure 6.1:** George jumps off the Brooklyn Bridge

into the East River. It is plausible to claim that if George had not plunged into the river, then he would not have jumped. People are inclined to find not jumping overwhelmingly the *most likely explanation* for George's hypothetical failure to plunge into the East River.[5] Lewis insists that counterfactual theorists need not be governed by such intuitions. Instead one needs to find an interpretation of similarity that will not imply that George's jumping is counterfactually dependent on his falling in the river, since his jumping is obviously not causally dependent on his falling into the river.

Let *w* in figure 6.1 be the actual world, in which George jumps from the Brooklyn Bridge and plunges into the East River. Among the causes of his plunging into the East River are his jumping and, let us suppose, a boat being downstream of the bridge rather than beneath him. Consider then the following three possible worlds (shown in figure 6.1) in which George does not plunge into the East River. In $w_1$ he does not jump. In $w_2$ he jumps, and the boat is beneath the bridge, so that he falls onto it. In $w_3$, he jumps, the boat stays put, but by some miracle he does not plunge into the river.

[5] In an incautious moment, Lewis himself writes that we should say that if a barometer had read higher, it would have been malfunctioning rather than that the pressure would have been higher. His reason is that "The barometer, being more localized and more delicate than the weather, is more vulnerable to slight departures from actuality" (1973a, p. 169). If one takes this literally, then Lewis is saying that the functioning of the barometer is counterfactually dependent on the reading, and one has a surprising example of backwards causation. Such remarks do not reappear in Lewis's later discussion in "Counterfactual Dependence and Time's Arrow," (1979) and I shall assume that they are a mistake.

Perhaps he is blown to shore. Most people would judge worlds like $w_1$ to be more similar to the actual world than are worlds like $w_2$ or $w_3$. If causal dependence is counterfactual dependence, Lewis had better not agree. So it had better not be the case that the miracle that moves the boat is, like the miracle that somehow keeps George out of the river in $w_3$, a "big" one, or that it has to begin earlier than the miracle that prevents the jumping. For then Lewis would have to agree that $w_1$ is more similar to $w$ than are $w_2$ or $w_3$, and his account would falsely imply that George's plunging into the river causes his jumping.

There are two ways to avoid this result. One is to hold that $w_3$ is more similar to $w$ than are $w_1$ or $w_2$. I would not do so, because I deny that worlds with miracles between cause and effect are more similar to the actual world than are worlds without such miracles. Lewis permits miracles between cause and effect, but I doubt that he would want to rely on one here, because the miracle that keeps George out of the water in $w_3$ does not seem to be small and isolated. The only other way to avoid the counterfactual dependence of the jumping on the falling into the water would be to maintain that there is some other possible world like $w_2$ that is at least as similar to $w$ as is $w_1$.[6] And to avoid making the position of the boat causally dependent on George's falling in the river, one would also have to maintain that worlds such as $w_1$ are at least as similar to $w$ as are worlds like $w_2$. This is the view to which my reformulation of Lewis's account is committed in general, and I think that in this case Lewis would accept it, too. He writes, "[W]e should sacrifice the independence of the immediate past to provide an orderly transition from actual past to counterfactual present and future. That is not to say, however, that the immediate past depends on the present in any very definite way" (1979, p. 40). Counterfactual theorists have a choice among implausibilities. They can say that $w_3$ is more similar to $w$ than is $w_1$ or they can say that there are ties between worlds like $w_1$ and $w_2$. The fact that my account is committed to the second alternative is no argument for favoring Lewis's formulation. I personally do not think that causation is counterfactual dependence. I am only arguing that my reformulation of a counterfactual theory is at least as plausible as Lewis's account.

The independence condition (I) implies that the causes of an event are not all causally connected to one another. If one assumes that counterfactual dependence implies causal connection, that is, that distinct events that are not causally connected are not counterfactually dependent on one another, it follows that not all the causes of an event will be counterfactually dependent on one another. The independence condition plus the assumption that counterfactual dependence implies causal connection also implies that

[6] I am indebted here to Horacio Arló-Costa.

effects of a common cause will each have their own causes that are not counterfactually independent on one another. Given the view that there will be ties between non-*b* possible worlds missing one or another of *b*'s independent causes, causes will not be counterfactually dependent on their effects and effects of a common cause will not be counterfactually dependent on one another. For a more rigorous presentation of this argument, see the proof of theorem 6.1, p. 135.

One can go on to deduce the sufficient condition **L** states (that counterfactual dependence implies causation) from the same premises plus the connection principle. If *b* counterfactually depends on *a*, then *a* and *b* must be causally connected. Since causes are not counterfactually dependent on their effects and effects of a common cause are not counterfactually dependent on one another, the only way that *a* and *b* can be causally connected is for *a* to be a cause of *b*.

It thus seems that one might defend the simplified version of Lewis's theory, **L**, with a different explanation for why individual causes are not causally dependent on their effects and why effects of a common cause are not causally dependent on one another. Given the independence condition, one need not rely on miracles, and one can ground the asymmetry of causation in the multiplicity of independent causes. There is however little reason to develop such a counterfactual theory as an *alternative* to the independence theory of causal priority, **CP**, because someone accepting the independence condition, the connection principle, and transitivity, is already committed to **CP**. By accepting **CP**, one can avoid the difficulties of analyzing counterfactuals and of defending an account of similarity among possible worlds.

Qualms about the independence condition are no reason to prefer Lewis's theory, because Lewis assumes that there is an asymmetry of overdetermination, which is closely related to the independence condition (see theorem 6.3, p. 136). Moreover, if one clings to Lewis's semantics and interprets miracles as interventions, then the claim that effects are not counterfactually dependent on their causes entails the strong independence condition (which in turn entails the independence condition **I**). Accepting Lewis's account is tantamount to assuming that there will always be something that prevents the effect and leaves the causes alone – that is, that there will always be something that is causally independent of the given causes.

## 6.4 Counterfactuals and Predictions

People sometimes reason counterfactually from present to past. Lewis maintains that in doing so, they interpret counterfactuals in a nonstandard way (1979, p. 34; see also Bennett 1984). Is this plausible? The alternative discussed above permits one to deny the counterfactual dependence of individual causes on their effects without invoking well-placed miracles, and it permits counterfactuals of the same kind forward and backward in time (see also Bennett 1984; Goggans 1992).

This seems to me a virtue, even though Lewis and others have argued that one cannot combine backward and forward counterfactuals lest one wind up maintaining inconsistently both "If he had jumped out of the window, he would have broken his neck" and "If he had jumped out of the window, he wouldn't have broken his neck" (because he would never have jumped without first installing a safety net). I am convinced by Jonathan Bennett's response, "Different standards for closeness to the actual world are *arbitrarily* associated with different temporal directions" (1984, p. 71). One can get the same inconsistencies, the same impossibility of finding a single way of determining closest or most similar possible worlds with respect to two forward conditionals. No single interpretation makes both counterfactuals true, because in endorsing one, we are holding fixed propositions that are inconsistent with those we hold fixed in endorsing the other. As pointed out in the above discussion of George jumping from the bridge, Lewis's own semantics for counterfactuals already requires some conditionals going backwards in time. Not only does the reformulation of Lewis's theory sketched in the previous section permit one to avoid ad hoc reliance on miracles, it also permits a unified treatment of counterfactuals both forward and backward in time.

There is another reason to favor the above account over Lewis's rules for comparing possible worlds. Counterfactual reasoning should permit one to work out the implications of counterfactual suppositions, so as to be prepared in case what one supposes actually happens. A child asks herself, "What will happen if I push that button?" and finds out by pushing the button and seeing what does happen. That can be a dangerous way to get an answer. It is handy to be able instead to pose counterfactual questions and to rely on their answers to predict what will happen. An analysis of counterfactuals ought to tie their truth closely to the truth of predictions concerning what will happen if . . . . A counterfactual of the form, "If I were to push the button, the alarm would go off," ought to license one to predict that the alarm will go off if one in fact pushes the button. Such a prediction may go astray, because of independent changes in other causes of the alarm's sounding, but such predictions must nevertheless be justifiable. Counterfactuals should satisfy a prediction condition.

Lewis's account of similarity among possible worlds implies that knowledge of counterfactuals will justify predictions about the results of *interventions*, which one can model as miracles. Consider the counterfactual: "If I were to push the button, the alarm would go off." On Lewis's account, in the closest possible worlds in which I push the button the other causal factors upon which the alarm sounding depends are unchanged, and my pushing the button is due to a minor difference between the laws of the actual world and the laws of this possible world. The alarm sounds in this possible world if and only if one can predict that it will sound in the real world when one *intervenes* and pushes the button. By definition, a button push that is brought about by an intervention is independent of changes in the values of any other variables, so the counterfactual is true if and only if the best prediction of what will actually happen if I push button is that the alarm sounds.

So far so good. But I would argue for a stronger prediction condition, which does not apply only in cases of intervention:

> **P** (*Prediction condition*) The knowledge that *b* would occur if *a* were to occur and that an event of kind **a** occurs taken by itself justifies the prediction that an event of kind **b** will occur.

By "taken by itself" I mean that one has no knowledge concerning the circumstances – that is, concerning whether other causes occur or other causal conditions obtain. Events of kind **a** instantiate the same causally relevant property that *a* does. The prediction condition says that knowledge of counterfactual dependence among tokens justifies the prediction that if a cause of the same kind occurs, then (ceteris paribus) an effect of the same kind will occur. **P** bears a vague resemblance to modus ponens: From "if *a* were to occur then *b* would occur" and "an event of kind **a** occurs" infer (fallibly) "an event of kind **b** occurs."

Counterfactuals should guide our thinking about nonexperimental as well as experimental situations. Regardless of whether one is concerned with the results of interventions or with the results of mere happenings, knowledge of the value of $x$ should provide more "guidance" concerning the value of $y$ when one knows that the value of $y$ counterfactually depends on the value of $x$ than when one does not know this. Knowledge of counterfactual dependencies should be reflected in our expectations about what will happen.

Consider, for example, the apparatus in figure 6.2, in which a salt solution flows through the pipes and mixing chambers. With constant flows, the salt concentration in chamber 5, $x_5$, is the average of the concentrations in chambers 3 and 4: $x_5 = .5(x_3 + x_4)$. This equation expresses a causal generalization: $x_3$ and $x_4$ influence $x_5$ in just this way. Paralleling the causal dependencies are apparently asymmetrical counterfactual dependencies. For example, the value of $x_5$ counterfactually depends on the value of $x_4$ and not vice versa. **P**, the prediction condition, says that knowing the counterfactual dependency of the value of $x_5$ on the value of $x_4$ justifies predicting that

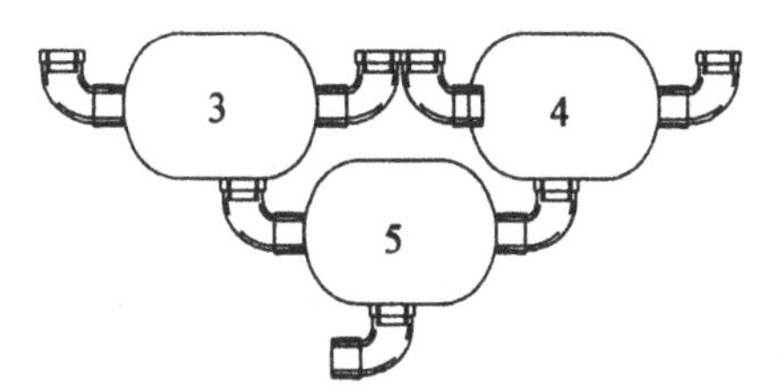


**Figure 6.2:** A simple causal structure

when one measures an increase in the concentration of basin 4, the concentration of basin 5 increases by half as much. Measuring an increased concentration in basin 5 does not, in contrast, justify any prediction concerning the concentration in basin 3 or the concentration in basin 4. It justifies only a prediction that one or both of these concentrations changes or that something changes in the mechanism carrying salt solution to basin 5. According to the prediction condition, counterfactual dependence makes this predictive difference.

*Lewis's account does not satisfy the prediction condition.* Consider the following four counterfactuals:

1. If *a* had occurred "miraculously" (such as via an intervention), then *b* would have occurred.
2. If *a* had occurred as a consequence of *d*, then *f* would have occurred.
3. If *a* had occurred, then *b* would have occurred.
4. If *a* had occurred, then *f* would have occurred.

According to Lewis's theory, if (1) is true, then (3) is true and (4) is false, regardless of whether or not (2) is true. One knows that (1) is true if and only if one knows that (3) is true. Suppose we know that (1) and (2) are both true and that some token of type **a** occurs. One cannot justifiably predict that a token of kind **b** will occur. Whether a token of kind **b** or of kind **f** occurs depends on what causes the token of kind **a**. The prediction condition says that if we know that (3), then a prediction that an event of kind **b** will occur will be justified (unless we have extra information concerning the state of other causal factors). Since the prediction is not justified and we do not have such extra information, we do not know that (3). By assumption we know that (1). Hence Lewis is mistaken to identify (1) and (3).

For example, engineers checking the design of a nuclear power plant may ask, "What would happen if that steam pipe were to burst?" They want their answer to match what one will observe in the event that the pipe actually does burst, though they hope never to make the observation. According to Lewis, they should consider a possible world exactly like ours until near the time of the pipe bursting, at which point some small miracle occurs, and the world evolves according to laws of nature like ours. In such a world, let us suppose that the reactor shuts down promptly.

The bursting of the pipe may have different consequences when it bursts because of an earthquake. The engineers will not and should not necessarily assume that the pipe burst because of a small miracle immediately preceding the bursting and they will not and should not predict that the consequences of the pipe's bursting will be that the reactor shuts down promptly. Knowledge that the pipe burst does not by itself justify *any* prediction about whether the reactor will shut down. The engineers need to do some backtracking and to say, "If the pipe were to burst, then either it was faulty, or a girder fell on it, or there was an earthquake, or there was sabotage, or the pressure became too great. The consequences of the bursting depend on which of these holds." *Responsible engineers must do such backtracking when the consequences of the pipe's bursting depend on what caused it to burst.* If the pipe burst because the pressure was too great, and the pressure was too great because the reactor was going out of control, then the consequences of the pipe bursting may be different than if it were caused by corrosion, a faulty weld, or a terrorist's bomb. In order to consider how the world would differ in the future in consequence of the bursting, the engineers must also think about how the world must have differed in order for the bursting to have occurred. Lewis's suggestion that the most similar possible worlds involve some small miracle just before the pipe bursts rules out the above reasoning, and it is for this reason mistaken. The counterfactual, "If the pipe were to burst, then the reactor would shut down safely" is *false*. No prediction is justified concerning the consequences of the bursting for whether the reactor shuts down until one specifies what caused the bursting.

The alternative view of counterfactuals sketched above (and developed in more detail in §6.2*) permits one to "hold fixed" laws of nature, not only in considering the consequences of *a*'s occurring, but also in considering what would have caused *a* to occur. It may not matter what caused some event to occur or not to occur, and then there is no harm in supposing it occurred in one specific way, such as by a miracle. But it may matter, and one will need to explore how the bursting could have followed lawfully from its causes (Bennett 1984, pp. 72–4). In considering what would happen if *a* occurred, one possibility is that *a* happened inexplicably, as if by intervention or miracle, but it would be irresponsible to suppose that that is the *only* way *a* might happen. If one accepts the prediction condition, then one should deny that worlds that differ in laws are more similar to the actual world than are worlds that differ in causal antecedents.

The quantitative example I introduced above underlines this point. In illustrating the asymmetry of counterfactual dependence and the content of the prediction condition, I assumed that the concentrations in basins 3 and

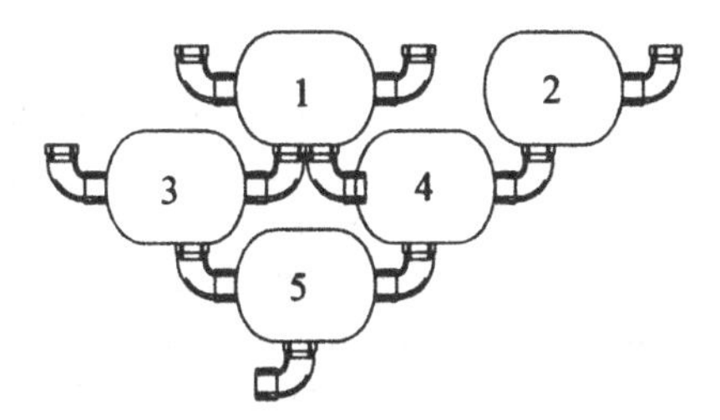


**Figure 6.3:** A structure with multiple connection

4 were independent of one another. But suppose $x_3$ and $x_4$ are not independent. Suppose that the apparatus is as shown in figure 6.3. The salt concentration in both chambers 3 and 4 depends on the salt concentration in chamber 1. Now one cannot justifiably predict what the concentration in basin 5 will be if the salt concentration in chamber 4 increases by $z^*$. If one insists on last-minute miracles, one will predict that the concentration in chamber 3 remains unchanged and the concentration in chamber 5 increases by $.5z^*$. But this prediction is not justified. When one measures an increase in the concentration in basin 4, the concentration in basin 3 need not remain unchanged, even if there are no independent changes. Nor need the concentration in basin 5 increase by $.5z^*$. These generalizations would be true if $x_3$ and $x_4$ were independent of one another, but they are not.

The consequences of the increase of the concentration in chamber 4 depend on its causes. If the increase in the concentration in basin 4 were due entirely to an increase in the concentration in basin 2, then the value of $x_5$ would increase by $.5z^*$. If the doubling were due to an increase in the concentration in basin 1, then the value of $x_5$ would increase by more than $.5z^*$. The concentration in $x_4$ might also result from an intervention such as the addition of salt through a hatch at the top of basin 4. A possible world in which an altered value of $x_4$ results from an intervention is no more similar to the actual world than is a possible world with a change in the values of $x_1$ or $x_2$ or both. The causes of the value of $x_4$ matter to its effects: in other words, $x_4$ does not screen off its causes from its effects. (An event *a* screens off its causes from a direct effect *b* if and only if *a* is independent of all other proximate causes of *b*.) Insisting on always postulating miracles at the last possible moment leads to a violation of the prediction condition. It is not the right way to prepare us to deal with actual happenings.

## 6.5 Critique of Lewis's Account of Similarity Among Possible Worlds

Should one conclude that Lewis is mistaken concerning similarities among possible worlds? In defense of Lewis, one can point out that only the state of the world at the moment when the pipe bursts or when there is a change in the salt concentration in basin 4 matters. Rather than inquiries into what caused the bursting or the change in salt concentration, one needs a correct specification of the values of all relevant *contemporary* variables. Once one specifies the value of $x_3$, there's no problem inferring the value of $x_5$ from counterfactual suppositions concerning the value of $x_4$. (Though with all the causes but one specified, there is equally little problem with the reverse inference.) Alternatively, a defender of Lewis's semantics can point out that one can consider counterfactuals with more complicated antecedents such as "If the pipe had been faulty and then had burst," or "If the pressure had become too great and the pipe had burst."

These maneuvers do not rescue Lewis's account. One is interested in the consequences of the pipe's bursting given the values of other relevant variables that one may actually encounter. And to determine what those values are, one needs to consider what might have caused the pipe's bursting, so that one can determine whether the other relevant variables depend on these causes. Similarly, one needs to backtrack to decide whether one needs to consider counterfactuals with more complicated antecedents and, if so, which ones one should consider. In determining the implications of a counterfactual supposition such as "What if this pipe burst?" one must backtrack.

A defender of Lewis's semantics could concede that backtracking is needed, yet maintain that backtracking plays only an epistemological role in the evaluation of counterfactuals. One backtracks to determine which counterfactual question to ask, not to give the answer. Backtracking helps one to transform counterfactual suppositions into counterfactual claims, not to determine whether counterfactual claims are true. Backtracking may be prevalent and important, but it has no role when one is stating the truth conditions of counterfactuals.

This response will not do. Suppose one tries to assess the counterfactual:

(C) If the value of $x_4$ were to increase by $z^*$, then the value of $x_5$ would increase by $.5z^*$.

Consider figure 6.4. The actual situation is depicted in figure 6.4a. $x_1 = x^*_1$, $x_2 = x^*_2$, $x_3 = x^*_3$, $x_4 = x^*_4$, $x_5 = x^*_5$ and the causal relations are as shown. Consider then the possible situations depicted in figure 6.4b, c, and d. In each of these, $x_4$, the concentration in basin 4, increases to $x^*_4 + z^*$. But the source of the increased concentration differs. In $w_1$ depicted in figure 6.4b, the increase is due to some difference between the laws of $w_1$ and the actual world (or to some unspecified mechanism or intervention), which leaves the values of $x_1$ and $x_2$ and $x_3$ and the causal structure relating $x_3$ and $x_4$ to $x_5$ unaffected. In $w_2$, shown in figure 6.4c, the increase in $x_4$ occurs because of an increase in $x_1$. In $w_3$, shown in figure 6.4d, the increase in $x_4$ is due to an

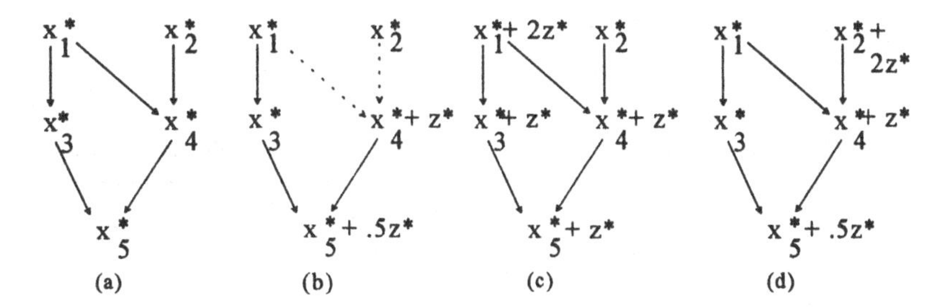


**Figure 6.4:** Which is closest?

increase in $x_2$. In $w_1$ and $w_3$ the value of $x_5$ is larger than it is in the actual world by $.5z^*$, while in $w_2$ the value of $x_5$ increases by $z^*$.

If we set aside questions suggested by Lewis's talk of "orderly transitions" (1979, p. 40), Lewis's semantics says that $w_1$ is closer to the real world than $w_2$ or $w_3$, and the counterfactual **C** is true. $w_1$ is closer to the actual world because the period of "perfect match" between $w_1$ and the actual world is longer than the period of perfect match between the actual world and either $w_2$ or $w_3$. This is a weak reason. Why should a few seconds of additional perfect match be decisive? Lewis answers in effect that unless a few seconds more match are decisive, one cannot account for our judgments of counterfactual and causal dependence. The alternative theory presented here undercuts this reason, because it accounts for our judgments concerning counterfactual and causal dependence without requiring that $w_1$ be more similar to the real world than is $w_2$ or $w_3$. All that remains is the dubious intuitive argument that any increase in the period of perfect match makes for greater similarity.

According to my account $w_2$ and $w_3$ are at least as similar to the actual situation as is $w_1$, and the counterfactual **C** is false. One needs to know how the increase in the salinity of basin 4 occurred before one can make any predictions or any true counterfactual claims concerning how much the concentration in basin 5 would increase. The falsity of the counterfactual **C** results from the "double-connection" between $x_1$ and $x_5$ – one connection via $x_3$ and one via $x_4$. If $x_1$ were not a cause of $x_3$ as well as $x_4$ – if $x_3$ and $x_4$ were causally independent of one another – there would be no need to bring $x_1$ or $x_2$ into the picture.

This account denies that non-*b* possible worlds involving miracles immediately before *b* are more similar to the actual world than non-*b* possible worlds without miracles right there. This account does not maintain that the most similar possible worlds will be free of miracles. It is plausible to maintain, as Lewis does, that possible worlds with completely different histories are very unlike the actual world. When there are no multiple connections between a cause and one of its effects, or, equivalently, when the causes of an event *b* do not matter to its effects, then one is free to suppose that *b* failed to occur by a miracle. There is no requirement that one keep backtracking endlessly. One stops backtracking when it no longer matters to the consequences of a supposition how the changes in causal ancestors came about, or when one comes to an event whose direct causes are all causally independent of one another. The context, one's purposes, and the causal relations enable one to isolate a "system" of interest in which miracles should be avoided.

### 6.6 Refutation of L and Defense of a Restricted Version

In §6.3 I showed how a view of comparative overall similarity that did not favor miracles, coupled with other plausible conditions, implies a simplified formulation of Lewis's sufficient condition for causation and permits one to deny that individual causes counterfactually depend on individual effects or that effects of a common cause counterfactually depend on one another. In the last sections I argued for this alternative view and against Lewis's account of similarity on the grounds that Lewis's account does not permit backtracking and that it implies the truth of counterfactuals that do not justify predictions. If one then accepts my alternative account, one must give up any hope of providing a noncircular counterfactual theory of causality, because similarity among possible worlds and the truth of counterfactuals would depend on explicitly causal facts.

One might, however, hope to defend a theory of the relations between causation and counterfactuals similar to the one sketched in Herbert Simon and Nicholas Rescher's essay, "Cause and Counterfactual" (1966). They link the asymmetry of causation to an asymmetry of counterfactual determinacy. Particular effects counterfactually depend on each of their causes, while particular causes do not counterfactually depend on any of their effects.

This hope cannot be sustained, because the necessary condition that **L** states is false. Causal dependence does not imply counterfactual dependence. The concentration in basin 5 causally depends on the concentration in basin 4, but the concentration in basin 5 does not counterfactually depend on the concentration in basin 4. Simon and Rescher's claim about the connections between causation and counterfactuals and Lewis's theory are both mistaken.

This difficulty does not derive from my suppression of Lewis's distinction between causal dependence and causation. Lewis's theory permits $x_4$ to be a cause of $x_5$ without any counterfactual dependence of $x_5$ on $x_4$. All that's needed is a chain of counterfactual dependence. But there is no chain of counterfactual dependence here. The salt concentration at the top of the pipe between basins 4 and 5 counterfactually depends on a change in concentration in basin 4, and the concentration at the bottom of the pipe counterfactually depends on the concentration at the top of the pipe. But the concentration in basin 5 does not counterfactually depend on the concentration at the bottom of the pipe. This is a case of causation without any chain of counterfactual dependence.

One might attempt to defend **L** as follows: "Regardless of how the change in salt concentration in basin 4 comes about, if the concentration in basin 4 were to change, then so would the concentration in basin 5. So there is after all no difficulty with **L**." This defense is unsatisfactory, and not only because Lewis also wants his account to apply to quantitative causal dependence (1973a, p. 166). Suppose in figure 6.3 that device number 1 is not a basin holding a salt solution, but some mechanism that generates exactly neutralizing quantities of an acid and an alkali, which flow through the pipes to basins 4 and 3, respectively. If the change in the acidity in basin 4 is due to a greater acid output from device 1, then there will be no change in the acidity of 5, because the greater acid output from 1 transmitted to 4 is neutralized by an increase in alkali output transmitted to 3. It would thus be false to say, "If the acidity of 4 were greater, then the acidity of 5 would be greater." If, for example, one had a switch which enabled one to affect the combined acid-base output rate of device 1, then when one moved the switch, the acidity of 4 would change without any change in the acidity of 5. One can give a similar qualitative example. It could be the case that whether the solution in 5 is acid at all does not depend counterfactually on whether there is any acid in 4. Yet the acidity of 5 causally depends on the acidity of 4. Causal dependence does not imply counterfactual dependence.

If one insists that causal dependence *must* be reflected in counterfactual dependence, one must take counterfactual suppositions, such as "if the acidity of 4 had been different" as suppositions that the differences came about via miracles. In that case, as already explained, one will be in the position of saying, "If *a* were to occur, *b* would occur, but one cannot predict whether events of kind **b** will occur when events of kind **a** occur." Second, the grounds for accepting Lewis's view of similarity among possible worlds would be one's knowledge of causal relations and one's desire to equate causal and counterfactual dependence. One would have tacitly abandoned the attempt to give a counterfactual theory of causation and in its place one would be offering a causal theory of counterfactuals – the theory of similarity among possible worlds would be grounded in an account of causation. **L** is false.

Let us say that there is a "multiple connection" between *a* and *b* if some cause *d* of *a* is or in the absence of *a* would be connected to *b* by a path that does not go through *a* (as $x_1$ is connected to $x_5$ via both $x_3$ and $x_4$). There will be a multiple connection between *a* and *b* if and only if controlling for events of kind **a** in these circumstances does not screen off *b*'s from the causes of *a*'s. If there is a multiple connection between *a* and *b*, then *b* will not counterfactually depend on *a* just as *a* does not counterfactually depend on *b*. One will have a case of causation without any chain of counterfactual dependence, and there will be no asymmetry of counterfactual dependence.

**L** can, however, be defended as an approximate truth, and a restricted form of **L** can be proven. When there are no multiple connections, **L** is true. Since situations involving multiple strong connections are rare, **L** is a good approximation. Given the connection principle and the revised account of similarity among possible worlds that is implicit in the discussion above, it can be proven that if there are no multiple connections between cause and effect, then causation implies counterfactual dependence. This claim is formulated more rigorously as theorem 6.4. Lewis's theory, restricted to circumstances in which there are no multiple connections, follows from a revised account of similarity among possible worlds, the claim that if distinct events are counterfactually dependent then they are causally connected, and the independence theory of causal priority presented in chapter 4.

## 6.7 What Does the Counterfactual Theory of Causation Teach Us?

The view of counterfactuals sketched in this chapter is not equivalent to:

> *A mistaken view*: If **P** were the case then **Q** would be the case if and only if **Q** is deducible from a nonredundant conjunction of statements including **P**, laws of nature, and specifications of the circumstances.

Consider the relationship between the salt concentrations in basins 1, 2, and 4 in figure 6.3, and suppose that the concentrations in basins 1 and 2 are causally independent of one another. If concentration in basin 2 is part of the "circumstances," one can deduce the concentration in basin 4 from the concentration in basin 1 or vice versa. If the concentration in basin 2 is not part of the circumstances, one can make no deductive inference in either direction. Deduction from laws and circumstances implies a *symmetrical* relationship between $x_1$ and $x_4$, but the counterfactual dependence here is asymmetrical, since there is (by assumption) no multiple connection. The concentration in basin 4 is counterfactually dependent on the concentration in basin 1, while the concentration in basin 1 is not counterfactually dependent on the concentration in basin 4. Where does the asymmetry come from?

The asymmetry arises from the fact that the values of $x_1$ and $x_2$ are causally connected to the value of $x_4$, but independent of one another. One has an "intransitive triplet" (Pearl and Verma 1994, p. 804). So when one supposes that $x_1$ has some other value, one "holds fixed" $x_2$, but when one supposes that $x_4$ has some other value, one does not hold fixed $x_2$, because the different value of $x_4$ may be due to a difference in the value of $x_2$. This is not a human quirk. If one measures different values of $x_1$ and predicts the value of $x_4$ on the assumption that the value of $x_2$ remains fixed, one will not necessarily be right, because the value of $x_2$ may vary independently. But if $x_1$ and $x_2$ are independent and one knows nothing about how $x_2$ may have varied, the calculated values of $x_4$ will be the best prediction of the actual values. This *fact* justifies making the prediction. If one measures different values of $x_4$ and calculates values of $x_1$ on the assumption that $x_2$ remains fixed, the calculated values will not be the best predictions of the actual values of $x_1$. On the contrary, the variations in the measured values of $x_4$ constitute evidence that the value of $x_2$ has changed. Thus statisticians prove that independence of error terms – of the other causes – is a necessary condition for unbiased estimation (Festa 1993, p. 39).

When there are no multiple connections, a causal intermediary *a* screens off its causes from its effects. It does not matter what caused *a* to occur. The other factors that contribute directly to the effects of *a* are causally independent of *a*. So one can hold them fixed, and the effects of *a* are individually counterfactually dependent on *a*. When there are multiple connections, as in the example of the basins, some of the causes of the given effect are not causally independent of one another (as, for example, the values of $x_3$ and $x_4$), and it is no longer the case that one should "hold fixed" the value of one when one supposes that the value of the other changes. It is the independence of causes that permits the counterfactual dependence of individual effects on individual causes and defeats the counterfactual dependence of individual causes on individual effects. But when there are multiple connections, not all the causes are independent, and individual effects do not counterfactually depend on individual causes.

These links between independence and counterfactual dependence establish a restricted version of **L**. **L** is thus not a tenable alternative account of causal priority. It is not tenable for the reasons already spelled out in this chapter. It is not an alternative, because it more or less presupposes the independence condition and, in deterministic single-connection circumstances, it follows from the independence theory, the claim that causal connection is a necessary condition for counterfactual dependence among distinct events, and a view of similarity among possible worlds that is at least as plausible as Lewis's.

Even if unacceptable as a theory of causation, a counterfactual theory can still tell us some things worth knowing about causation. Although there are cases in which *a* is a cause of *b*, even though *b* does not depend counterfactually on *a*, such cases are infrequent unless the causal connection between *a* and *b* is remote. It is seldom the case that there are multiple connections in which more than one of the connections is strong enough to worry about. Consequently, it is a good first approximation to say that there is an asymmetry of counterfactual dependence. It is worth noting the connections between causation and counterfactual dependence, even if one cannot defend a counterfactual theory of causation.

# 6*

# Independence and Counterfactual Dependence

After briefly discussing two other counterfactual theories of causation, this chapter formulates the account of similarity among possible worlds employed in chapter 6 and proves the claims it makes.

## 6.1* Mackie's Counterfactual Theory

John Mackie argued,

> if on a particular occasion *A*'s doing *X* is causally related to *B*'s doing *Y*, and if they had not been so related but things had otherwise been as far as possible as they were, *A* would still have been doing *X* but *B* would (or might) not have been doing *Y*, then *A*'s doing *X* is conditionally and causally prior to *B*'s doing *Y*. (1979, p. 24)

For example, suppose one breaks the connection between a car's engine and its wheels: the engine continues turning while the wheels stop.

Let us call the event of *A*'s doing *X* on the particular occasion "*a*" and the event of *B*'s doing *Y* on the occasion "*b*." Suppose that in the closest possible world in which some event *c* failed to occur, *a* would not be causally connected to *b*. If the causal connection between *a* and *b* depends on the existence of *c*, then *c* must be a cause of at least one of *a* or *b*. On Mackie's view *a* is causally prior to *b* if and only if, in the absence of some minimal difference *c*, *a* still occurs, but *b* does not (*B* does not do *Y*). So *c* causes *b* only. Mackie thus indirectly assumes that there are causes of *b* that are not causes of *a*, and this assumption is, of course, an immediate implication of both **CP** and agency views. One can thus capture Mackie's intentions without introducing the further complications raised by such counterfactuals. These complications are nevertheless enlightening. Mackie has pointed out a further feature of causal priority, which is neatly explained by **CP** and $\mathbf{AT_g}$.

## 6.2* Swain's Theory of Causal Asymmetry

Although invoking last-minute miracles neatly permits one to deny the counterfactual dependence of causes on effects, Marshall Swain suggests that invoking miracles also leads one mistakenly to deny the counterfactual dependence of effects on causes. A world with a "pure deletion miracle" which "deletes" a cause "but which leaves the *rest* of the actual world entirely unchanged" would apparently be extremely similar to the actual world (1978, p. 9). If such a world is more similar than one in which both the cause and its effects are missing, then effects do not depend on their causes.[1]

Swain proposes the following alternative to Lewis's theory:

**D7′** Where *c* and *e* are specific events that occurred, *c* is a cause of *e* iff:

1. there is a chain of occurrent events from *c* to *e*;
2. where $w_1$ is a world in which *c* occurs and *e* does not occur, and *w* is the actual world, $w_1$ would only. . . have to have been different from *w* in the following respect: some cause *a* (other than *c*) which occurs in *w* and upon which *e* depends causally in *w* fails to occur in $w_1$;
3. where $w_2$ is a world in which *e* occurs and *c* does not occur, and *w* is the actual world, $w_2$ would have to be different from *w* in at least the following respects: (1) some event *f* (other than *e*) which occurs in *w* and upon which *c* depends causally in *w* fails to occur in $w_2$; and (2) some event *g* occurs in $w_2$ such that *e* is not causally dependent upon *g* in *w* but *e* is causally dependent upon *g* in $w_2$.[2]

In Swain's account, the laws of similar possible worlds such as $w_1$ and $w_2$ are the laws of the actual world, *w*. So for events to fail to occur or for new events to occur, the past might have to be drastically different. Swain accepts Lewis's notion of causal dependence: *e* causally depends on *c* if and only if *c* and *e* are distinct and if *c* had not occurred, then *e* would not have occurred. But he denies that causal dependence is generally asymmetric. If, as Swain holds, the laws of the most similar possible worlds are the laws of the actual world, then causes will often causally depend on their effects. Thus on Swain's account, the asymmetry of causation does not derive from an asymmetry of causal dependence. It rests instead on the fact that worlds in which *c* occurs without *e* require only that some other cause or causal condition of *e* not occur, while to have *e* without *c*, it must be the case both that some cause *f* of *c* fails to occur and that some other event *g* occurs, which causes *e* in the absence of *c*.

Swain's account supposes that whenever *c* causes *e* there is some distinct cause *a* of *e* that is not itself causally dependent on *c*. Otherwise there would be no easy way to have *c* without *e*. Similarly when Swain discusses effects of a common cause, he requires that each have its own distinct and inde-

[1] This criticism of Lewis preceded the publication of Lewis's "Counterfactual Dependence and Time's Arrow" (1979) and seems to be answered by Lewis's discussion there of the asymmetry of miracles and overdetermination. For other criticism of Swain's view see Davis (1980).

[2] 1978, p. 11. Swain's footnote (after "only") is omitted. The last two occurrences of "*g*" in the quotation are misprinted as "*f*" in the printed text.

pendent causes (1978, p. 12). So Swain's account presupposes something like the independence condition **I**, and unless Swain would take issue with the connection principle or transitivity, Swain is thus committed to the independence theory, **CP**. Once committed to **CP**, there is little reason to pursue a counterfactual theory as an *alternative*.

Swain's revision of Lewis's theory is problematic. There is no reason why the causal structure might not be such that the absence of *f* required in the second clause of **D7′** in order for *c* to be absent might not itself be sufficient to bring about *e*.[3] More generally, one might state the following worry: **D7′** replaces Lewis's claims that non-*e* possible worlds with *c* are more similar to the actual world than non-*e* possible worlds without *c*. **D7′** compares numbers of changes in the immediate vicinity of *c* and *e* rather than overall similarity. Second, rather than comparing non-*e* possible worlds, **D7′** compares *e*-and-non-*c* possible worlds to *c*-and-non-*e* possible worlds. These two changes raise two questions: What does the number of different occurrences or nonoccurrences in the immediate vicinity of *c* and *e* have to do with causal asymmetry? Second, what is the relevance of a comparison between *c*-and-non-*e* possible worlds and *e*-and-non-*c* possible worlds? Until these two question are answered, the account seems arbitrary.

## 6.3* An Alternative Account of Similarity Among Possible Worlds

The argument in chapter 6 relies on the following general principle concerning the similarity among possible worlds:

> **SIM** (*Similarity among possible worlds*)
>
> 1. *Worlds with miracles are not the most similar.* For any event *b* there are non-*b* possible worlds without at least one of *b*'s causes that are at least as close to the actual world as are any non-*b* possible worlds in which all of *b*'s causes occur.
>
> 2. *It doesn't matter which cause is responsible.* For any event *b*, if *a* and *c* are any two causes of *b* that are causally and counterfactually independent of one another, there will be non-*b* possible worlds in which *a* does not occur and *c* does occur that are just as close to the actual world as are any non-*b* possible worlds with *a* and without *c*, and there will be non-*b* possible worlds without *a* and with *c* that are just as close to the actual world as are any non-*b* possible worlds without both *a* and *c*.[4]

[3] Such a case would involve a multiple connection between *f* and *c* – one connection via *c* and another via some other chain of consequences of *f* not appearing. As we will see below, multiple connections create problems for Lewis's theory, too.

[4] On the grounds that the fewer the differences, the more similar the worlds, one might question whether non-*b* worlds without both *a* and *c* can be just as similar to the actual world as are non-*b* worlds without just one of these causes. Those who find this plausible can change this clause of **SIM**. The arguments in this chapter, including the proofs of the three theorems employing **SIM** as a premise, go through just the same. If one accepted the reasoning that the fewer the differences, the

3. *The fewer the irrelevant differences in events, the more similar the world.* For any event $b$, consider two possible events $e$ and $f$ that are not causally connected to $b$, where $e$ occurs and $f$ does not. Then there are non-$b$ possible worlds with $e$ that are more similar to the actual world than any non-$b$ possible worlds without $e$, and there are non-$b$ worlds without $f$ that are more similar to the actual world than any non-$b$ worlds with $f$.

4. *The fewer the irrelevant differences in laws, the more similar the world.* For any event $b$ there is a non-$b$ possible world in which all other laws, apart from those relating $b$ to its causes or the causes of $b$ to one another, are the same as the actual world that is more similar to the actual world than is any non-$b$ possible world which differs with respect to some such laws.

The first clause in **SIM** denies that possible worlds with miracles between $b$ and its causes are more similar to the actual world than are possible worlds where the causal relations between $b$ and its causes hold and the miracle, if any, comes earlier. It does not insist that laws must be held sacrosanct. Lewis rejects (1), but his main reason seems to be that he cannot account for the asymmetry of causation unless he locates miracles as late as possible. The second clause maintains that it is equally easy to get rid of $b$ by getting rid of any combination of its independent causes. This clause is neither asserted nor denied by Lewis. The third clause says that unrelated changes detract from similarity, while the fourth says that additional differences in laws detract from similarity. Lewis would endorse the last two clauses of **SIM**. **SIM** does not offer comprehensive rules for comparing the closeness of possible worlds. Only the first clause of **SIM** is incompatible with Lewis's account.

Unlike the first, third, and fourth clauses of **SIM**, the second is not plausible. Its difficulties are discussed above in connection with the example of George jumping off the Brooklyn Bridge (p. 116). The difficulties concerning similarity are just as serious for Lewis's version of the theory as they are for the revision I am exploring. I think that difficulties in specifying a plausible relation of comparative overall similarity among possible worlds give one good reason to avoid counterfactual theories altogether.

## 6.4* Independence and Counterfactual Dependence

To derive implications concerning counterfactuals from the independence theory of chapter 4, one needs one further condition:

**CDCC** (*Counterfactual dependence implies causal connection*) If $a$ and $b$ are distinct events and $b$ counterfactually depends on $a$, then $a$ and $b$ are causally connected.

more similar the worlds, one would have to predict in figure 6.4 that $x_5$ has one of two possible values instead of any value between the two values. It is for this reason that I prefer the present formulation.

**CDCC** states the metaphysical principle that counterfactual dependence among distinct events is always causal or nomological. This principle is not obvious, but it is widely accepted, and Lewis can have no objection to it, since it is implied by his theory. Given the links between causal connection and probabilistic dependence, **CDCC** implies that counterfactual dependencies between individual events will typically be reflected in probabilistic dependencies in the circumstances between events of the relevant kind. One can then prove:

**Theorem 6.1: SIM, CDCC,** and **I** imply that individual causes will not be counterfactually dependent on individual effects and effects of a common cause will not be counterfactually dependent on one another.

Proof: **I** says that for every cause $a$ of every event $b$ there will be some other cause $c$ of $b$ causally independent of $a$. Given **CDCC**, $a$ and $c$ will also be counterfactually independent. Clause 2 of **SIM** then implies that for any non-$b$ world without $a$, there will be a non-$b$ world without $c$ that is at least as similar to the actual world. So individual causes will not be counterfactually dependent on their effects. **I** also implies that if $b_1$ and $b_2$ are effects of a common cause $a$, then $b_1$ has a cause $a_1$ that is independent of any cause of $b_2$, and $b_2$ has a cause $a_2$ that is independent of any cause of $b_1$. Given **CDCC** and **SIM** there will be non-$b_1$ possible worlds without $a_1$ that are just as close to the actual world as are non-$b_1$ possible worlds without $a$, and in those possible worlds $b_2$ will still occur. So $b_2$ does not counterfactually depend on $b_1$. Given **SIM** there will also be non-$b_2$ possible worlds without $a_2$ that are just as close to the actual world as are non-$b_2$ possible worlds without $a$, and in those worlds $b_1$ will still occur. So $b_1$ is not counterfactually dependent on $b_2$.

**Theorem 6.2: CDCC, SIM, I,** and **CC** imply that if $b$ counterfactually depends on $a$, then $a$ causes $b$.

Proof: Suppose $b$ counterfactually depends on $a$. Then (by **CDCC**) $a$ and $b$ are causally connected. **I** and **SIM** imply that $b$ does not cause $a$ and that $a$ and $b$ are not causally connected only as effects of a common cause. By **CC** it follows that $a$ causes $b$.

Artificial event fusions create difficulties for the claim that individual causes are not counterfactually dependent on their effects. Suppose that determinism is true and that $a_1, \ldots, a_n$ cause $b$. If $b$ were not to occur, while $a_1, \ldots, a_{n-1}$ occurred, then if miracles between the cause and effect under consideration are not allowed, the other cause, $a_n$, must have failed to occur. This does not, of course, show that if $b$ had failed to occur $a_n$ would not have occurred. It shows instead that if $a_1$ through $a_{n-1}$ occurred and $b$ had failed to occur, then $a_n$ would not have occurred. Let $e$ be the event that occurs whenever $a_1$ through $a_{n-1}$ and $b$ all occur. So when $a_1$ through $a_{n-1}$ occur and $b$ does not, then $e$ does not occur. Then it might appear that $a_n$ is counterfactually dependent on $e$, and if counterfactual dependence is sufficient for causation, one will be led to the false conclusion that $e$ causes $a_n$. This claim is problematic, since there are other ways that $e$ might fail to occur, but to give a general answer to this objection requires that one say something about artificial events such as *e*. In chapter 13 I shall attempt to say something about when event fusions can stand in causal relations and how their causal relations depend on the relations among natural tropes. At this point, however, I am concerned only with relations among natural events or tropes, and *e* is not a natural event.

### 6.5* The Asymmetry of Overdetermination

Qualms about **I** do not give one reason to prefer Lewis's version of his own theory, because **I** is implicit in the view that worlds with small miracles are closest, and because it is hard to explain the asymmetry of overdetermination, which Lewis's theory requires, if **I** is not true. Ordinary overdetermination, like that discussed in §3.5 and in §13.1, involves the existence of multiple minimal sufficient conditions. The overdetermination that Lewis is concerned with involves individual conjuncts. Let us then say that a trope *e* is *determined* by *f* if and only if *f* is sufficient in the circumstances for *e*. *f* is sufficient in the circumstances for *e* if *e* is necessary in the circumstances for *f*. Necessity in the circumstances is defined by **DC** (p. 43). *e* is overdetermined by *f* and *g* if and only if *f* and *g* are both sufficient in the circumstances for *f*.

Consider then the following strong restatement of Lewis's asymmetry of overdetermination claim:

> **AOD** (*The asymmetry of overdetermination*) If causation is deterministic, then (1) events will be determined by a great many of their (natural) effects, and (2) events will be not be determined by any of their (natural) causes.

The following theorem shows that **AOD** follows from **I** when causation is deterministic and there is no preemption or ordinary overdetermination.

**Theorem 6.3:** If there is no preemption or ordinary overdetermination (overdetermination by conjunctions of natural causes), then **DC** and **I** entail **AOD**.

Proof: Given **DC** and the absence of ordinary preemption or overdetermination, the conjunction of (the properties of) the cause tropes is necessary in the circumstances for the effect trope, and so each cause is determined by each of its effects. Suppose that *a* causes *b*. By **I** there will another cause *f* of *b* that is causally independent of *a*. Since *a* will not be necessary in the circumstances for *f* nor vice versa, *a* by itself – that is, without *f* – will not be sufficient in the circumstances for *b*. Effects will not be determined by any of their individual causes.

Individual causes are (given the absence of alternative sufficient conditions) necessary, but not sufficient for individual effects, so individual effects are sufficient for individual causes. Since there are cases of preemption and ordinary determination, it is not true that events are always overdetermined by their effects, but since preemption and ordinary overdetermination are rare, it is usually the case that events are overdetermined by their effects, while they are never overdetermined by their causes. **I** is only sufficient for the asymmetry of overdetermination, not necessary. Only the much weaker condition that not all of the causes of an event determine one another is necessary. If **I** were false, the separate causes of an event would come closer and closer to overdetermining it as the lawful connections among the separate causes grew tighter. Although **I** is not necessary for the asymmetry of overdetermination, it is difficult to see why the asymmetry of overdetermination should be true if **I** were not true.

## 6.6* Proof of a Restricted Version of L

As argued in §6.6, causation does not imply counterfactual dependence when there are multiple connections between cause and effect. So the necessary condition that **L** states and consequently **L** itself are false. But one can prove:

**Theorem 6.4:** Given **CC**, **DC**, **SIM**, and no multiple connections – that is, if *a* had not occurred, no cause of *a* would have been a cause of *b* –, if *a* causes *b*, then *b* is counterfactually dependent on *a*.

Proof: Suppose that *a* causes *b* and there is no preemption or overdetermination. Then, by **DC** *a* is necessary in the circumstances for *b*. Suppose (counterfactually) that *a* does not occur, and consider the possible worlds in which *b* occurs anyway. In some worlds *b* occurs miraculously, but these will be very unlike the actual world. In others *b* occurs because of some causes. Those causes cannot be preempted actual causes, because by assumption there is no preemption. They cannot be a causes of *a*, since there are no multiple connections. So whatever causes *b* must be some new occurrence or nonoccurrence of something that does not cause *a* or *b* in the actual world, but which, as a consequence of some new law, causes *b* in this possible world. By clauses 3 and 4 of **SIM**, possible worlds with *b* will be less similar to the actual world than are some possible worlds without *b*. So *b* is counterfactually dependent on *a*.

Theorems 6.2 and 6.4 obviously imply:

**Theorem 6.5: CC, I, DC, SIM,** and **CDCC** entail **L** restricted to circumstances in which there are no multiple connections.

I have argued that **SIM** is more plausible than Lewis's account of similarity among possible worlds. **CDCC** is as plausible as any claim I know of relating causation to counterfactuals and unobjectionable to counterfactual theorists such as Lewis. **DC** merely restricts the argument to deterministic relations between cause and effect. So theorem 6.5 shows that the independence theory of chapter 4 explains what is true about the counterfactual theory.

Given the precise formulation of **SIM**, **L** (restricted to circumstances without multiple connections) does not directly imply or presuppose the truth of **I**. Suppose some event *e* has only one cause *c* or only two causes $c_1$ and $c_2$ that are causally connected. One can still deny the counterfactuals "if *e* had not occurred *c* would not have occurred" or "if *e* had not occurred $c_1$ would not have occurred" on the grounds that possible worlds in which *e* fails to occur as the result of a miracle need be no less similar than are possible worlds in which the causes of *e* failed to occur. If, on the other hand, one maintained that worlds with the same laws relating events to their causes as the actual world are *more similar* (not merely no less similar) to the actual world than are worlds with miracles, then **L** would imply **I**. One also needs something very like **I** in order to defend Lewis's asymmetries of miracles and of overdetermination.

# 7

# Counterfactuals, Agency, and Independence

In the previous two chapters I examined the relations among three theories of causation: the independence theory (**CP**), Lewis's counterfactual theory (**L**), and one version of an agency theory ($\mathbf{AT_g}$). That discussion left many loose ends, which this chapter takes up. In particular I need to say something about a token-level formulation of agency theory and about how agency theories and counterfactual theories relate to one another. This chapter also addresses a criticism of the independence condition, which one defender of an agency view has formulated as a criticism of Lewis's theory. Examining that criticism will show the centrality of independence to causation. This chapter ends by pulling together the defense of the independence theory scattered through previous chapters and summarizing the relations between time, agency, counterfactuals, and independence.

## 7.1 Agency, Counterfactuals, and Independence

Agency theories can be formulated as counterfactual token-level theories rather than (as in chapter 5) as type-level theories. Furthermore, the notion of an intervention helps counterfactual theories avoid counterexamples. One might wonder whether **L** might be derivable from some version of agency theory. Agency and counterfactual theorists might profitably join forces.

The following token-level formulation of agency theory is analogous to the type-level formulation of chapter 5:

> **AT** *(Agency theory)* *a* causes *b* if and only if *a* and *b* are distinct, and if *a* had come about as a result of a direct manipulation, then that intervention would have been a cause of *b*.

One might want the results of interventions to provide truth conditions for a nontransitive relation of causal dependence and then define causation, as Lewis does, in terms of chains of causal dependence. But as in the last chapter, I shall omit this complication and postpone grappling with the problems of preemption until chapter 13. **AT** can be formulated to require only that the intervention be nomically connected to *b*. It is natural to think of an intervention counterfactually, but awkward to formulate agency theory as a counterfactual theory of token causation, because the "lever" that one "wiggles" is the variable or property, not a token event.

A token event *a* comes about via a human intervention – that is, as a result of a direct manipulation *i* – if and only if the human action *i* is a direct cause of *a* that has no causal connections to any other events of interest except those that follow from its being a direct cause of *a*. In a causal graph in which edges represent direct causal relations, there is no edge into the vertex representing *i* and only one arrow out of it pointing toward the vertex representing *a*. Abstract interventions are events satisfying these conditions whether or not they are human actions.

As in the case of the type theory, agency theorists must assume that interventions are possible, and **AT** (like $\mathbf{AT_g}$) follows from the definition of an intervention and the assumption that interventions are possible. But the derivation of **CP** from **AT** (which would be analogous to the derivation of $\mathbf{CP_g}$ from $\mathbf{AT_g}$) does not go through. **CP**, the independence theory, concerns the actual situation, while **AT** concerns what would happen if events came about via interventions.

Consider the following variant of **CP**:

> **CP*** (*Modal independence theory*) *a* causes *b* if and only if *a* and *b* are causally connected and for all possible events *c* distinct from *b*, if *c* were causally connected to *a*, then *c* would be causally connected to *b*.

**CP*** states a logically stronger necessary and sufficient condition for "*a* causes *b*." It must be true in the closest possible worlds as well as in the actual world that nothing nomically connected to *a* is independent of *b*. If the connection principle and transitivity hold in all causally possible situations, **CC** and **T** together with the semantics for counterfactuals presented in chapters 6 and 6* imply the necessary condition **CP*** states.

At the same time **CP*** states a logically *weaker* necessary and sufficient condition for "*a* does not cause *b*." According to **CP*** *a* may fail to be a cause of *b* even though there is nothing in the actual situation distinct from *b* that is causally connected to *a* and not to *b*. **CP*** only requires that some possible world containing an event distinct from *b* that is causally connected to *a* and not to *b* is more similar to the actual world than any world without such an event. So **CP*** does not entail **I**, and **CP*** rather than **CP** is deducible from the token version of agency theory. **CP*** rather than **CP** thus has a claim to be the objective correlate of a token-level agency theory. Since it derives from the assumption that (token) interventions are possible, it will be more plausible than **CP** if that assumption is more plausible than is the independence condition and if one is willing to face the difficulties attached to counterfactual theories.

Why did the type-level derivation go through? What enables $\mathbf{CP}_g$ (unlike **CP**) to link up so nicely to agency theories is that $\mathbf{AT}_g$ derives from the assumption not merely that interventions are *possible*, but that intervention variables or types *exist*. I argued (p. 94), however, that intervention variables exist if and only if (token) interventions are possible. Whenever token interventions are possible, there must be some open line of causal "attack." If an intervention is like moving some lever, there must be some lever to be moved. Interventions don't install such levers, they move them. The levers must still be there when no intervention takes place. (And when interventions install levers, there must be something to connect them to.) I am not sure about the status of this assertion. It is, I suggest, implicit in the conception of a *causal structure* that is embodied in causal relations among events. If it is the case that interventions are possible if and only if there are intervention variables, then the theorems of chapter 5* capture the links between agency and independence. With the addition of this premise – that interventions are possible if and only if there are intervention variables – **AT** entails **CP**, and the possibility of intervention implies independence (see theorems 7.3 and 7.4, p. 153).

## 7.2 Agency and Counterfactuals

Agency theorists generally suppose that when *a* causes *b*, one could have prevented *b* by intervening to prevent *a*. This suggests a simpler formulation of agency theory:

> **AT′** (*Counterfactual agency theory*) *a* causes *b* if and only if *a* and *b* are distinct, and if one intervened and prevented *a*, then *b* would not occur.

As in the presentation of Lewis's theory, **L**, I am taking the claim that *a* and *b* are distinct events as implying that they exist. There are two reasons to define **AT′**. First, it is worth demonstrating that the relations I have established between the independence view of chapter 4 and agency views do not depend on a gerry-rigged formulation of agency theory. Theorems 7.5 and 7.6 in chapter 7* accomplish this. This version of agency theory, like **AT**, follows from the definition of an intervention and implies **CP***.

Second, **AT′** enables one to explore the relations between agency theory and **L** – that is, Lewis's counterfactual theory. The basic result (theorem 7.7) is that **AT′** and **L** are equivalent if there are no multiple connections and interventions are possible. A defender of **AT′** could argue that Lewis's theory conceals the agency aspect of causation in an account of similarity among possible worlds. Doing so hides the relevance of agency to the understanding of causation and mistakenly conflates counterfactual and causal dependency. A proponent of Lewis's theory could argue that **AT′**

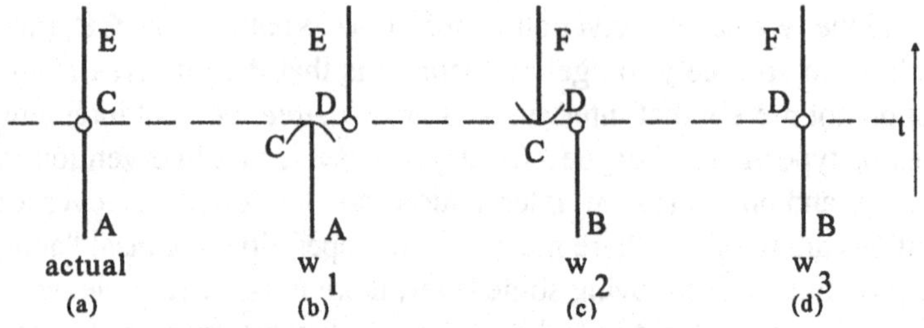

Figure 7.1: No asymmetry of overdetermination

mistakenly takes an aspect of similarity among possible worlds to be a feature of human agency. The previous two chapters contain my arguments denying that there is a good case for either **L** or **AT′** as theories of causal asymmetry, but more remains to be said.

### 7.3 Price Contra Lewis

A recent criticism of Lewis's theory by Huw Price helps to adjudicate among theories of causal asymmetry. Price believes that his criticism also applies to theories like the one proposed in chapter 4, but that it does not apply to his version of agency theory. Price argues that "the asymmetry of overdetermination is a product of macroscopic statistical considerations" (1992, p. 506) that do not obtain at the microscopic level. These macroscopic statistical considerations derive from the fact that the world is not in thermodynamic equilibrium. Initial low entropy states coupled with microchaos leads to overdetermination of causes by their effects. In microscopic interactions among a small number of entities, no asymmetry of overdetermination obtains. Effects are no less overdetermined by their causes than are causes by their effects. There is also no asymmetry of independence or of screening off.

Consider figure 7.1, which derives from Price's diagram 2 (1992, p. 509). Figure 7.1a represents the actual situation. A particle travels from *A* to *E* without interacting with anything else and produces some effect *e* at *E*. Figure 7.1b represents a possible world $w_1$, which differs from the actual world by a miraculous shift of location of the particle just before time *t* from *C* to *D*. Figure 7.1c represents a second possible world $w_2$ in which the particle travels from *B* to *D* and then just after *t* there is (from the perspective of the laws of *w*) a miraculous shift to *C*. If the event *e* at *E* causally depends on *c* – the particle being at *C* at time *t* –, then, according to Lewis's theory of causation, the following counterfactual must be true: "If the particle had not been at *C* at time *t*, then *e* would not have occurred." In order for this counterfactual to be true, $w_1$ must be more similar to the actual world than is $w_2$. But Lewis's account of similarity among possible worlds implies that $w_1$ and $w_2$ are both equally similar to $w$. $w_1$ and $w$ are identical until slightly before $t$, while $w_2$ and $w$ are identical beginning slightly after $t$, and so $w_1$ and $w_2$ are on a par with respect to the extent of period of perfect match with $w$. And the miracle in $w_1$ is no "smaller" than the miracle in $w_2$. So causal dependence does not imply counterfactual dependence, even though there are no multiple connections.

Price argues that one has to deny that there are causal relations in much of the microscopic realm, or one has to surrender accounts of causal asymmetry that depend on agent-independent asymmetries, such as Lewis's asymmetry of overdetermination. Since physicists do speak of causal relations in the microscopic realm, one should conclude, Price argues, "that the apparent asymmetry of overdetermination is insufficiently general to account for the asymmetry of causation" (1992, p. 512). Although the relationship between macroscopic means and ends derives from physical asymmetries such as the asymmetry of independence, one can extend the notion that one thing might serve as a means for controlling another to the microscopic realm where these asymmetries do not exist. Agency theories are not subject to the same limitations as are other theories because "agents are essentially macroscopic, and depend on the very thermodynamic asymmetry which is the source of the various physical asymmetries to which writers such as Lewis, Hausman and Papineau appeal" (1992, p. 516).

Price's objection does not bear in quite the same way on the revised counterfactual theory of causation present in chapter 6 as it does on Lewis's theory. On the view of similarity among possible worlds sketched above, $w_1$ and $w_3$ may be equally similar to $w$, but since $e$ does not occur in either $w_1$ or $w_3$, this creates no problem for the claim that $e$ counterfactually depends on $c$. $w_2$ involves both a miracle and a difference in a cause that is not needed to account for the failure of $c$ to occur. Thus my account holds that $w_2$ is less similar to $w$ than are $w_1$ or $w_3$. (Clause 1 of **SIM**, p. 133, implies that $w_1$ and $w_3$ are equally similar to $w$, while clause 3 implies that $w$ is less similar to $w_2$ than to $w_3$.) Because a world $w_4$, which is like $w_1$ except that the shift from $C$ to $D$ occurs slightly later (so that $c$ occurs), may be as close to the actual world as is $w_3$, in which $c$ does not occur, I can still deny the counterfactual, "if $e$ had not occurred, $c$ would not have occurred."

These remarks do not answer Price's objection. My view takes $w_1$ to be closer to the actual world than $w_2$, only because it supposes that we already know the direction of causation. My account of similarity among possible worlds is indefensible if the distinction between causes and effects that it presupposes cannot be defended. Price's challenge still stands: What grounds could there be for maintaining that $c$ causes $e$ and that $e$ does not cause $c$? Apart from the time order, everything is symmetrical. In particular, it appears that everything nomically connected to *c* is nomically connected to *e*, and everything nomically connected to *e* is nomically connected to *c*.

## 7.4 Irreversibility and Independence

What happens when the only difference between two situations is time order? For example, consider two situations *S* and *U*. These involve interactions at different times between the same two colliding billiard balls on the same table (see Reichenbach 1956, §6). Suppose that a movie taken of the collision of the balls in *S* is, when run backward, indistinguishable from a movie of the collisions in situation *U*. Notice that *S* and *U* are not possible worlds. They are situations in our world. Someone thinking only of *U* might find that **CP** appears to be satisfied because the initial velocities of the two balls are independent of one another, the masses, and other factors responsible for the final velocities. Since the film of *S* run backward is the same as the film of *U* run forward, one may, however, start feeling queasy, for the final velocities in *S* are simply the opposites of the initial velocities in *U*. How can one of these pairs be independent, while the other is causally connected?

There are, of course, nonreversible features of the collisions that differ in backward *S* and forward *U*. Since the balls are not perfectly elastic, they will warm up slightly and rapidly as a result of the collision, and so the balls in the film of *S* run backward are slowly getting warmer as they apparently head for the collision, while the balls in the forward film of *U* are constant in temperature before the collision. But it is not obvious that there will always be irreversible features to save the day. Price argues that such differences will not be available when one is considering microscopic interactions among small numbers of entities. Can one maintain that in the history of our world there will never be two causal interactions such that one is the perfect temporal reverse of the other? To pursue the central issue, let us suppose something that is not consistent with the laws of nature: that there are no physical facts about anything shown in the film which distinguish backward *S* from forward *U*.

In that case, one has two physically indistinguishable situations, which differ locally only in the direction of time and, if there are causal relations within *S* and *U*, in the direction of causation. If there are causal relations within *S* and *U*, no analysis of causation that relies on local physical differences between causes and effects other than time can be correct. **CP** and $\mathbf{CP}_g$ apparently rely on a physical difference: the independence of causes and the connectedness of effects. So either they are incorrect, or the physical difference they rely upon is not local, or there are no causal relations within *S* and *U*. In both *S* and *U* the final velocities must be causally connected, as effects of a common cause, while the initial velocities are independent. Yet *U* is physically indistinguishable from *S* run backward in time. Since the causal relations are reversed when the film of *S* is run backward, even though what happens within the film is identical to what happens in *U*, it seems that the only possible difference between cause and effect is their position in time. Even this difference appears to be merely perspectival: There is nothing in the films of *S* backward and *U* forward that distinguishes "forward" in time from "backward."

Since *S* and *U* are not separate universes but specific parts of this universe, the difference in time is also a difference in their connections to events that are outside of *S* and *U*. The independence theory does not require *local* differences between cause and effect. The independence condition states that if *a* causes *b* or *a* and *b* are causally connected only as effects of a common cause, then *b* has a cause that is causally independent of *a*. This cause need not be a proximate cause (see figure 4.4, p. 64). The direction of causation may differ in *S* run backward and *U* run forward not because of what happens *within* these situations, but because of the independence of more distant causes.

To offer this defense of the independence theory is to concede that if the whole world were like *S* or *U*, either **CP** would be false, or there would be no causal relations that are not relativized to some open subsystem. Independence is tied to openness: there are, I suggest, no (asymmetrical) causal relations in closed systems – although there may be such relations in open subsystems. One has here a rationale for the initially disquieting fact discussed in chapter 4 (p. 66–7) that the independence condition cannot be satisfied in closed systems.[1]

## 7.5 Reversibility, Intervention, and Independence

An agency theorist might argue that there is another *local* difference between *S* backward and *U* forward. One can distinguish *U* forward from *S* backward by claiming that the final velocities in *U*, unlike the initial velocities in *S*, depend on the value of an intervention variable. This intervention variable is not to be found by a more powerful microscope. The difference is modal or, if one prefers, it is a matter of an *absence*. If one had reached out and grabbed a ball headed toward a collision in a film of *U* run forward, no collision would occur, while if one reached out and grabbed a ball headed toward a collision in the film of *S* run backward, the collision would still occur. This counterfactual difference, which might at first glance appear to vindicate Price, can, however, be restated without mentioning human agency.

[1] I am indebted here to Leslie Graves.

In *S* and *U* these interferences do not occur. Forward *U* is by assumption identical with backward *S* with respect to all the variables measured by science. But if interventions are possible, other variables were "there" just the same, and their located "off" values are other independent causes. These vindicate **I** and $\mathbf{I_g}$ (and hence **CP** and $\mathbf{CP_g}$) directly, and they also explain how it can be that the "initial" velocities in backward *S* – that is, the reverse of the final velocities in *S* – are causally connected while the initial velocities in forward *U* are independent. At the end of the time period *S* occupies, the values of the intervention variables for *S* are "fixed." At the beginning of the time period *U* occupies, the values of the interventions variables for *U* are not fixed. The possibility of an intervention has not yet been closed. There is no causal dependence of the values of the initial velocities on the values of the masses and final velocities because of the possibility of "bilking": Any causal dependence would be disrupted by an intervention, which remains a possibility. Since there is no backward causal dependence, the initial velocities are not causally connected. The logical deduction of the initial velocities in *U* may be identical (apart from the signs) to the logical deduction of the final velocities in *S*, but the possibility of intervention shows that there is no backward causal dependence and no causal connection between the initial velocities.

If interventions presuppose the existence of intervention variables, then one of the causal factors involved in occurrences in which interventions are possible but do not take place is the intervention variables having at that place and time their "off" or nonintervention values. The lever that an intervention moves must have some position when there is no intervention. One can in this way argue that located "off-values" will have independent causes that vindicate the independence condition. Whenever Price's own agency theory gets the causal direction right, so will the independence theory: As shown in chapter 5, the truth of **I** is a consequence of the possibility of interventions. Notice that although the existence of interventions vindicates **I**, **I** does not require that interventions be possible or that there be any local difference between forward *U* and backward *S*.

Price's criticisms thus have less force against the independence theory than they do against his own agency theory. An agency theory gives the right answer with respect to the one-particle situation in figure 7.1a only if interventions are possible. Such manipulations are possible only if there are intervention variables, and *e* depends not only on *c* but also on these intervention variables having their "off" values. If Price's case does not count against his own agency theory, then interventions must be possible. And if interventions are possible, then there are also the independent causes required by $\mathbf{CP_g}$ and **CP**.

This response is open to two serious criticisms. First, as Reichenbach argued (1956, p. 45), the notion of an intervention relies implicitly on a temporal asymmetry. It is the openness of future possibilities as against the determinateness of the past that provides independent intervention or noninterference causes and, as just argued, secures the independence of other causes. This openness of the future as opposed to determinateness of the past is obviously a temporal difference, but it is not *only* a temporal difference, because if *S* and *U* were closed systems so that the only difference between *S* run backward and *U* forward were a difference in time, there would be no possibility for intervention. Time matters so much in causation – causes typically precede their effects – because time order is built into the notion of an intervention. The independent causes that interventions constitute or derive from are independent causes of *future* events.

Second, an agency theorist might object that our perspective as agents enables us to impose a causal direction even in closed systems. As we watch them unfolding, we can *imagine* interventions that would influence only what comes later, and so we take *a* as causing *b* when there is a lawful connection between *a* and *b* and when *a* comes before *b*. But in this story we only imagine interventions. By assumption they are not possible. Our perspective as agents could lead us mistakenly to believe that there are causal relations within deterministic closed systems, when in fact there are not. Not only is it impossible to *determine* what causes what, when the independence condition is not satisfied (see chapter 12), but a central feature of causation, which is closely related to features of causal *explanation*, is missing. What happens is not the result of separable factors that might be rearranged and recombined. In closed deterministic systems, happenings may still be determined (symmetrically by past or future), but they cannot be *caused*.

It is worth briefly comparing this account to John Mackie's "fixity" account (1966, 1980, ch. 7). Here is the second of the two versions Mackie gives:

> [Suppose that *a* and *b* are causally connected] Then despite this, *a* was not causally prior to *b* if there was a time at which *b* was fixed while *a* was unfixed. If on the other hand, *a* was fixed at a time when *b* was unfixed, then *a* was causally prior to *b*. Again, if *a* was not fixed until it occurred, then even if *b* also was fixed as soon as *a* occurred . . . *a* was causally prior to *b*.[2]

Mackie's fixity account makes reference to time, but it does not require that causes always precede their effects. However, as Mackie notes, this fixity

[2] 1980, p. 190. A further clause is omitted. Mackie wrote "*X*" where I wrote "*a*" and "*Y*" where I wrote "*b*." Mackie's fixity account has some problems I shall not discuss. As J. A. Foster (1975) points out, *a* may be sufficient for some indirect effect *d*, which is thus fixed at the time when *a* is fixed, yet not sufficient for one or another of two causal intermediaries *b* and *c*, which both lead to *d* and which are accordingly fixed later than *a* or *d*. See also Sanford (1976, p. 194; 1984, pp. 53–6).

account implies that determinism is inconsistent with the existence of causal priority. Similarly, on the assumption that everything is determined in one big closed system, the value of an intervention variable is in truth no more open in the future than it is in the past. Even if, as seems likely, determinism is in fact false, causal asymmetry does not exist only because determinism fails. If $S$ and $U$ were closed deterministic systems in which the causal relations in one are reversed in the other, then since the only physical difference between them is temporal, the only physical difference between cause and effect would be temporal, too. Either there is no nontemporal difference between cause and effect or there are no causal relations within closed deterministic systems.

If one regards causation as that relation among events that holds when citing one can be used to give a causal explanation of the other, then Mackie's fixity view can be taken as a condition on what sorts of systems permit causal explanation. If systems are closed, then there is no room for specifically causal explanation. If they are open, there is. *Even if the universe as a whole is a closed deterministic system, causal explanations and the identification of causal relations remains possible in open subsystems.* If the universe were a closed deterministic system, there would be no place for causal relations in a theory of the universe as a whole, and one might hold that in the objects – that is, apart from the questions, classifications, and systems in which we are interested, there are no causal relations. Rather than finding fault with the theory (whether it be Mackie's fixity theory or my independence theory) one should surrender the insistence that causation be purely objective. If instead causal relations are relativized to certain systems or fields (as we have already found reason to require), then it is no objection to a theory of causation that it implies that there are no causal relations in closed deterministic systems. Even if the universe were a closed deterministic system, there could be causal relations within all of its myriad open subsystems.

## 7.6 Conclusions: Independence, Agency, and Counterfactuals

When $a$ causes $b$, there will be many probabilistic dependencies among the effects of $a$ and few probabilistic dependencies among the causes of $a$. What enables the perspective of an agent to get a grip on the world are objective (but messy) relations between causation and probabilities. It is because of these objective relations that when one "wiggles" a cause, an effect wiggles, but not vice versa. But for the approximate asymmetry of probabilistic independence, which follows from the asymmetry of causal independence (which in turn constitutes the asymmetry of causation), it would not be a good approximation to say that effects are counterfactually dependent on their causes.

The independence condition **I** can be interpreted as a rough generalization, which is presupposed by the applicability of counterfactual and agency views, and which breaks down in closed deterministic systems involving reversible processes, or **I** can be regarded as going to the very heart of causation and as giving us reason to deny that there are causal relations in such systems. In closed deterministic systems there is no room for agency, no asymmetry of overdetermination, and no asymmetry of counterfactual dependence. Might one not then conclude that there is also no causal asymmetry and no causation? The notion of tracing out the ways in which independent factors come together to produce an outcome, which is arguably central to a causal explanation, is out of place. One can imaginatively extend the notion of causation to such systems. One can suppose falsely that interventions are still possible, that there are intervention variables "there" with "off" values. But one is then supposing that one does not have a closed system after all. When one believes that an intervention is possible, one is committed to an independence condition. Insofar as one's belief in the possibility of intervention stems from an (agent's) image of the future as open and the past fixed, then an agency view takes for granted a temporal asymmetry of causation, and a commitment to independence that relies on the possibilities of intervention also takes for granted a temporal asymmetry of causation.

There is no temporal or anthropomorphic presupposition involved in the claim that there are typically probabilistic dependencies among effects of a given type of cause and that there are not typically such dependencies among causes of a given type of effect. Such an asymmetry may be *explicable* in terms of some temporal story about the universe. But regardless of its explanation, the objective asymmetry of independence permits one to make significant generalizations concerning the direction of causation that do not make reference to time order. And that asymmetry also permits one to tie together the partial asymmetries of agency and counterfactual dependencies.

The status of the token-level formulation of agency theory resembles that of the type formulation. Both complement accounts such as the independence and Humean views, and both are dubious as alternatives to these accounts. One learns important things about causation when one recognizes the links between causation and agency. But in elucidating those links, one is committed to independence. For the same reason, coupling agency with counterfactuals does nothing to resuscitate agency theory or a counterfactual theory of causation.

This chapter shows how interconnected are the four accounts of causal asymmetry discussed so far. Lewis's counterfactual theory will not do, but there is a crucial counterfactual element that is linked to considerations of agency, independence, and time. Agency theories won't do, but a more abstract notion of an intervention ties in with considerations of independence, counterfactuals, the openness of the future, and (as we shall see in chapter 8) explanation. The independence theory links up with agency views, asymmetries in explanation, and partial probabilistic asymmetries, and it explains the successes and failures of counterfactual views. Hume's view won't do, but temporal elements turn out to be central in agency, counterfactual, and independence theories. When these theories are juxtaposed and their relations clearly delineated, one begins to comprehend causal asymmetry. But one critical piece of the puzzle is missing: an account of causal explanation. We will turn to that task in chapter 8.

# 7*

# Agency, Counterfactuals, and Independence

## 7.1* Independence and a Token Version of Agency Theory

Analogous to the conditions in chapter 5 let us define:

> **AT** *(Agency theory)* *a* causes *b* if and only if *a* and *b* are distinct, and with respect to some set of (actual) events *E*, if *a* had come about as a result of a direct manipulation, then that intervention would be a cause of *b*.

If one wants to make the circularity less blatant, one can replace the last nine words with "then that intervention would be nomically connected to *b*," and all the same results hold. As was the case with $\mathbf{AT}_g$, the quantification over a set of events is necessary, because the notion of a direct manipulation is relativized to a set of events.

> **DHI** *(Definition of a human intervention)* *i* is a direct human intervention or manipulation that brings about *a* only (with respect to a set of events *E* of interest) if and only if (1) *i* is distinct from every event in *E*, (2) *i* is a direct cause of *a*, (3) the structure of the causal relations between *a* and its effects is the same when *a* is caused by *i* as when there is no intervention, (4) *i* has no causal connections to any other events in *E* except those that follow from its being a direct cause of *a*, and (5) *i* is a human action.

**DI**, the token-level definition of an (abstract) intervention, is **DHI** without the last clause. Condition (3) has no analogue in the case of $\mathbf{DHI}_g$, because it is implicit in the notion of a variable that may have different values that the causal graph does not itself change with changes in the values of variables, at least within some range of values. **DHI** defines a weaker notion of a direct intervention than that assumed by many agency theorists, because a direct manipulation of *a* need not break all connections between *a* and its other causes.

> **PI** (*Possibility of intervention*) Given the set of events *E* of interest, for each event *e* in *E*, there is a possible situation with the same causal relations between *e* and its effects in which *e* and only *e* comes about (or is prevented) via a direct manipulation.

**PI** is, of course, false, if interventions are construed as human interventions. It is a boundary condition to the validity of **AT**. One can then prove:

**Theorem 7.1: DHI** (or **DI**), **PI, T,** and **PP** imply **AT**.

Proof: Suppose that *a* causes *b*. By **PI** it is possible for there to have been an intervention that brought about *a*. By **DHI** that intervention would cause *a* and would be distinct from *b*. By **T** that intervention would thus be a cause of *b*.[1] To prove the converse, suppose that every direct manipulation that would have brought about *a* would be a cause of *b*. By **PI** it is possible for there to be such direct manipulations of *a*. By **DHI**, these interventions cannot cause *b* except via directly manipulating *a*, and by the path principle *a* must be a cause of *b*.

**PP** is the token-level path principle: If *a* causes *b*, there is a causal path from *a* to *b*.

**DHI** (or **DI**), **T, CC,** and **AT** do not entail **CP**. The analogues to theorems 5.2 and 5.4 are invalid. If one attempts to prove that **DHI, T, CC,** and **AT** entail **CP**, one must prove:

1. if (if *a* causes *b* and if *a* had come about via a direct manipulation, then that manipulation would have caused *b*), then everything distinct from *b* and causally connected to *a* is causally connected to *b*, and
2. if *a* does not cause *b* and if *a* had come about via a direct manipulation of *a* that might not cause *b*, then there is something distinct from *b* and causally connected to *a* that is not causally connected to *b*.

The proof of (1) goes through as before, but **DHI, T,** and **CC** do not imply (2). The *possibility* that *a* had come about via a direct manipulation that might not have caused *b* does not imply that anything causally connected to *a* and not to *b* actually *exists*. If **CP** said that *a* causes *b* if and only if anything distinct from *b* that would be causally connected to *a* would be causally connected to *b*, then the derivation would go through. So consider the following variant of **CP**:

> **CP*** (*Modal independence theory*) a causes *b* if and only if *a* and *b* are causally connected and if for all possible events *c* distinct from *b*, if *c* were causally connected to *a*, it would be causally connected to *b*.

If one assumes that **CC** and **T** hold in all causally possible situations, the proof of theorem 4.1 establishes:

**Theorem 7.2: CC, T,** and **I** entail **CP***.

But there is no analogue to theorem 4.7. Conjoined with the other conditions, **CP*** does not entail **I**. One can, however, now prove the analogues to theorems 5.2, 5.4, and 5.5.

[1] Assuming that **T** applies to causal relations in all causally possible situations.

**Theorem 7.3: CC, T, DHI (or DI), and AT entail CP*.**

Proof: Part 1 of the proof is trivial from theorem 4.1. What remains to be proven is that if *a* does not cause *b* and if *a* might have come about via a direct manipulation that did not cause *b*, then it is not the case that for all *c* distinct from *b* if *c* were causally connected to *a*, then it would be causally connected to *b*. The consequent is equivalent to the claim that there might be something distinct from *b* and causally connected to *a* but not causally connected to *b*. Since by assumption there might be a direct manipulation that brings about *a* without causing *b*, and since a direct manipulation of *a* cannot have any other kind of causal connection to *b*, the result follows.

One can formulate the independence condition implicit in **CP*** and a modal analogue to the strong independence condition as follows:

**I*** (*Modal independence condition*) If *a* causes *b* or *a* and *b* are causally connected only as effects of a common cause, then *b* might have a cause that would be causally independent of *a*.

**$I_s$*** (*Strong modal independence condition*) Every event *b* that has at least one cause might have another independent cause, *f*, such that for all events *a*

1. If *a* were not caused by *b* and there were a causal path that did not go through *f* from *a* to *b* or from some cause of *a* to *b*, then *a* and *f* would be causally independent.
2. If *b* were to cause *a*, then there would be no path that does not pass through *b* from *f* to *a* or from any cause of *f* to *a*.

With the help of one more condition, one can state and prove the analogue to theorem 5.6:

**NIC** (*Nonintervention causes*) If *a* can be caused by an intervention, then *a* also has causes that are not interventions.

**Theorem 7.4: CC, NIC, DI, and PI entail $I_s$*.**

Proof: Given **NIC**, any event *b* that can be caused by an intervention has another cause *a*. By **PI**, *b* could be caused by an intervention *i*. For any event *e* causally connected to *b* and not an effect of *b*, **DI** implies that *i* and *e* are independent. For any effect *g* of *b*, **DI** implies that there is no path to *g* from *i* or from any cause of *i* that does not pass through *b*. So **$I_s$*** holds.

Since **$I_s$***entails **I***, one can conclude that (given **NIC** and **CC**) **DI** and **PI** entail **I***.

Suppose we accept the following additional principle:

**SPI** (*Structural presupposition of interventions*) Interventions are possible if and only if intervention variables exist.

Given **SPI**, **AT** will hold among tokens if and only if **$AT_g$** holds among the relevant types and **$CP_g$** will be true if and only if **CP**. The analogues to theorems 5.2 and 5.4 then follow immediately. Given **SPI**, the possibility of a token intervention implies the existence of an intervention variable, which in turn (by theorem 5.6) implies **$I_{gs}$**, the type-level strong independence condition, which in turn implies $\mathbf{I_s}$. So given **SPI**, the analogues to theorems 5.6 and 5.7 will also hold.

## 7.2* AT′, CP, and L

Consider:

> **AT′** (*Counterfactual agency theory*) *a* causes *b* if and only if *a* and *b* are distinct, and if one intervened and prevented *a*, then *b* would not occur.

One intervenes and prevents *a* if there is an intervention that causes it to be the case that *a* does not occur. One can demonstrate that the central claims I made about **AT** are true of **AT′** as well.

**Theorem 7.5: DC, SIM, DI, PI, CDCC, and PP entail AT′.**

Proof: Suppose *a* causes *b*, and suppose that an intervention *i* occurs that prevents *a*. If causes are, as **DC** says, necessary in the circumstances, then there exists a trope, the absence of which would prevent *a*. **PI** says that an intervention to prevent this trope is possible. By **DC** a possible world in which *a* fails to occur will lack *b* unless some other sufficient condition for *b* occurs. Since there is, by assumption, no preemption or overdetermination, and since (by **DI**) *i* has no causal connection to *b* except possibly via *a*, *b* will fail to occur unless there is some *independent* occurrence or nonoccurrence to provide an alternative sufficient condition. By clause 3 of **SIM** such worlds will be less similar to the actual world than are worlds without such independent occurrences or nonoccurrences, and so if an intervention prevents *a*, *b* will not occur. Conversely, suppose that an intervention *i* occurs that prevents *a*. By assumption, in the closest possible worlds *b* does not occur. *b* is thus (negatively) counterfactually dependent on *i*, and by **CDCC** (p.134), *i* and *b* are causally connected. By **DI** *b* cannot cause *i* and *i* and *b* cannot be causally connected only as effects of a common cause. So *i* prevents *b*, and by **PP** *a* causes *b*.

Theorem 7.5 shows that my claim that agency theory is little more than a definition of an intervention and an assertion of the possibility of interventions is not just an artifact of a peculiar formulation of **AT**. As the following theorem shows, neither is the derivation of independence from agency views:

**Theorem 7.6: DI, CDCC, CC, T, DC, and AT′ entail CP*.**

Proof: Since the necessary condition in **CP*** follows from **CC** and **T** (if these are interpreted as applying to all possible worlds), it suffices to prove that if one were to intervene to prevent *a*, and *b* might still occur, then it is not the case that everything distinct from *b* and causally connected to *a* is causally connected to *b*. Since *b* might still occur given an intervention that prevents *a* (even without preemption or overdetermination), *A* is not an INUS condition for *B* and hence (by **DC**) *a* does not cause *b*. By **DI**, an intervention that prevents *a* can be causally connected to *b* only if *a* causes *b*. So the intervention is not causally connected to *b*. So it is not the case that everything distinct from *b* and causally connected to *a* is causally connected to *b*.

One can also prove that **AT′** and **L** are conditionally equivalent, provided there are no multiple connections and interventions are possible.

**Theorem 7.7:** Given **SIM, DI, PI,** and the absence of multiple connections, **AT′** entails and is entailed by **L**.

Proof: It suffices to prove that *e* is counterfactually dependent on *c* if and only if, if an intervention prevented *c*, *e* would not occur. If *e* is counterfactually dependent on *c*, then whether by intervention or by some other difference, if *c* were not to occur, then *e* would not occur. So if an intervention were to prevent *c*, *e* would not occur. To prove the converse, note that if there are no multiple connections, then the consequences of *c*'s not occurring are independent of whether *c* fails to occur because of an intervention or some other way. So if *e* does not occur in worlds in which *c* is prevented by intervention, non-*c* worlds without *e* are closer to the actual world than are non-*c* worlds with *e*, and *e* is counterfactually dependent on *c*.

Unsurprisingly, one can also establish relations between **L** and **AT** and between **AT′** and **AT**.

**Theorem 7.8** Given **DI, CDCC, DC, SIM, T, PP,** and the absence of multiple connections and of preemption or overdetermination, **L** entails and is entailed by **AT**.

Proof: It suffices to prove that in circumstances without multiple connections, *b* is counterfactually dependent on *a* if and only if, if *a* had come about as a result of a direct manipulation, then that intervention would have been a cause of *b*. Suppose first that *b* is counterfactually dependent on *a*. Consider one of the closest possible worlds without *a* and *b* and suppose some intervention *i* were to occur that caused *a*. Since *b* is counterfactually dependent on *a*, *b* would occur and hence *b* would be counterfactually dependent on *i*. By **CDCC** *i* and *b* would be causally connected and by **DI** *i* would be a cause of *b*.

Conversely assume that if *a* had come about as a result of a direct manipulation *i*, then *i* would be a cause of *b*. By **PP** *a* must cause *b*. By **DC** *a* must be an INUS condition of *b*. Given the absence of preemption, overdetermination, and multiple connection, clauses 3 and 4 of **SIM** imply that *b* must be counterfactually dependent on *a*.

**Theorem 7.9:** Given **DI, CDCC, DC, SIM, T, PP,** and the absence of multiple connections or of preemption or overdetermination, **AT′** entails and is entailed by **AT**.

Theorem 7.9 follows trivially from theorems 7.7 and 7.8.

# 8

# Causation, Explanation, and Laws

Several features of the causal relation suggest a close connection between causation and explanation.

1. The relevant aspects of cause and effect make reference to properties and do not appear to be fully concrete entities.
2. The relations between causation and laws and the scientific centrality of type-causal claims force one to emphasize the role of properties in the causal relation.
3. The importance of absences and nonoccurrences in causal relations and the fact that the same pair of substances can be the locus of causal relations in either direction, depending on context, all call into question a view of causation as a purely physical relation of the sort suggested by transfer theorists.
4. The relativization of causal relations to particular fields or systems points to links between causation and explanation.
5. A connection between independence and explanation is implicit in the thought explored at the ends of chapters 5 and 7 that independence should be regarded as specifying the circumstances in which causal relations obtain – or in which causal explanations are appropriate.

So something must be said about the intimate relations between explanation and causation. For millennia, citing a cause has been taken to be explaining. When Aristotle presents his doctrine of the four kinds of *causes*, he is developing an account of four different kinds of *explanations* one may give. Material causes of objects are explanations of their properties in terms of the material of which they consist. Formal causes are explanations in terms of structure. Final causes are explanations in terms of ends. Proximate causes – causes in our modern sense – are apparently linked to explanation as well.

A theory of explanation is needed, but chapter 8 of this long book on causal asymmetries is hardly the place to present a full-scale theory of explanation – even if I were capable of doing so. So shortcuts will have to be made and some hard questions begged.

## 8.1 The Deductive-Nomological Model and its Sufficiency

Within the broadly Humean tradition of twentieth-century analytical philosophy, explanation has been analyzed in terms of logical or probabilistic relations among true propositions, some of them natural laws. Neo-Humean analyses of causation are similar but somewhat more complicated, since causation apparently involves concrete events and their spatiotemporal relations. These analyses unify our understanding of causality and of explanation and take much of the mystery out of the notion of causation. They spare the analysis of scientific explanation the difficult problems involved in clarifying what causal relata are. They allow for the possibility of explanations that are not causal.

I shall focus my discussion on the deductive-nomological (DN) model of explanation (Hempel 1965), because the DN model is so well articulated and well known. The problems with which I shall be concerned are, however, problems for all accounts of scientific explanation. Let us say that a "deductive-nomological (DN) *argument*" is any sound argument that includes essentially as a premise some law of nature. (A sound argument is a logically valid argument with all true premises.) Hempel's DN model of explanation maintains that deterministic explanations, including causal explanations, are DN arguments. Many have argued that not all scientific explanations are DN arguments and that not all DN arguments are scientific explanations. I shall be concerned with objections to the *sufficiency* of the DN model – that is, objections maintaining that some DN arguments are not scientific explanations. Answering these objections requires one to consider what scientific explanations require in addition to the logical and semantic conditions Hempel defends.

There are two main classes of DN arguments that are not scientific explanations. The first involve failures of *causal relevance*. Examples here are Salmon's case of "explaining" a man's failure to get pregnant in terms of his faithful consumption of birth control pills (1971, p. 34) or Kyburg's case of "explaining" the dissolving of salt in terms of its having had a dissolving spell cast on it and the law that salt that has had such a spell cast on it dissolves in water (1965). The second class involve failures of causal *direction* – so-called explanatory asymmetries. Examples include Bromberger's case of deriving the height of the Empire State Building (or of a flagpole) from the length of its shadow and the angle of elevation of the sun (1966, pp. 92–3) or Jeffrey's "explanations" of storms in terms of abrupt drops in barometer readings (1971, p. 21). A good theory of scientific explanation ought not to count these as scientific explanations.

The most plausible diagnosis of these cases of DN arguments that are not explanations is that the premises in these arguments fail to focus on the *causes* of the phenomena described in their conclusions.[1] The consumption

[1] This is not to deny that there might also be asymmetries in explanations that do not cite causes. Peter Railton (1980, pp. 385–410) and James Woodward (1984) have argued convincingly that there are different kinds of explanatory asymmetries.

of birth control pills does not explain a man's failure to get pregnant, because it is causally irrelevant, as is the hexing of the salt. The length of the shadow of a flagpole does not explain its height, because the length of the shadow is not a causal condition of its height. Drops in barometer readings are effects of the atmospheric phenomena that cause storms, not themselves causes.

If one accepts this diagnosis, then one must either abandon the attempt to analyze causal explanation in terms of truth, logic, and laws, or one must attempt to analyze causation in those terms. Since the latter appears impossible, it seems that the theory of explanation must incorporate other elements, such as those that have figured in the accounts of causation discussed above. This is a very significant conclusion, and it ought not to be accepted lightly. I consider the arguments for it and against it in §8.1* and §8.2*. If explanation cannot be analyzed with concepts like those Hempel employs, then it may be – as this diagnosis of explanatory asymmetries suggests – that explanation needs to be analyzed in terms of causation or elements of theories of causation. Perhaps explanation of events simply consists in citing their causes, and that's all there is to say. On the other hand, perhaps there are differences in the character of derivations from causes as opposed to derivations from effects that account for why the one and not the other is explanatory. This chapter will pursue this possibility.

One possible moral, which some contemporary philosophers have drawn, is that one should abandon the whole project of providing a general theory of scientific explanation, including the attempt to account for the explanatory importance of causal relevance and causal direction. One learns more about scientific explanation by examining the constraints and ideals concerning explanation accepted by practitioners of particular disciplines. To notice that explanations of economic phenomena should be in terms of the constrained maximizing efforts of individuals tells one more about explanation in neoclassical economics than does the DN model.

Even though most interesting truths about scientific explanation depend on discipline-specific constraints and ideals, a philosophical inquiry into explanatory asymmetries is not misconceived. The problems have nothing to do with the peculiarities of particular disciplines. Even if discipline-specific constraints resolved them (which seems unlikely), their general exploration might still tell us something important about causation and explanation.

### 8.2 Explanatory Asymmetries

As just noted, many explanations – both causal and noncausal – show asymmetries that are not reflected in their logic. In a case such as the flagpole, there are three variables and a law of nature that enable one to derive the value of any one of the variables from the values of the other two. Yet only one of the three derivations is explanatory. The salt basins discussed in chapter 6 show a similar explanatory asymmetry.

The standard diagnosis of explanatory asymmetries is that explanations must cite causes of their explananda (Salmon 1978; Beauchamp and Rosenberg 1981, p. 313). This diagnosis only accounts for asymmetries in causal explanations and rarely amounts to a precise formulation of a causal requirement (which is no mean task; see Railton 1980, pp. 213f). Even those who emphasize the variety of explanations typically accept this diagnosis of the asymmetry of specifically causal explanations (Glymour 1980, pp. 48–9; Railton 1980, pp. 191–224; Woodward 1984, p. 436).

These intuitions concerning explanatory asymmetries have been questioned. Carl Hempel argues that there are pragmatic reasons why people mistakenly deny that the derivation of the voltage in a circuit from its current and the resistance is a scientific explanation (1965, p. 352–3). Hempel's claim is hard to accept. The intuitions he contests are strong and widespread among ordinary people and scientists as well as philosophers. Arguing about intuitions is not, however, very profitable. A more effective answer to Hempel's skepticism about explanatory asymmetries demonstrates that derivations that cite effects or other effects of a common cause lack crucial explanatory virtues possessed by derivations that cite causes.

The derivation of the voltage in a circuit from Ohm's law and the values of the current and resistance seems not to explain the voltage. Intuitively the reason is that the voltage does not causally depend on the current and resistance. *But why should citing causes explain effects, while citing effects does not explain causes or other effects of a common cause?* If causation and explanation are distinct from one another, some account is needed of how causation matters to explanation. If causation and explanation are not distinct, and causation is merely as a species of explanation, one still needs to explain how explanatory asymmetries are related to the other asymmetries of causation. What do the various asymmetries of the relation have to do with one another? Why should they match up? How can one *justify* the assertion that the causal relation is an asymmetric explanatory relation?

According to David Lewis, "to explain an event is to provide some information about its causal history" (1986a, p. 217). Wesley Salmon similarly maintains that one conception of explanation simply *is* to cite causes. Given such views of explanation, it is obvious why citing causes explains, while citing effects doesn't. But the difficulty remains and merely needs to be restated. *Why* does giving information concerning causal history satisfy the aims of explanation, while giving information concerning causal consequences does not? How are the aims of explanation linked to the asymmetrical features of causation considered in the previous chapters?

Regardless of how one defines one's terms, there is a real problem here. I prefer to formulate it as: Why do *arguments* that cite causes explain, while *arguments* that cite effects or effects of a common cause do not explain? The focus on arguments may seem to be evidence of an unhealthy preoccupation with the deductive-nomological model. As its critics have correctly pointed out, explanations are often not arguments. But they can usually be restated as arguments, and one learns a good deal about explanations by thinking about the contrast between arguments that cite causes and arguments that cite effects or effects of a common cause. The underlying problem could, however, just as well be formulated as: What are the connections between the asymmetries of causation and the aims of explanation?

To address this problem, however it is formulated, one needs to have some initial, pretheoretic notion of explanation to which features of causation can be linked. I suggest that the root notion of an explanation involves *showing that something had to be as it is*. I hope that this suggestion will resonate with you, my readers. There is little I can do to justify it, but here are some comments to motivate it: One explains mathematical truths by proving them from axioms that one can regard as in some sense necessary. One explains scientific regularities by showing that they must be true, given more fundamental truths about nature. One explains why events occur by showing that they had to happen given their causes. If asked to explain why citing causes explains effects, while citing effects does not explain causes, one might be inclined to say something like, "Causes explain because they state what *makes* phenomena occur. They explain by showing why the phenomenon had to be as it is." By the standards of contemporary analytical philosophy, this is less an answer than a restatement of the problem. What it is for a cause to *make* its effect happen (see Honderich 1982)? In what sense does an explanation show its explanandum to be *determined*? One needs to clarify the relations between the "making" of effects and the "determination" of explananda.

A satisfactory answer to the question, "why does citing causes explain while citing effects does not?" must do three things. First, the core idea of showing how an explanandum is "determined" or "necessitated" needs clarification. Second, the notion that causes make their effects happen must be linked to some more clearly understood difference between causes and effects. Finally, given a precise specification of the aims of explanation and of some difference between causes and effects, one needs to show what these have to do with one another.

It is not obvious how the facts about causal asymmetry discussed in previous chapters could account for the fact that citing effects does not explain their causes. On Hume's view, for example, the constant conjunction between cause and effect can be equally constant from either direction. There seems to be no way to capture the notion that causes *make* their effects happen, apart from Hume's sketchy psychological theory of necessary connection. According to Hume the tendency of the mind to pass from the impression of the cause to the idea of the effect is, however, no stronger than the tendency of the mind to pass from the impression of the effect to the idea of the cause. The only difference between causes and effects is that causes precede effects, and it is hard to see what temporal precedence has to do with explanation. At the end of chapter 5, I suggested that the thought that causes necessitate their effects might be a metaphysical exaggeration of the fact that agents can make things happen by means of their causes. In any case, we need an account of the relevance of causal asymmetries to the aims of explanation.[2]

## 8.3 The Aims of Scientific Explanations

Let us start by considering what the aims of scientific explanations might be. In clarifying what scientific explanations should accomplish, perhaps one will see why citing causes explains while citing effects does not. Apart from the position that explanations should reveal causes, which is unhelpful in this context, the philosophical literature on scientific explanation contains four major views of the aims of explanation. Each interprets the core idea that to explain is to reveal how explananda are *made* to be as they are, and each points to additional explanatory goals.

1. A good explanation of a phenomenon should lead one to have expected it to occur (Hempel 1965). Hempel domesticates the notion that an explanation makes its explanandum necessary as the thesis that an explanation shows its explanandum statement to be a logical consequence of the initial conditions and relevant laws. The explanandum is an instance of a law, not a pure contingency. Before the explanation, the explanandum was happenstance. After the explanation there is order. One could have predicted that the explanandum event would happen, if one had known enough beforehand.

On this view, one should often be able to explain causes by citing their effects. The angle of elevation of the sun is as predictable from knowledge of the height of a flagpole and of the length of its shadow as is the shadow's length from knowledge of the height and angle of elevation. If the only virtue of scientific explanations is nomic expectability, then intuitions

[2] Henry Byerly (1979, 1990) offers another account that I shall not discuss here. He argues that the asymmetry comes from the fact that the causes, unlike the effects, determine what the system is. Thus the latter cannot explain the former. I do not know what to make of this suggestion.

concerning explanatory asymmetries are mistaken (Hempel 1965, pp. 352–4).

2. According to Wesley Salmon, to explain is to provide a "complex of objective facts" concerning the "causes of the explanandum, causal processes that connect the causes to their effects, and causal regularities that govern the causal mechanisms involved in the explanans" (1985, p. 274). He gives this claim content by offering a theory of causal processes and causal interactions, which has been superseded by the one discussed in chapter 1*. A scientific explanation places the phenomenon within the causal nexus and brings in view the causal processes and interactions. Deterministic explanations show why the phenomena had to be as they are by revealing the causal mechanisms.

This "ontic" conception of explanation provides a persuasive account of what is wrong with deductive-nomological arguments in which the purported explanans is not causally *relevant*. To deduce that a man who consumes birth control pills will not get pregnant from the law that nobody who consumes the pills faithfully gets pregnant does not explain why the man does not get pregnant, because it cites an inoperative mechanism. Explanations should tell one which mechanisms are relevant.

Despite Salmon's explicit concern with explanatory asymmetries, he never makes clear how his ontic theory of explanation accounts for them, and, as we saw in chapter 1*, it is difficult to link process theories to concerns about explanation. Recall the example of ice cubes in water. One can explain both why the water got colder and why the ice cubes melted in terms of the transfer of heat from the water to the ice cubes, but the explanation for the water getting colder appears to be in terms of its effects (as a process or transference theorist would see things) rather than its causes. Is this a case where citing an effect explains a cause, or is this a case in which (as we would say in ordinary language) the ice cubes caused the water to get colder? Why do the causal interactions between sunlight and a flagpole explain the length of the shadow but not the height of the flagpole? Salmon's answer seems to be that the processes and interactions are not responsible for height of the flagpole. But what is the "is responsible for" relation? If it is a causal relation, then it seems that Salmon has reiterated the requirement that explanations cite causes without justifying it. How is the "responsible for" relation relevant to scientific understanding?

3. According to Michael Friedman (1974) and Philip Kitcher (1981, 1989), a scientific explanation *unifies*. Subsuming is explaining, because subsumption reduces the number of independent phenomena. Explanations may reveal mechanisms or ground predictions, but their explanatory worth lies in unifying human knowledge. Unification provides an appearance of necessity; the phenomena to be explained belong to a larger pattern. More

specifically, Kitcher argues that explanations of causes in terms of effects will lessen the unifying power of a whole set of explanations by introducing redundant explanatory patterns that do not explain much (i.e., accounts of heights of towers in terms of the shadows they cast). Causes are simply those factors that are genuinely explanatory (1989, p. 477). This account matches our intuitions well (though Barnes 1992 disagrees), but it does not address the question of why the factors that permit unification also precede, permit manipulation, are independent, and so forth.

4. A scientific explanation answers why-questions of a particular kind (Bromberger 1966, van Fraassen 1980, Achinstein 1983, Miller 1987). To say what scientific explanations are requires attention to the occasions when people ask "Why?" and to the information they seek. Explanation is pragmatic. Explanations resolve curiosity. Making a phenomenon nomically expectable, unifying phenomena, or revealing mechanisms may sometimes answer why-questions, but one needs to consider the contexts in which why-questions are asked, the interests that prompt them, and the contextually determined standards that answers must satisfy. This pragmatic view of explanation may be linked to a pragmatic view of necessity to elucidate the claim that explanations show their explananda to be necessary: They show what could or could not be done about their explananda. According to a pragmatic view of explanation, people want to know the causes when they ask for explanations. For this reason, only citing causes will explain.

## 8.4 Causation and Why We Ask, "Why?"

According to Bas van Fraassen, ". . . the asymmetries of explanation result from a contextually determined relation of relevance. . ." and thus it should "be possible to account for specific asymmetries in terms of the interests of questioner and audience that determine this relevance" (1980, p. 130). Such a pragmatic view of explanation links causal and explanatory asymmetries: Human interests in knowing causes explain why citing causes explains while citing effects does not.

The best developed and most plausible pragmatic theory of explanation is, in my opinion, Richard Miller's (1987). According to Miller, explanation is "adequate causal description" (1987, Part I, esp. ch. 2). "Adequacy" depends both on context-independent features, such as whether the cause was actually present, and on context-dependent features, such as whether the description conforms to a "standard causal pattern." On Miller's account, a why-question asks, "What are the salient causes?" Obviously information concerning effects will not answer such questions.

This connection between causation and explanation is superficial. Why are why-questions requests for information about causes? Why does information about causes satisfy human interests better than information concerning effects? How do differences between cause and effect that are independent of human interests mesh with human interests? Miller, like Bromberger, van Fraassen, and Achinstein, says little about what the differences between causes and effects are, how they *explain* why why-questions are requests for information concerning causes, and whether the differences *justify* this human habit. Why are why-questions requests for information about causes?

The connections between causal asymmetry and agency explored in chapters 5 and 7 explain why when people ask "Why?" they particularly want to know the causes. An explanation that states the causes may tell one how to *do something* about the kind of phenomenon that one seeks to explain. An explanation that states only the effects (or another effect of a common cause) offers no help in controlling what one seeks to explain. Given the practical interests of human actors, people thus ask "Why?" to find out about causes, not effects. Causal explanations are not always of practical use, but a desire to know the causes when one cannot act might merely generalize the desire to know the causes when one can. Since information about disease-causing agents is of particular practical importance to those concerned with the treatment and prevention of infectious illnesses, the standard causal pattern for the explanation of infectious diseases in medicine involves identifying the relevant infectious agent (Miller 1987, pp. 89–90).

This connection between causation and explanation seems vulgar. Surely humans are not only concerned to know the causes when they are about to act. The account needs to be broadened. One way to do this is to point out that agents may want to know in how things *might be* changed, even when they have no practical interest in changing anything. People want to know "What if. . . ?" Knowledge of causes, unlike knowledge of effects helps to answer "What if. . . ?" questions. For example, suppose one wants to know why the resistance of a particular aluminum wire is $r^*$ ohms. Consider two deductive-nomological arguments (arguments that meet the conditions Hempel imposes on nonprobabilistic explanations) with the conclusion that the resistance is $r^*$ ohms: the first in terms of Ohm's law and the current and voltage in the circuit, the second in terms of facts about the material, length, and diameter of the wire. Consider then the two questions, "What if the wire had been made of copper instead of aluminum?" and "What if the current in the circuit had been twice as great?" Their counterfactual answers appear respectively to be "The resistance would have been lower," and "The resistance would have to have been lower or the voltage of the battery would have to have been greater" (see Simon and Rescher 1966, pp. 125–30). The impossibility of drawing counterfactual conclusions concerning particular causes discussed in §6.2 means that a counterfactual "What if . . . ?" question concerning how the causes of some phenomenon would have been will never have a definite answer.

Knowledge of a specific cause is not always useful in answering a "What if. . . ?" question, because, as shown in §6.4 and §6.5, when there are multiple connections, one does not know how effects would differ were a cause to differ. But cases involving significant multiple connections are infrequent, and multiple connections do not interfere with inferences concerning consequences of *interventions*. Knowing the causes, unlike knowing the effects of an event, typically tells us about the possibilities of bringing about or averting events of that kind. The approximate asymmetry of counterfactual dependency matters a great deal, because, as active beings, people want to know why phenomena *must* or *might* occur (where "must" and "might" are to be understood in terms of the counterfactuals just discussed). The desire to know the causes need not follow *immediately* from any practical interest. It may derive instead from an interest in understanding what might happen. This general interest in turn follows from practical interests in action. In this way practical interests lead humans to seek information about causes and to regard causes as explaining their effects but not vice versa.

If explanations give people the information they seek, and people seek information about causes because they want to know what actions are possible, then one can explain why citing effects or citing other effects of a common cause fails to explain. It is hard to see how else causation could be relevant to explanation. Since Humean, independence, and counterfactual theories of causation can be linked to the agency view, one can take this as a general strategy for explaining why one cannot explain an event by citing its effects or by citing other effects of its causes.

Two aspects of this account are unsatisfactory. First, the account may exaggerate the extent to which explanation depends upon interests in action. If only practical interests ground a preference for derivations that cite causes, why then not conclude that in *pure* science "explanations" in terms of effects *should* be just as valuable as explanations in terms of causes? Does this pragmatic account *justify* as well as *explain* the importance of causation to explanation? There seems too little difference between demonstrating the pragmatic relevance of causation to explanation, as van Fraassen, Achinstein, and Miller do, and exposing the fact that the perspective of human actors causes people *mistakenly* to reject perfectly good explanations, as Hempel does. In espousing a pragmatic account, has one not simply put a conciliatory "spin" on the dismissal of the explanatory asymmetries? Though I shall make some arguments that justify the importance of causal direction to explanation, I suspect that a satisfactory account of the relevance of causation to explanation must refer to human interests and action.

Second, the pragmatic account does not specify what non-interest-dependent differences between knowledge of causes and knowledge of effects enable the first to serve human interests better than the second. Human interests can justifiably give rise to explanatory asymmetries only if knowledge of causes serves human interests differently than knowledge of effects. One way that knowledge of causes can serve human interests better than knowledge of effects is if causes precede their effects in time. Since one cannot act to change the past, knowledge of effects, unlike knowledge of causes, will not serve interests in action. Perhaps the discussion in chapters 5 and 7 of the relations between independence and intervention will illuminate further the practical differences between knowledge of causes and knowledge of effects.

## 8.5 Causation and Independent Alterability

Consider the following two arguments:

| (C) | (E) |
|---|---|
| 1. $h = h^*$ | 1′. $s = s^*$ |
| 2. $a = a^*$ | 2. $a = a^*$ |
| 3. $h/s = \tan(a)$ | 3. $h/s = \tan(a)$ |
| ∴ 4. $s = s^*$ | ∴ 4′. $h = h^*$ |

I called the first "(C)" because it is a derivation that cites only laws and causes of what is to be explained and the second "(E)" because it is a derivation that cites an effect of what is to be explained. $h$ is the height of a flagpole, $a$ the angle of elevation of the sun, and $s$ the length of the shadow. $h^*$, $a^*$, and $s^*$ are particular values of these variables. Equations 1, 1′, and 2 "specify" the values of variables, and I shall call them "specifying equations." Equation 3 derives from the law that light travels in straight lines. On the assumption that premises 1 and 2 are true, (C) appears to explain why the pole casts a shadow of length $s^*$. (E) does not appear to explain why the pole has height $h^*$. Why are arguments like (C) explanations, and why aren't arguments like (E) explanations?

The answer does not lie in logic or mathematics. Both (C) and (E) are sound arguments, and all the premises are essential. There are, however, other differences:

1. There is the practical difference we have just been discussing. One can change the length of the shadow by changing the height, but not vice versa.
2. There is an implicit temporal difference. The length of the shadow at $t$ is a function of the height of the flagpole slightly before $t$.

3. There is an asymmetry of independence. Of the three pairs of variables – (*h*, *s*), (*a*, *s*), and (*h*, *a*) – only the values of *h* and *a* are causally independent.
4. As a consequence of this last asymmetry, there is an asymmetry of counterfactual dependence and, given the prediction condition of chapter 6 (p. 120), an asymmetry of prediction. The length of the shadow is counterfactually dependent on the height of the pole and not vice versa, and if the height of the flagpole were to change, one could "plug in" the new value of the height in derivation (C) and make the best prediction of the new length of the shadow. But if one were to plug in a new shadow length in derivation (E) and calculate a new value for *h*, one would not be justified in predicting that *h* would have that value, for the change in the shadow length might have been due to a change in the angle of elevation of the sun.

This last asymmetry apparently bears on the aims of explanation. (C) provides one with a better "grip" on the phenomenon than does (E). Causal independence between *h* and *a* is reflected in a freedom to substitute new values of *h* and *a* and, barring independent changes, get correct predictions. However, as we saw in chapter 6, the asymmetry of counterfactual dependence and any related asymmetry of justifiable prediction is only approximate. Although (given **I**) causes are never counterfactually dependent on their effects, effects are not always not counterfactually dependent on their causes. Whether an event that one is explaining is counterfactually dependent on the causes one cites does not, however, matter to the worth of the explanation. A derivation of $x_5$, the salt concentration in basin 5 in figure 6.3, in terms of the concentrations in basins 3 and 4, $x_3$ and $x_4$, appears to be a satisfactory explanation, even though $x_5$ is not counterfactually dependent on $x_3$ or $x_4$.

Independence is linked to other differences between derivations (C) and (E). First the specified variables in derivation (C), unlike those in derivation (E), are pairwise counterfactually independent – that is, neither of the values is counterfactually dependent on the value of the other. The counterfactual, "if the value specified for *a* were to change, then the value specified for *h* would change" is false. Second, the variables in (C) unlike the variables in (E) are "independently alterable":

> **IA** (*Independent alterability*) For every pair of variables, *x* and *y*, whose values are specified in a derivation, if the value of *x* were changed by intervention, then the value of *y* would be unchanged.

A derivation satisfies **IA** if any intervention that changes the values specified for some of the variables would leave the values specified for the others unchanged. If causation is deterministic, then the definition of an intervention and principles governing counterfactuals imply that independent alterability holds if and only if the variables whose values are specified in a derivation are not related as cause and effect (theorems 8.2 and 8.3). The specified variables in (C) (*h* and *a*) are both counterfactually independent and independently alterable. The specified variables in (E) ($s$ and $a$) are neither counterfactually independent nor independently alterable. $s$ is counterfactually dependent on $a$, and if one were to alter the value of $a$ by intervention, it is not the case that $s$ would remain unchanged. In this case $a$ and $h$ are also causally independent, but this need not hold in general. $x_3$ and $x_4$, the concentrations in basins 3 and 4 in figure 6.3, are both counterfactually independent and independently alterable, even though they are not causally independent.

Counterfactual independence and independent alterability are coextensive in these examples, but not generally. Independent alterability is a stronger condition that implies pairwise counterfactual independence but is not implied by it (theorem 8.1). An equation system in which the values of the salt concentration in basins 4 and 5 ($x_4$ and $x_5$) were both specified provides a case of counterfactual independence without independent alterability. Neither $x_4$ nor $x_5$ is counterfactually dependent on the other, but the value of $x_4$ is not independently alterable. I shall focus on independent alterability, because it is stronger and appears more closely connected to the aims of explanation.

What I call "simple derivations" – that is, derivations that do not specify the value of two variables that are related as cause and effect – satisfy **IA**. Derivations from exclusively exogenous variables – that is, variables that are either not causally related to any of the other variables under consideration or are related to them as causes only – are always simple derivations. To clarify the notion of a simple derivation, consider following derivation of $x_5$, the salt concentration in basin 5, in figure 6.3:

1. $x_2 = 8$ mg/ml
2. $x_1 = 4$ mg/ml
3. $x_4 = .5(x_1 + x_2)$
4. $x_3 = 4$ mg/ml
5. $x_5 = .5(x_3 + x_4)$

$\therefore$ 6. $x_5 = 5$ mg/ml

This derivation, which explains why $x_5 = 5$ mg/ml, is not a simple derivation, because the value of $x_3$ and of a variable upon which it causally depends, $x_1$, are both specified. Since $x_3$ is counterfactually dependent on $x_1$, and independent alterability implies counterfactual independence, the value of $x_1$ is not independently alterable.

Given determinism, one can always transform derivations that cite only the values of causes of what is to be explained into simple causal derivations. In deterministic circumstances if a causal derivation specifies the value of a variable $y$ and the value of variables upon which $y$ depends, delete the equation specifying the value of $y$, add equations relating $y$ to the variables upon which it depends, and add equations specifying the values of any variables upon which $y$ depends that were not previously mentioned. In this way, causal derivations can be transformed into simple causal derivations. Theorem 8.2 shows that simple causal derivations satisfy the independent alterability condition.

Derivations that specify the values of effects of the variables whose values are to be calculated can sometimes be simple derivations, and they will thus satisfy **IA**. But when this happens, the nonspecifying equations will not hold up. Suppose, for example, that sunlight approaching a flagpole is split by a system of mirrors into two beams that strike the flagpole from different directions and create two shadows. The angle of one of these beams with the horizon is twice that of the other. One can then calculate the height of the pole from the following system of equations:

1. $s_1 = s_1*$
2. $s_2 = s_2*$
3. $a_1 = 2a_2$
4. $\tan a_1/\tan a_2 = s_2/s_1$
5. $h = s_1 \tan a_1$

where $s_1$ is, of course, the shorter of the two shadows. The lengths of the two shadows are independently alterable. By an adjustment of the mirrors or by providing another source of illumination, one can alter the length of one of the shadows without the other changing. But when one intervenes in this way, equation 3 will not in general continue to hold, and the values for $h$, $a_1$, and $a_2$ one calculates will not in general be correct.

In a simple causal derivation, unlike a derivation – whether simple or not – that cites effects of what is derived, the variables are independently alterable, *and* the other equations continue to express correct relations when the values of the specified variables change. If $E$ expresses a law or mechanism, then its truth should not depend on the value of any of the variables specified in a simple causal derivation. This assumption seems to me part metaphysics and part methodology. If a *relationship* between variables changes when the value of a variable changes, one concludes that the relationship is misspecified or that one is outside of the relevant range of the values of the variables. When the nonspecifying equations do not depend on the values of the specified variables in a derivation, let us say that they are "insensitive."

> **IC** (*Insensitivity condition*) If the value specified in a derivation for any variable were changed by intervention, then the nonspecifying equations in the derivation would continue to hold.

The definition of an intervention and the semantics for counterfactuals entail that simple causal derivations typically satisfy both **IA** and **IC**.[3] Furthermore, derivations of values of variables from values of their effects or values of effects of common causes will not satisfy both **IA** and **IC** (theorem 8.5).

Given independent variability and insensitivity, if one intervened and changed the value specified for one of the variables, then one would get the correct values for all the other variables by substituting that new value into the equation system. Theorems 8.2 and 8.4 imply that if $x$ causes $y$, then in a simple causal derivation of the value of $y$ from the value of $x$, if the value of $x$ were changed by intervention and the new value were substituted into the derivation, the value of $y$ would be as calculated. This is equivalent to the necessary condition of the modal invariance claim (**MI**) discussed in chapter 11. Given in addition the prediction condition, one can conclude that the value $y$ *would have* if the value of $x$ were changed by intervention is the best prediction of what $y$'s value *will be* if one intervenes and changes the value of $x$. The value of $y$ one calculates may not be correct, because there may be some independent change in the system or in the value of other variables, but it is the best prediction nevertheless. Since simple causal derivations satisfy independent alterability and insensitivity, one can predict what will happen when one changes the value specified for a variable: Simply plug in the new value. In derivations from effects, on the other hand, one cannot do this.

## 8.6 Independent Alterability and Explanation

The definition of an intervention implies that simple causal derivations satisfy the independent variability and insensitivity conditions and that derivations that cite effects do not. It is easy to see why a derivation that does not satisfy **IA** and **IC** is not entirely satisfactory: One cannot substitute a new value for one of the independent variables and get a justifiable prediction of new values of the dependent variables.

What does this deficiency have to do with explanation? After all, when there are multiple connections, derivations that satisfy **IA** and **IC** may be almost as unsatisfactory. In derivations of the salt concentration in basin 5 $x_5$ from the concentration in basins 3 and 4, $x_3$ and $x_4$, the variables satisfy **IA** and **IC**, but one can calculate a new value of $x_5$ from a new value of $x_4$ only when the new concentration results from an intervention. What do independent alterability and insensitivity have to do with explanation?

[3] See theorem 8.4, p. 183. The reason for the qualification, "typically," is that there is a further condition. Suppose a causal derivation is simple because it omits some variable $z$ that an independent variable $x$ depends on and that has an independent effect on the dependent variable $y$. Such a derivation will not satisfy **IC**. See also the discussion of modal invariance in §11.3.

I shall argue that they are necessary for explanation. I cannot *prove* this, because I do not know what is the correct theory of explanation. The most I can argue is that shared intuitions concerning scientific explanations justify the claim that explanations require independent alterability and insensitivity. Actually I shall make a more cautious claim. When giving explanations, people often do not dot all the i's and cross all the t's, and the actual utterances people take to be explanations may not satisfy **IA** or **IC**. I contend only that independent alterability and insensitivity are required in fully articulated explanations. Here are five arguments for this contention:

1. People often seek explanations when they are puzzled. Explanations mitigate puzzlement and reduce contingency. Without both independent alterability and insensitivity, a derivation does not mitigate puzzlement. Suppose, for example, one wants to know why the length of a shadow is $s^*$ feet. This why-question can be answered by citing the height of the pole, the angle of elevation of the sun, and the law that light travels in straight lines. The explanation can be formulated as a derivation such as (C) above. Given the premises, there is nothing puzzling: $s$ "had to be" $s^*$. Of course, plenty remains unexplained. One has not explained the angle of elevation of the sun, the height of the pole, or why light travels in straight lines. A great deal always remains unexplained.

Consider in contrast an attempt to explain the height in terms of the shadow's length and the angle of elevation of the sun. Although the derivation has the same number of givens as the explanation of the shadow's length, it is more difficult to offer additional explanations of them. How is one to explain the shadow length without citing the facts that the shadow length supposedly explained? The fact that **IA** and **IC** do not hold also creates *a new mystery*. The derivation of $h$ is correct only if $a$ and $s$ have just the right values. If one intervened and used mirrors to change the angle from which light falls on the pole to $a^{*\prime}$, then (barring coincidence or the action of an unspecified mechanism) the height one derives from $a^{*\prime}$ and the old value of $s$ would be incorrect. There is a new coincidence and a new puzzle: why do $s$ and $a$ have just the right values to permit a correct calculation of the height? In an aberrant case of a flagpole with height or shadow adjustment mechanisms,[4] this question will have an informative answer, but in the case of an ordinary flagpole, all one can say is that they have the right values because $s$ depends on $h$, and because nobody intervened. The purported explanation generates a puzzle whose solution depends on the fact that was to have been explained (Jobe 1976, 1985; Woodward 1984).

Nothing turns on the details of the example. The new puzzle – why do the supposedly explanatory variables simultaneously have just the values they

[4] Such as was, I have heard, given to Carl Hempel on his eightieth birthday.

must have if the derivation of the explanandum is to be correct? – will arise whenever independent alterability or insensitivity fails. To reduce contingency, derivations must satisfy **IA** and **IC**.

2. Perhaps something more can be made of the notion of subsumption under law as a construal of contingency reduction. Hempel articulates this notion as nomic expectability (1965), and on his account, knowledge of effects apparently reduces contingency just as well as knowledge of causes. But this is, I suggest, a myopic misconstrual of the underlying intuition. If there are deterministic and calculable relations among values of the variables, then obviously the direction of the causal arrows is irrelevant to the inference of the values of some variables from the values of others. But the direction of the causal arrows is not irrelevant to inferences about the consequences of *changes* in the values of variables. When **IA** and **IC** hold, one can base nomic expectations on knowledge that the value of a single variable has been changed by an intervention. Independent alterability and insensitivity permit one to trace phenomena to *separate explanatory factors* and to show the possibilities of control. Deductions from causal variables thereby avoid the *fragility* of deductions of values of causal variables from values of their effects, and they guide action. Reducing contingency and subsuming relations between variables under laws of nature requires that the relations not be fragile in this way. What one seeks is not nomic expectability as defined by Hempel, but the ability to ground expectations on knowledge of changes in values of individual variables. In the relevant sense, the value of $v$ is not expectable, if one cannot predict what the value of $v$ will be, if one intervenes and changes the value of one of the explanatory variables.[5] Independent alterability and insensitivity are necessary for genuine subsumption and nomic expectability.

3. According to Wesley Salmon (1985), to explain is to reveal the operative causal mechanisms, while according to David Lewis (1986a), explanation presents information concerning the causal history. Since causes rather than effects enter into causal histories and into relevant causal mechanisms, only causes explain. But Lewis needs to say why information concerning causal histories rather than future causal trajectories explains, and Salmon needs to clarify why the relevant mechanisms involve the causes of the phenomena to be explained rather than their effects.

The fact that only simple causal derivations satisfy both **IA** and **IC** could be cited by Lewis or Salmon to complete their accounts. Suppose one mistakenly took some mechanism involving sunlight to be responsible for the height of a flagpole, $h = h^*$. A deduction of $h = h^*$ that relies on the law that light travels in straight lines will be correct only if the values of $a$ and

[5] Compare Woodward's (1979) remarks on "functional interdependence."

*s* are just right, and they will be just right only through the action of an unspecified mechanism or by coincidence, and only in the absence of interventions. Interventions that change the values of *a* or *s* will, ceteris paribus, lead to a faulty deduction. In a derivation that cites effects, a crucial mechanism has been omitted. If one takes the law that light travels in straight lines instead to be part of the mechanism responsible for shadows, the derivation will satisfy **IA** and **IC**, and no mechanism will be missing.

The separability of different factors that is implicit in the requirement of independent alterability is implicit in the notion of a mechanism. When factors do not interact, the causal capacity of each should remain unchanged by an intervention that changes one of the specified values. Causal relationships between dependent and independent variables should be independent of changes in the values of independent variables. This kind of explanation will only be possible when derivations satisfy **IA** and **IC**.[6] One does not have a mechanical explanation, and one has not correctly specified all the relevant features of the causal map, unless independent alterability and insensitivity hold.

4. Without independent alterability and insensitivity, one faces a puzzle and a problem. The puzzle is: Why do the purported explanatory variables simultaneously have the values that guarantee a correct deduction of the explanandum? If one intervenes and changes the value of a supposedly explanatory variable that is an effect rather than a cause, then the deduction will, ceteris paribus, have a false conclusion. In ordinary cases, the only solution to the puzzle would require using the values of the explanandum variables to explain the values of the variables in the explanans and citing the fact that there were no interventions. This solution constitutes the problem: Such explanatory loops cut these systems out of broader causal networks. There would be no way to concatenate such "explanations" into wider accounts and to provide a unified and compact theoretical picture of the world. Such "explanations" rest on sand and collapse in the face of interventions. Insofar as explanations are supposed to be resilient and to unify our knowledge, they require independent alterability and insensitivity.

5. The links between causation, counterfactual dependence, intervention, **IA**, and **IC** complement the pragmatic account of explanatory asymmetries. Because of their insensitivity and independent alterability, causal explanations serve human purposes much better than answers to why-questions that cite effects of what is to be explained. Without **IA** and **IC**, one cannot answer "What if?" questions, and one cannot predict the consequences of intervening to change the value of some variable. The discussion of inde-

[6] Independent alterability and insensitivity do not, however, imply that causal capacities will be invariant across a wide range of contexts. See p. 191.

pendent alterability and insensitivity clarifies why, when they ask, "Why?" people want to know causes. Explanatory asymmetries depend on human interests, but they equally depend on a non-interest-dependent consequence of strong independence, which is that derivations satisfy **IA** and **IC**.

### 8.7 Explanation, Intervention, Time, and Independence

One cannot explain why something happens by citing its effects or other effects of its causes, because derivations from effects or from other effects of causes do not satisfy **IA** and **IC**. Without independent alterability and insensitivity, arguments do not reduce contingency, they do not show the phenomena to be explained to be instances of deeper regularities, they do not unify our knowledge, and they do not capture all the relevant mechanisms. Without **IA** and **IC**, putative explanatory arguments cannot help agents to control their environment, and they do not answer "What if?" questions. They do not provide the answers to why-questions that people want. Independent alterability and insensitivity are essential. The notion of an intervention (which presupposes independence) and the account of counterfactuals defended in chapter 6 imply that there will be independent alterability and insensitivity in simple causal derivations and that independent alterability or insensitivity fails in derivations from effects.

The exploration of explanation in this chapter has wider implications. The importance of independent alterability and of insensitivity suggests that causal explanation is tied to human interests in the possibility of manipulation and control. One particularly important kind of explanation involves uncovering separate elements linked by laws of nature to the phenomena to be explained and independently alterable by intervention. This kind of explanation is extremely important, because tracing phenomena to independently manipulable factors responds to human interests in control. When one has an explanation of this kind, then one can control when one can intervene.

Consider the following speculative anthropology. Suppose that the variables $x$, $y$, and $z$ bear to one another the nonprobabilistic lawlike relation, $F(x, y, z) = 0$. Given the values of two of the variables, one can calculate the value of the third, and $F(x, y, z) = 0$ can be rewritten as $x = f(y, z)$; $y = g(x, z)$; or $z = h(x, y)$. Why distinguish causes and effects? Indeed why introduce an asymmetrical notion of causation at all? Suppose that the values of $x$, $y$, and $z$ are initially $x^*$, $y^*$, and $z^*$, and suppose that if the value of $x$ were altered by an intervention to $x^{*\prime}$, then the value of $z$ would be unchanged at $z^*$, and $y$ would equal $g(x^{*\prime}, z^*)$. Suppose, on the other hand, that if the value of $y$ were altered by an intervention, the value of $z$ would be unchanged but that x would not be $f(y^{*\prime}, z^*)$. There is thus an important asymmetry in the relationship between *x* and *y*. *x* is as it were a lever by which one can manipulate *y*. Although one can calculate the value of *x* from the values of *y* and *z*, the values of the latter two variables are not "responsible" for the value of *x*, because no agent could make *x* have that value by setting the values of *y* and *z*. Causal explanation is knowledge of the recipe whereby the value of a variable could be made what it is, and causation consists in the asymmetrical relations that permit agents to make variables have their values.

Because of limits to their muscle or mental power, people often cannot intervene, even though events or variables have the structure of causal and explanatory relations. The agency theory of explanation and causation sketched in the previous paragraph is thus only a first approximation. A theory like Hume's might seem tempting, because it will almost always be the case that the causes come earlier in time. But such an account is also only an approximation, because of the problem of epiphenomena. What makes possible explanation and what constitutes causation is instead a certain *structure*. A structure of counterfactual relationships given abstractly possible (though not necessarily feasible) interventions constitutes causation. This structure of counterfactual relationships presupposes noncounterfactual relationships. In particular, the possibility of abstract intervention implies the strong independence condition, and the asymmetries among the consequences of abstract interventions derive from transitivity: When *x* causes *y*, an abstract intervention with respect to *x* will not be independent of *y*.

The independence condition, **I**, is a weaker necessary condition. When it is not satisfied, there can be no recipes. It is not possible separately to add a little more salt. A certain kind of explanation for the flavor of the soup cannot be given. Without independence, there is no basis for defining asymmetrical relations among nomological connections and no asymmetrical relations to be defined. Causes are simply independent factors. Our search for causes and our conception of causes derives from our explanatory interests, and these in turn derive from our interests as agents. Whether *a* causes *b* does not, however, depend on our interests. Whether *a* causes *b* depends on whether there are events lawfully connected to *b* and not to *a*.

# 8*

# Causation, Explanation, and Independent Alterability

The first two sections consider whether there is any way to strengthen the DN model to deal with the problem cases without relying on the notion of causation or on elements of a nonconditional analysis of causation. The third section provides proofs underlying the conclusions drawn in §8.5.

## 8.1* Nomic Sufficient *Conditions* and Explanations

In his essay "The Direction of Causation and the Direction of Conditionship" David Sanford offers an account of causal priority in terms of a nonsymmetrical relation among propositions, "is a causal condition of." If his account succeeds, then he has provided the core of a purely conditional analysis of causation. Sanford also implicitly offers an account of scientific explanation. Sanford does not claim to have done so much, because his account also relies on an unanalyzed notion of "causal connection in a direct line."[1]

Let us, following Sanford, say that "**A** is nomically sufficient for **B** in circumstances **C**" (or "**B** is nomically necessary for **A** in circumstances **C**") if and only if "**A&~B&C&L**" is logically inconsistent where **A**, **B**, **C**, and **L** are all propositions, the conjunction is strongly nonredundant, and **L** is a (conjunction of) law(s) of nature. **A** and **B** state that events with properties *A* and *B* occur (1976, pp. 200–1). The exposition differs here from Sanford's in four significant ways: (1) Sanford's exposition is more general and applies to other sorts of necessity. (2) Sanford uses the notion of a logical impossibility rather than the narrower one of logical inconsistency. (3) In the case of causal relations, Sanford requires that **L** consist of causal laws

[1] I comment at some length on Sanford's account because of both its merits and its influence. John Mackie, for example, argues, "It seems, then, that we cannot derive a direction of conditionality from necessity and sufficiency, whether or not we add 'in the circumstances.' . . . I have previously, on this account, despaired of finding an analyzable direction of conditionality. But an article by David Sanford has persuaded me that it is possible, . . ." (Mackie 1979, p. 23). Beauchamp and Rosenberg criticize Sanford's views at length (1981, pp. 227–36) but also write, ". . . we consider Sanford's attack on this problem [of causal directionality] the most impressive and incisive of the alternatives to Hume's supposed views" (1981, p. 235).

and speaks of causal rather than nomic necessity and sufficiency. But since the distinction between causal and noncausal laws is problematic – with apparently noncausal laws such as the perfect gas law or the law of the pendulum perfectly capable of serving in causal explanations, I think the account will only be strengthened by this generalization. (4) Sanford does not mention explicitly the properties exemplified by the events said to occur by propositions **A** and **B**.

Both the notions of nonredundancy and of laws of nature are problematic, and like Sanford I shall not attempt to analyze them. The nonredundancy clause is crucial, because it renders necessity in the circumstances and sufficiency in the circumstances nontransitive. **A** may be necessary in the circumstances for **B** and **B** necessary in the circumstances for **D**, yet **A** not necessary in the circumstances for **D**. It may be the case that **D&~A&C&L** is redundant, while **B&~A&C&L** and **D&~B&C&L** are not.

Sanford then defines, "**A** is a nomically sufficient (necessary) *condition* for **B** (in circumstances **C**)" as **A** is nomically sufficient (necessary) for **B**, and there are admissible circumstances in which everything nomically necessary for **A** except **B** itself is nomically necessary for **B** (1985, p. 225; see also 1976, p. 205). Let us postpone discussing what circumstances are admissible except to note that the actual circumstances are always admissible. Sanford seeks to define notions of necessary and sufficient *conditions* such that **A** may be a sufficient (or necessary) condition of **B** without **B** being a necessary (or sufficient) condition of **A**. The requirement that everything necessary for **A** be necessary for **B** is supposed to capture the intuition that conditions are in some sense operative and responsible for what they are conditions for.[2]

One might then propose the following account of a scientific explanation: If **F** is the explanandum statement and **E** is a proposed explanans, <**E**, **F**> is a deterministic scientific explanation for **F** if and only if all of the propositions in **E** are true and some proposition (or conjunct in a proposition) in **E** is a nomic sufficient condition for **F**. All of the conditions of the DN model apart from the truth of the explanans and explanandum are already captured in the account of nomic sufficiency.

Adding the insistence that some proposition in the explanans be a nomically sufficient *condition* for the explanandum appears to solve the problems of causal *relevance* presented before. Even if a man's consumption of birth control pills is sufficient in the circumstances to avoid becom-

[2]As a general account of necessary and sufficient *conditions*, I do not think that Sanford's theory is acceptable. It has the disturbing implication that all logically sufficient conditions are also logically necessary conditions. If **P** is a sufficient condition of **Q**, then on Sanford's theory everything necessary for **P** must be necessary for **Q**. In other words **Q** is sufficient for everything necessary for **P**. Since the conjunction of everything necessary for **P** will be sufficient for **P**, **Q** must be sufficient for **P**. So if **P** is logically sufficient for **Q**, it must also be logically necessary for **Q**.

ing pregnant, it is not necessary in the circumstances, and not everything necessary for taking the pills will be necessary for not becoming pregnant. Hence taking the pills fails to explain the man's not becoming pregnant. Nor is hexing salt necessary in the circumstances for its dissolving.

Yet as a conditional analysis of causal priority or as a theory of explanation, Sanford's account does not fare very well. Consider an example Sanford discusses:

> A pendulum's having a certain length is a causally necessary and sufficient condition, in the circumstances, of its having that period. Everything necessary in the circumstances for its having that length is necessary in the same circumstances for its having that period. But something necessary in the circumstances for its having that period, for example, the amount of gravitational force acting upon it, is not necessary in any admissible circumstances for its having that length. (1976, p. 206)

To say that the proposition, ***G***,[3] which states the value of the gravitational force, is necessary in the circumstances for ***T***, the proposition stating the period, requires that one include in the circumstances the proposition ***L***, stating the length of the pendulum. Only given ***L*** is ***G*** nomically necessary for ***T***. To deny that ***G*** is necessary for ***L*** requires on the other hand that one *not* include ***T*** in the circumstances. For given ***T*** and the law of the pendulum, ***G*** can be deduced from ***L*** and is hence necessary for ***L***. So either Sanford's account is false or he must supply grounds for including ***L*** but excluding ***T*** from the circumstances. Intuitively what Sanford is driving at seems to be captured better by the independence account of chapter 4. Although the length of a pendulum and the strength of the gravitational fields are not independent of one another (since the gravitational force stretches the string holding the bob), there are physical properties of the bob upon which the frequency of oscillation depends that are causally independent of the gravitational force. But neither facts about the construction of the pendulum nor about the gravitational force are independent of its frequency of oscillation, and hence the latter is the effect. It is, by the way, best to take the effect to be the frequency of oscillation rather than the *period*, since it is not clear that the period at a time, as an instantiation of a dispositional property, is a distinct trope (see Railton 1980).

Given the law of the pendulum, and taking one of ***G***, ***T***, and ***L*** as a description of the circumstances, the other two propositions are nomically necessary and sufficient for one another in these circumstances. Moreover, everything nomically necessary in such circumstances for any one of these propositions will be necessary in the circumstances for the other (provided that the nonredundancy condition is not violated). But the derivation of ***T***

[3] The propositions are written using both boldface and italics to avoid confusing these names with the conditions G, T, and L.

from *L* and *G* is explanatory, while the derivations of *G* from *T* and *L* and of *L* from *T* and *G* are not explanatory, for the length and the gravitational force are (as a first approximation) not dependent on one another or on the period. Logically and mathematically the three propositions or the variables whose values these propositions state are perfectly symmetrical, but from the perspective of causation or explanation (and, as we saw in chapter 6, counterfactual dependence) they are not. It seems that no account of explanation in terms of the logical relations among propositions (including reference to the circumstances) is adequate.

Can this conclusion be avoided? Sanford maintains that *L* is not explained by *T*, because some proposition necessary in the circumstances for *T* is not necessary in the circumstances for *L*. Although *G* will not do the trick, can one find some other such proposition? Consider, for example, any proposition *L′* that is entailed by *L* and logically independent of *T* and *G*. Since *T* is nomically sufficient in conditions *G* for *L*, it will be nomically sufficient for *L′*, and *L′* will be nomically necessary for *T*. But *L′* is logically and hence not nomically necessary for *L*. So there is after all something nomically necessary for the effect that is not nomically necessary for the cause. Unfortunately, in just the same way one can always find propositions that are nomically necessarily for the cause, but not for the effect.

One must revise Sanford's proposal to say something like: **A** is a nomic sufficient condition of **B** in the circumstances if and only if **A** is nomically sufficient for **B** and every proposition that is logically independent of **A**, **B** and the circumstances **C** and that is nomically necessary for **A** in the circumstances is nomically necessary for **B** in the circumstances. But having revised Sanford's account in this way, one has ruled out precisely the conditions in which, owing to redundancy, necessity in the circumstances fails to be transitive. So **A** will be a nomic sufficient condition for **B** if and only if **B** is a nomically necessary condition for **A**, and Sanford's account of causal asymmetry fails.

## 8.2* Some Reformulations

Perhaps I have missed what Sanford is after. Consider the following passage:

> In the pendulum example above, the construction or adjustment of the pendulum is sufficient for its having a certain length. This is independent of the gravitational force and thus independent of the period of the pendulum. A pendulum does not have its particular length as a consequence of its having its particular period, because there is always something else, independent of its particular period, which is sufficient for its having its particular length (1976, p. 206).

These remarks suggest an alternative approach. Let us then say that **E** explains **F** only if (1) **E** is a nomic sufficient condition of **F** in the circumstances and (2) anything nomically sufficiently for **F** in the circumstances is, without reference to the circumstances, nomically necessary or sufficient for **E**. ***L*** explains ***T*** because ***L*** is a nomic sufficient condition for ***T*** in the circumstances and because there is no other independent nomic sufficient condition for ***T*** in the circumstances. Even though ***T*** is a nomic sufficient condition for ***L*** in the circumstances, it does not explain ***L*** because there is another nomic sufficient condition for ***L***, that is neither necessary nor sufficient for ***T***.

This revised account seems plausible, but there are complications. First, there are problems involving preemption and overdetermination, which I am bracketing until chapter 13. Second, and more seriously, if one judges that the derivation of ***L*** from ***T*** is not explanatory on the grounds that there is another independent derivation of ***L***, then one must also judge, for the same reason, that the other derivation of ***L*** is not explanatory. It would appear that Sanford's account implies that the proposition **A** describing "the construction or adjustment of the pendulum" cannot explain ***L***, because "there is always something else, independent of" **A** – namely ***T*** – "which is sufficient [in the circumstances] for its having its particular length." Sanford is suggesting a deterministic analogue to Salmon's no-screening-off condition (1971, pp. 55f). And, as is well known, without further restrictions, perfectly good explanatory factors may be screened off by spurious ones. (For another instance of the same problem, see p. 49 above.) The same difficulty is also fatal for Bromberger's account of explanation in his well-known essay, "Why Questions" (1966). For he insists that in order for **B** to be an answer to the question, "Why **A**?" there must, in effect, be no other (mutually exclusive) sufficient condition for **A** (1966, pp. 100, 105–7).

Can we capture what Sanford is after yet avoid this unfortunate implication? Can one make anything of the apparent fact that there are independent alternative sufficient conditions for the voltage and the resistance of a circuit, but not for the current? One possibility might be to say that when there are multiple independent derivations of some proposition **P**, the explanatory derivation is the one remaining after one subtracts all the derivations that explain some other proposition. But this suggestion appears hopeless. It is unworkable in practice. It assumes, falsely, that a derivation cannot explain more than one proposition. It supposes that one will always have explanatory derivations for all variables.

Evan Jobe has a different proposal that seems more promising (1976). Define "**B** is *explanatorily dependent* on **A**" as "**A** is a nomic sufficient condition for **B** in the circumstances, and there is no other nomic sufficient condition for **B** in the circumstances that is independent of **A**." (This is not

Jobe's formulation.) One might then suggest, as a first approximation, that a necessary condition for **B** to explain **A** is that **B** not be explanatorily dependent on **A**.

Even when refined, this account will not do. It lacks any compelling rationale, and it apparently accepts spurious explanations of one effect of a common cause by another (Glymour 1978). This last difficulty may be reparable (Jobe 1985), but the account still arbitrarily rules out the possibility of mutual dependence (Woodward 1984). The amount of resistance in a circuit is in fact to some extent dependent on the amount of current flowing. One may, prima facie, reasonably cite the amount of current in an explanation of the amount of resistance. Furthermore, suppose that the frequency of oscillation of a particular pendulum explains in the circumstances some other proposition **P**. Then there may be another independent sufficient condition for ***T***. In that case ***T*** would not be explanatorily dependent on ***L***, and one would not then be able to reject as nonexplanatory the derivation of ***L*** from ***T***.

Consider one other related approach. In *Determinism* (1971) Bernard Berofsky proposes that one look beyond the particular purported explanation and consider whether the laws in purported explanations can be subsumed in deductively systemized theories of phenomena like those for which the supposed explanation is offered. Consider a law like the law of the pendulum, which relates propositions such as ***T***, ***L***, and ***G***, and consider our most comprehensive theories of phenomena such as periods of pendula, lengths of pendula, and gravitational fields. ***T*** here is explained by ***L*** and ***G***, because the law of the pendulum is deductively subsumed into the most comprehensive theory of the periods of pendula, and it is not subsumed into the most comprehensive theories of gravitational fields or lengths of pendula. Similarly, the height of a flagpole explains the length of its shadow and not vice versa because the law that light travels in straight lines is deductively systematizable in a theory of phenomena such as shadows, while there are more comprehensive theories of phenomena such as the height of flagpoles than any which include the law that light travels in straight lines. Facts about shadows contribute nothing *in general* to the understanding of phenomena such as the height of flagpole, just as the law of the pendulum contributes nothing to the understanding of gravitational fields or strings with bobs at their ends.

Berofsky's account places a great burden on ways of classifying phenomena. It seems that one can cook up classifications of phenomena so as to make laws of optics crucial in understanding phenomena such as the height of flagpoles. Second, Berofsky's account appears to face counterexamples. Isn't Ohm's law a part of a theory of voltage or of resistance? Third, Berofsky's account may not qualify as one that relies only on logical relations among true propositions including laws, because the notion of a state description, upon which the account relies, makes essential reference to time. Fourth, Berofsky's reliance on the notion that statements provide a "theory of" some phenomena appears to be question-begging. He comes close to claiming that ***T*** is explained by ***G*** and ***L*** because ***G*** and ***L*** enter into the explanation of phenomena like ***T***.

The project of analyzing explanation in terms of a logical relation among propositions, including essentially propositions expressing natural laws, cannot be successfully carried out. Something more is needed. One must rely on additional nonlogical conditions on explanations. The possible resources of logical relations among **A**, **B**, and background conditions have been exhausted. Insisting that an explanatory factor **A** be necessary in the circumstances for the explanandum **B** has independent intuitive appeal, and it solves the problems of causal relevance, but it exacerbates the directional problems: If **A** is nomically necessary and sufficient for **B**, then **B** is nomically necessary and sufficient for **A**. No combination of necessity or sufficiency in the circumstances will do in cases in which the failure is at least in part directional.

## 8.3* Some Proofs

The two conditions satisfied by the derivation on p. 166 of the value of an effect from a value of a cause (C) and not by the derivation of the value of a cause from a value of an effect (E) are:

> **CI** (*Pairwise counterfactual independence*) If the values of two variables $x$ and $y$ are specified in a derivation, then $y$ is not counterfactually dependent on $x$ and $x$ is not counterfactually dependent on $y$.

> **IA** (*Independent alterability*) For every pair of variables, $x$ and $y$, whose values are specified in a derivation, if the value of $x$ were changed by intervention, then the value of $y$ would be unchanged.

Suppose that the value of a variable $y$ is $f(x^*, x^*_1, \ldots, x^*_n)$. I shall say that $y$ is counterfactually dependent on $x$ if and only if for all $x^{*\prime}$ within some specified range, if the value of $x$ were $x^{*\prime}$, the value of $y$ would be $f(x^{*\prime}, x^*_1, \ldots, x^*_n)$. Although this form fits quantitative relations most naturally, the variables and their values need not be quantitative. The specified variables in a derivation are counterfactually independent if none are counterfactually dependent. The specified variables in a derivation are independently alterable if an intervention that changes the values of some leaves the others unchanged. Counterfactual independence and independent alterability are not coextensive.

**Theorem 8.1: SIM** and **DI$_g$** (the definition of an abstract intervention) imply that if the specified variables in a derivation are independently alterable, then they are counterfactually independent.

Proof: I will prove the contrapositive. Suppose that counterfactual independence fails and that the values of $x$ and $y$ are both specified and that the value of $y$ counterfactually depends on the value of $x$. By clause (1) of **SIM** if $y$ counterfactually depends on $x$ then $y$ counterfactually depends on interventions that change the value of $x$. So the value of $y$ would not remain unchanged.

A counterexample to the converse is provided by the salt basins. The concentration in basin 5, $x_5$ is not counterfactually dependent on $x_4$ nor, of course, is $x_4$ counterfactually dependent on $x_5$. So $x_4$ and $x_5$ are pairwise counterfactually independent, but the value of $x_4$ is not independently alterable in the equation system.

**Theorem 8.2:** Given $\mathbf{CC_g}$, $\mathbf{DI_g}$, and **CDCC**, if the variables whose values are specified in a derivation are not related as cause and effect, then the derivation satisfies **IA**.

Proof: Suppose that $x$ is among the variables whose values are specified. Given $\mathbf{CC_g}$, $\mathbf{DI_g}$, and the assumption that the specified variables are not related as cause and effect, none of the variables except $x$ is causally connected to an intervention that changes the value of $x$. Hence, by **CDCC** none is counterfactually dependent on the intervention.

**Theorem 8.3:** Given **SIM**, $\mathbf{DI_g}$, $\mathbf{T_g}$, $\mathbf{CC_g}$, and **DC**, if a derivation satisfies **IA**, then none of the variables whose values are specified are related as cause and effect.

Proof: Suppose that $x$ and $y$ are among the variables whose values are specified and that $x$ causes $y$. If the value of $x$ depends on an intervention variable $z$, then by $\mathbf{DI_g}$ and $\mathbf{T_g}$, $z$ causes $y$. By $\mathbf{DI_g}$ there is no multiple connection in the causal relationship between $z$ and $y$, and so by the proof of theorem 6.4, $y$ is counterfactually dependent on $z$, and so the derivation does not satisfy **IA**.

Simple causal derivations do not necessarily satisfy **IC** unless one further condition is met:

**SUF** (*Sufficiency*) For all variables $x$ in the set $V$, $V$ contains all direct causes of $x$ that are direct or indirect causes of any other variable in $V$ by paths that do not go through $x$.

"Direct" here means merely that there are no causal intermediaries among the variables of interest. The set $V$ of variables includes at least all the independent and dependent variables in the explanatory derivation. Given **SUF**, one can then state and prove:

**Theorem 8.4:** $\mathbf{DI_g}$, **SIM**, and **SUF** entail that simple causal derivations satisfy **IA** and **IC**.

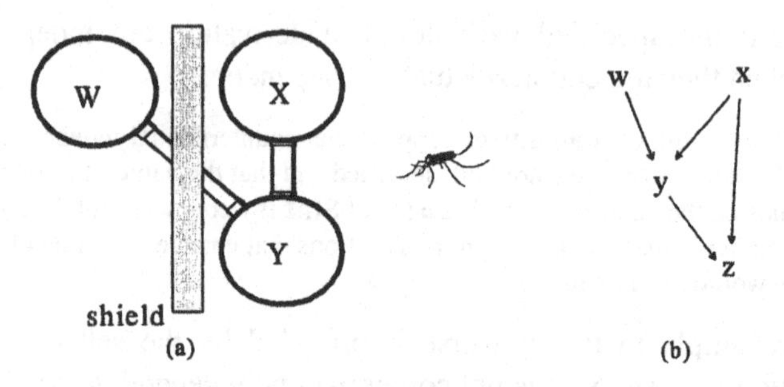


**Figure 8.1:** Segmenting causal derivations

Proof: Suppose that the value of one specified variable $x$ is changed by intervention. By assumption the values of the other specified variables do not causally depend on $x$ and by theorem 8.2 they would remain unchanged. By assumption those variables that are nonzero functions of $x$ are effects of $x$ and do not depend on causes of $x$ by any paths that do not go through $x$. By **DI$_g$** and clauses 3 and 4 of **SIM**, the nonspecifying equations in the derivation would continue to hold.

**Theorem 8.5:** **CDCC** and **DI$_g$** entail that derivations of values of variables from values of their effects or from values of effects of common causes do not satisfy both **IC** and **IA**.

Proof: Suppose (a) that $x$ causes $y$ or that $x$ and $y$ are causally connected only as effects of some variable $z$, (b) that given the values of all other variables the value of $x$ is a nonzero function of the value of $y$, $F(y)$, and (c) that the value of $y$ is specified in a derivation of the value of $x$. By **DI$_g$** an intervention with respect to $y$ that changes its value to $y^{*\prime}$ is not causally connected to $x$, and so by **CDCC** $x$ is not counterfactually dependent on it. So it is not the case that if an intervention were to change the value of $y$ to $y^{*\prime}$, then the value of $x$ would be $F(y^{*\prime})$.

Given determinism, one can always transform derivations that cite only the values of causes of what is to be explained into simple causal derivations. In an indeterministic case, there are further complications. Suppose, there are, as shown in figure 8.1a three basins of radioactive liquid, $W$, $X$, and $Y$, and a mosquito. The radioactive liquids in $W$ and $X$ flow into $Y$, and the mosquito is exposed to the radiation from basins $X$ and $Y$. The values of $w$, $x$, and $y$ are respectively the radiation levels in basins $W$, $X$, and $Y$, while $z$ is a variable measuring the health of the mosquito. $y$ is probabilistically dependent on $w$ and $x$, and $z$ is probabilistically dependent on $x$ and $y$. The causal graph is shown in figure 8.1b. One can save a probabilistic version of independent alterability by replacing the relationship between $z$ and $y$ with a relationship between $z$ and $y^+$, the difference between $y$'s actual and expected value. In a probabilistic derivation of $z$ from $w$, $x$, and $y^+$, the variables are independently alterable.

# 9

# Probabilistic Causation

There are two main motivations for developing a theory of probabilistic causation. The first is esoteric. According to contemporary physics, many occurrences are not determined. A complete specification of the state of some ensemble only determines a set of probabilities. Despite the temptation to maintain that what happens by chance is not caused, causal questions remain. When one sends a photon through a slit and illuminates a particular square on a grid behind the slit, one wants to say that sending the photon through the slit caused the square to be illuminated, even when the transmission of the photon created only a small probability that the particular square would be illuminated. There are also probabilistic phenomena involving large numbers of individuals, like those in statistical mechanics, that apparently call for probabilistic *explanation*, even though the relations among the individuals may be deterministic.

The second motivation is more prosaic. As is argued with special vividness by Elizabeth Anscombe (1971), causal attributions in everyday life often appear nondeterministic. People say that punishments deter theft even though they do not believe that the deterrence is perfect. People say that dropping a glass caused it to break, even though they are aware of similar circumstances in which similar glasses were dropped and did not break (see also Rosen 1982).

These prosaic cases do not force one to conclude that the causal relation is not deterministic. Many causal factors influence thefts in addition to the threat of punishment. One can *conjecture* that the threat of punishment completely deters theft given certain arrays of these other factors, and that it does not deter theft given other arrays. Indeed there may even be circumstances in which the threat of punishment causes thefts. By specifying all the circumstances precisely, one might restore determination to causation.

This response to the prosaic indeterminism of causation seems to be based on faith rather than on evidence. Why should one believe that every glass exactly like this one dropped in exactly the same circumstances breaks? Esoteric knowledge of the microscopic indeterminism that lies "beneath" the macroscopic phenomena of daily life enhances these doubts. Furthermore, those who would dismiss indeterministic and probabilistic causation apparently shirk an important philosophical task. Suppose that causation is always a deterministic relation and that it turns out that smoking is a deterministic cause of lung cancer in some circumstances, irrelevant in others, and a deterministic preventative in still other circumstances. If causation is a deterministic relation, then presumably there are circumstances of all three kinds. (Those who doubt that there are circumstances of the third kind should think of those people whose physiology is such that if they smoke, they will die soon from a heart attack. For such people, smoking is an effective, though unattractive way to prevent lung cancer.) What then can the surgeon general's claim that smoking causes lung cancer mean? Presumably it does not imply that smoking is never a preventative, and it could hardly imply that everyone who smokes gets lung cancer. Whether or not causation is a deterministic relation, an interpretation of everyday causal claims is needed.

The prosaic cases that count for an indeterministic view of causation do not necessarily count for a specifically probabilistic theory of causation. For it is no easier to say what the precise probability is that dropping a glass will cause it to break than to formulate deterministic generalizations about when glasses will break. Many theories of probabilistic causation do not, however, require that one be able to state precise probabilities. They require only that causes make their effects more likely.

Despite these reasons to develop a theory of probabilistic causation, I shall argue that one should deny that events are caused probabilistically. Indeterminism requires that one should recognize that some things are not caused. Instead they happen by chance. Sometimes only the probabilities of events are caused. Purportedly probabilistic causation should be regarded as deterministic causation of propensities. Furthermore, I shall argue that the problems of interpreting claims such as the surgeon general's have little bearing on the metaphysics of causation. Speaking in defense of this view are its simplicity and coherence with accounts of deterministic causation. This chapter argues in addition that the internal dialectic of philosophical discussions of probabilistic causation points to the same conclusion.

### 9.1 Causes as Increasing the Probability of Their Effects

The theories of probabilistic causality discussed in this chapter distinguish causes from effects by temporal order and have little to say about causal asymmetry. Probabilities play their part instead in the theory of causal connection. These theories imply that the types **a** and **b** are nomically connected if and only if some probabilistic relationship obtains. What is that relationship?

One possible answer, suggested by Hempel's theory of statistical explanation (1965, pp. 376–411), is that the types **a** and **b** are causally connected (in certain circumstances *K*) when it follows from laws of nature that $\Pr(B/A\&K)$ is very high. Although this account could apparently justify the claim that slashing tires causes them to deflate, it will not justify or explain how sending a photon through a slit could cause a particular square to be illuminated.

More plausibly, one can say that **a** and **b** are causally connected in circumstances *K* when $\Pr(B\&A/K) \neq \Pr(B/K)\cdot\Pr(A/K)$.[1] When $\Pr(B\&A/K) > \Pr(B/K)\cdot\Pr(A/K)$, **a** is positively causally relevant to **b**. When $\Pr(B\&A/K) < \Pr(B/K)\cdot\Pr(A/K)$, the causal relevance is negative. Causes do not necessarily make their effects probable, and negative causes or preventatives do not necessarily make their effects improbable. Causes and preventatives only make their effects more or less probable than they otherwise would have been. One can then say that *a*'s cause *b*'s (in the circumstances) when *a*'s precede *b*'s and $\Pr(B/A\&K) > \Pr(B/{\sim}A\&K)$ or $\Pr(B/A\&K) > \Pr(B/K)$. Just as in the case of deterministic causation, one has to insist that these probabilities hold as a consequence of laws of nature (though this may be built into the interpretation of probability).

Even with the addition of a nomological requirement, such account of probabilistic causality, like Hume's theory, runs squarely into the problem of spurious causation. It may be that **a** is a prima facie cause of **b** – that is, that tokens of **a** precede tokens of **b** and $\Pr(B/A\&K) > \Pr(B/{\sim}A\&K)$, yet **a** and **b** are effects of a common cause. One cure is to say that **a** is a cause of **b** when it is a prima facie cause and there is no partition of the events (given the circumstances *K*) prior to occurrences of **a** that screens off **b** and **a**. This is essentially the theory Patrick Suppes defended.[2]

Both the claim that causes must be prima facie causes and the requirement that genuine causes are not screened off are unsatisfactory. *a*'s may cause *b*'s in circumstances *K* even though $\Pr(B/A\&K) \not> \Pr(B/K)$ because, for example, some cause of *a*'s occurring has an independent negative causal influence on the occurrence of *b*'s or because *a*'s positive influence on *b*'s by one causal path is canceled out or outweighed by its negative influence by another path (Hesslow 1976). Birth control pills can cause thrombosis even though those who take birth control pills have a *lower* incidence of thrombosis, because the pills prevent pregnancy, which is a

[1] This is a type-level restatement of the operationalizing assumption (**OA**) of chapter 4. Here its role will be to provide definite truth conditions rather than merely an approximate linkage.

[2] 1970, esp. p. 28; 1984, pp. 151–3. For a fine brief discussion of Suppes's theory, see Spohn (1983b, pp. 75–82). See also Hesslow (1976), Rosen (1978), and Otte (1981). Suppes does not distinguish token and type levels and does not explicitly cast his theory as a theory of causal relations among types.

stronger cause of thrombosis. Furthermore, a genuine cause may be screened off by some partition of prior factors, even though it is not screened off by a maximum conjunction of all (causally) *relevant* background factors (Skyrms 1980, p. 108).

The most common way to respond to these difficulties has been to say that *a*'s causes *b*'s in circumstances *K* if and only if for all *j* $Pr(B/A\&S_j\&K) > Pr(B/S_j\&K)$, where $\{S_j\}$ is a partition of all the relevant factors apart from *A* and *K*. That is, let $\mathbf{c}_1, \ldots, \mathbf{c}_m$ be all the relevant factors, and suppose that each of these factors is dichotomous – it is either present or absent. Then there will be $2^k$ different combinations of these factors. Some of these combinations will not be possible, and in some the probability of **b** will be one or zero regardless of **a**. Those combinations that remain define the "test situations," $S_j$ – that is, the cells in the partition.[3]

A scheme like this is defended by Nancy Cartwright:

**CCC** **a** causes **b** if and only if $Pr(B/A\&K_j) > Pr(B/K_j)$ for all state descriptions $K_j$ over the set $\{\mathbf{c}_i\}$ where $\{\mathbf{c}_i\}$ satisfies:

1. If $\mathbf{c}_i$ is in $\{\mathbf{c}_i\}$, then $\mathbf{c}_i$ causes **b** or ~**b**.
2. **a** is not in $\{\mathbf{c}_i\}$.
3. For all **d**, if **d** causes **b** or **d** causes ~**b**, then **d** = **a** or **d** is in $\{\mathbf{c}_i\}$.
4. If $\mathbf{c}_i$ is in $\{\mathbf{c}_i\}$, then **a** is not a cause of $\mathbf{c}_i$.[4]

Analyses similar to Cartwright's have been endorsed by many others.[5]

In **CCC** the relevant factors consist of all and only the factors that **b** or ~**b** depend on, apart from **a**, the factors specified in the circumstances *K*, and the factors that depend on **a**. The first condition has been dubbed by John Dupré the requirement of "contextual unanimity" (1984, p. 170). Causes must raise the probability of their effects in all relevant contexts. Cartwright argues that the fourth condition in **CCC** is not correct. What one should do instead is to hold fixed all causal factors apart from **a**, *K*, and those that are in fact on the particular occasion caused by the token event *a* rather than

[3] It is controversial whether one should include in the partition those cells in which Pr(B) is one or zero regardless of **a**. In such cells **a** is causally irrelevant to **b**. Such facts should not be ignored. For a discussion of these issues and of the problems that arise when the conditional probabilities are undefined, see Ray (1992, pp. 231–40).

[4] 1979, p. 26. I have changed Cartwright's lettering, and in order to avoid confusing this condition with my connection principle (CC), and I have changed its name from "CC" to "CCC." Where I use the word "causes," Cartwright uses instead a special notation to make clear that she is concerned with causal generalizations or relations among types rather than with token causation.

[5] Some influential texts are Skyrms (1980), Eells and Sober (1983), and Eells (1991). Skyrms suggests that it is enough that for some *j* $Pr(B/A\&S_j\&K) > Pr(B/S_j\&K)$ and for no *j* $Pr(B/A\&S_j\&K) < Pr(B/S_j\&K)$. Eells argues that this suggestion is harmless, but that it is better to regard the causal relevance of **a** to **b** in such cases as "mixed" (1991, pp. 95–7). In some contexts **b** depends on **a**, while in others **b** does not.

allowing all factors that depend on **a** in circumstances *K* freely to vary (1979, p. 30). One should hold fixed everything distinct from **a** upon which **b** depends except the actual (token) causal intermediaries. Cartwright holds that condition 4 is an unsatisfactory approximation and that no purely type-level formulation is correct.[6] This account does not purport to reduce causation to probabilities, since one has to know what else **b** depends on and whether those things in turn depend on **a** in order to specify the test situations in which one checks to see whether Pr(B/A) > Pr(B).

As is apparent from inspecting **CCC**, Cartwright (1979) does not explicitly relativize causal claims to background circumstances. Most theorists of probabilistic causality follow her in this regard. (Eells is a partial exception, but he stresses *populations* rather than *circumstances*.) One might think that this is harmless, since the relevant features of the circumstances *K* will be incorporated into the partitioning. Suppressing the reference to *K* is, however, misleading, because one may then forget that causal relationships may change from circumstance to circumstance. Suppose that *K* were a single property, and that there were but one other relevant property, *D*. If one did not relativize the causal claim to circumstances *K*, then one must consider the causal relevance of **a** to **b** in the four situations of *D&K*, *~D&K*, *D&~K*, and *~D&~K*. But one may be concerned with the causal relevance of **a** to **b** only in circumstances *K* and thus only with the first two of these. The so-called problem of causal interaction – the fact that the probabilistic impact of **a** on **b** may differ in direction and magnitude from circumstance to circumstance – arises only if one expects the causal relations between types or variables to be same in all test situations. Although the knowledge that *a*'s cause *b*'s will not be useful unless *a*'s increase the probability of *b*'s in a variety of circumstances, it is unrealistic to hope to identify causal relations that are not relativized to some background.

An alternative way to cope with circumstantial character of causal relations or with the problem of interaction is to enrich one's description of the cause, rather than to relativize causation to the circumstances (Cartwright 1979, p. 31; Humphreys 1989, esp. §25). If **a** is positively relevant to **b** given **d**, but not given ~**d**, then one should regard **a&d** rather than **a** by itself as a cause of **b**. Any variation from context to context in the relevance of **a** to **b** would, on this approach, demonstrate that **a** is but a part of a cause of **b** rather than a cause itself. On this view, what people call causes are only parts of causes: Striking a match would not count as a cause of its lighting,

[6] Eells and Sober (1983), on the other hand, deny that condition 4 is a compromise, and Eells argues at length that a modification of 4 is correct (1991, ch. 4). They argue that partitioning too finely may lead to small and unrepresentative samples and thus to *misestimates* of probabilities, but if a factor is truly irrelevant, conditioning on it should make no difference to the genuine probabilities. See also Spohn (1983b, p. 83).

but merely a part of a cause, because striking a match does not always increase the probability of its lighting. In a deterministic context, this is a view of causes as minimal sufficient conditions, rather than at least INUS conditions. Such a "combined-factors" approach (in Eells's terminology, 1991, p. 131) runs into difficulties in comparing Pr(B/A&D) and Pr(B/~(A&D), because *~(A&D)* is a disjunction, *~A&D* or *A&~D* or *~A&~D* (Eells 1991, ch. 3), and it may be that Pr(B/A&D) > Pr(B/~A&D) while Pr(B/A&D) < Pr(B/A&~D) or Pr(B/A&D) < Pr(B/~A&~D). Furthermore the quantitative contribution of *A&D* would appear to be undefined. Humphreys (who is mainly concerned with causal relations among quantitative variables) addresses these problems by proposing that one compare $Pr(z/xy)$ to $Pr(z/x_0y_0)$, where $x_0$ and $y_0$ are "zero" levels of the variables $x$ and $y$ (1989, §15). But it is questionable whether zero levels can always be defined nonarbitrarily.[7] Eells's and Cartwright's "revised-contexts" approach avoids having to define such zero levels and, like Mackie's view of deterministic causes as INUS conditions, it is consistent with the causal language of everyday life and of science. For these reasons the revised-contexts approach seems preferable to Humphreys's combined-factors approach.

One might insist that there be quantitative invariance across test situations and that merely qualitative contextual unanimity demands too little. Suppose that Pr(B/A&C) > Pr(B/~A&C) and (B/A&~C) > Pr(B/~A&~C), but that Pr(B/A&C) - Pr(B/~A&C) ≠ Pr(B/A&~C) - Pr(B/~A&~C). Then the causal relation between **a** and **b** differs depending on whether tokens of **c** or **~c** occur, and one omits relevant information if one merely notes that **a** is positively relevant to **b** in both cases. Calling "watch out" may increase the probability that people crossing a street toward you avoid getting run over by a truck whether the pedestrians are deaf or not, but if they are deaf, the probability that they will read your lips and notice your warning is small. Such unanimity across different contexts seems an uninteresting consequence of the significant causal facts, which concern the consequences of calling "watch out" in the different situations. Later (pp. 199–200), I shall consider and reject an argument of Paul Humphreys for the conclusion that merely qualitative facts about the signs of these differences in different test situations have a role in the explanation of undetermined outcomes.

When contextual unanimity holds, the factor **a** has an invariant influence on **b**. Cartwright argues that insisting on contextual unanimity involves attributing a causal capacity to **a** (1989, p. 145; see also Woodward 1993). In Cartwright's view, an analogous invariance is crucial in nonprobabilistic

[7] For a deeper critique, see Hitchcock (1993, pp. 344–5). As Hitchcock points out, the problem of disjunctive causal factors creates problems for Eells's and Cartwright's approaches, too.

settings as well. To understand gravitation is to understand a causal capacity that not only results in a certain downward acceleration in the absence of other forces, but issues in an invariant tendency toward such an acceleration even when other forces are present. If contextual unanimity or the analogous invariance illustrated by the last example never held across any but the narrowest range of contexts, the world might be equally law-governed, but, according to Cartwright (and Woodward 1993, 1995), causal language would lose its point, which is precisely to isolate separate factors that (except when there are interactions) act in invariant ways. Woodward's views are discussed in §11.2 – §11.4.

One should not link causation to invariance of effect, because interaction is ubiquitous. The failure of analogues to contextual unanimity and of contextual unanimity itself in cases of interaction points to no deficiency in an explanation. Contextual unanimity is an *additivity* condition rather than an *independence* condition, and explanation requires independence.

One should not, however, jump to the conclusion that, because the probabilities within the individuals cells of the partition are fundamental, contextual unanimity has no importance. Although contextual unanimity and the invariance it exemplifies are not essential to causation, causal knowledge would be fragile and causal explanations would be shallow if we were never able to generalize across contexts.

## 9.2 Type and Token Causation Revisited

Cartwright's influential analysis gives rise to many difficulties. One problem, which was noted earlier by I. J. Good, is that apparently causes sometimes *lower* the probability of their effects! Good gives the following example:

> Sherlock Holmes is at the foot of a cliff. At the top of the cliff, directly overhead, are Dr. Watson, Professor Moriarty, and a loose boulder. Watson, knowing Moriarty's intentions, realises that the best chance of saving Holmes's life is to push the boulder over the edge of the cliff, doing his best to give it enough horizontal momentum to miss Holmes. If he does not push the boulder, Moriarty will do so in such a way that it will be nearly certain to kill Holmes. Watson then makes the decision (event F) to push the boulder, but his skill fails him and the boulder falls on Holmes and kills him (event E). (1961, p. 318)

It seems that Watson's decision causes Holmes's death and lowers its probability. Good argues that one should distinguish between the "tendency" of the decision and its degree of causal influence. Eells and Sober (1983) interpret Good as distinguishing between type and token causation. In their view, Watson's decision was a token cause and a type preventative of Holmes's death.

One might dispute Eells and Sober's view as follows: Assume that there are no significant purely stochastic relations between Watson's decision and its consequences. Let *K** be a specification of *exactly* the circumstances in which Watson made the decision to push the stone. Let *K* be Good's less detailed specification of the circumstances. Let **f** be the decision to push the stone and **e** be Holmes's death. The token causal claim that Watson's decision to push the stone caused Holmes's death implies that $Pr(E/K^*\&F) > Pr(E/K^*\&{\sim}F)$.[8] This is a type-level claim. In circumstances of just this kind, decisions like Watson's tend to cause death. The token causal claim will fail to imply such a type-causal claim only in indeterministic circumstances when the token cause lowers the probability of its effect in exactly the circumstances in which it leads to its effect.

$Pr(E/K^*\&F) > Pr(E/K^*\&{\sim}F)$ and $Pr(E/K\&F) < Pr(E/K\&{\sim}F)$ are consistent. So it looks as if the push is a type cause in one set of circumstances and a type preventative in another. But this is a misreading of the facts. $Pr(E/K\&F) < Pr(E/K\&{\sim}F)$ does not imply that **f** is a type-level preventative. *K* is equivalent to a disjunction of precise state descriptions, of which *K** is one, and **f** does not decrease the probability of **e** across all homogeneous circumstances truly described by *K*. **f** is not a type preventative of **e** in circumstances *K*. As argued in chapter 5, type-level causal claims are generalizations of token causal claims, and so the two will only diverge due to random elements or to a difference between the circumstances specified in the generalization and the circumstances found in the token occurrence. The claim that Watson's push lowered the probability of Holmes's death is either mistaken or it makes reference to *different* circumstances. Watson's push did not tend in the circumstances to prevent Holmes's death. Watson believed it would and intended it to, but it did not.

Such a response cannot always be given, and one may want to rely on Eells and Sober's distinction between token and type causation or between singular causal claims and causal generalizations to defend a theory like Cartwright's from counterexamples in genuinely indeterministic circumstances. Suppose, as in figure 9.1, that an indeterministic system initially in state *S* can wind up in one of three different stable states, *X*, *Y*, or *Z*. *S* has a 20% chance of going directly to state *X* (that is, of an event of kind **a** occurring), a 40% chance of going to intermediate state *M* (of an event of kind **b** occurring), and a 40% chance of going to intermediate state *N* (of an event of kind **c** occurring). Once in *M* the system has a 90% chance of ending up in state *Y* and a 10% chance of winding up in state *Z*, while in *N* it has a 90% chance of winding up in *Z* and a 10% chance of winding up in

[8] This is both plausible and true on Eells's own account of causation (1991, ch. 6) just after the time the stone is pushed.

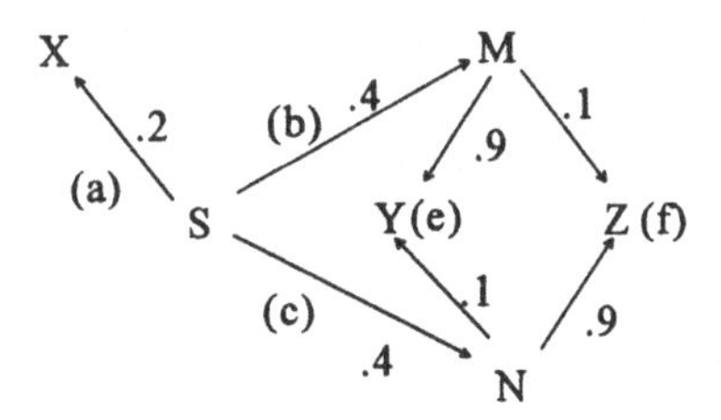


**Figure 9.1:** Token causation and type prevention

state *Y*. Let **e** and **f** be the event types of ending up in states *Y* and *Z*, respectively. So Pr(E) and Pr(F) = .4, while Pr(E/B) and Pr(F/C) = .90, and Pr(E/C) and Pr(F/B) = .10. Suppose the system winds up in state *Y* via the intermediary state *N*, and that one regards the transitions as causal (which Suppes 1984, p. 161, for one denies). Then it appears that *c* (entering intermediate state *N*) is a token cause of *e* (winding up in state *Y*) even though **c** is a type-level preventative of **e**.[9]

Does the possibility that *c* is a token-level cause and that **c** is a type-level preventative force a reconsideration of the conclusions of §5.2*? There I argued that type-level causal claims are generalizations of token causal claims. I answered Eells arguments for the independence of token and type causation by relying on the differences between the circumstances in token causal relations and the circumstances specified in causal generalizations. In the example considered in the previous paragraph, on the other hand, the circumstances are just the same. In exactly the same circumstances, it seems that *c* is a token cause and that **c** is a type preventative. Must one not conclude that Eells is right to maintain that claims about causation at the type level are not reducible to generalizations concerning token causation?

The conclusion is unavoidable *if* one grants that *c* is a token cause of *e* and that **c** is a type preventative of **e**. But one does not have to grant this assumption, and one of the main goals of this chapter is to defend the conclusion that *c* is *not* a token cause of *e* and that **c** is not a type preventative of **e**. *Tokens of* **e** *are not caused.* Given the occurrence of *b* or of *c*, whether one winds up in state *Y* or state *Z* is a matter of chance. *c* or **c** only causes Pr(E), and it does this in just the same way at the type and token level. If probabilistic causation is deterministic causation of probabilities, then type causation is reducible to token causation.

[9] One might resist this conclusion on the grounds that *e* occurs (the system winds up in state *Y*) *despite* the occurrence of *c*, rather than because of it. Eells's theory of token causation (1991, ch. 6) attempts to regiment these intuitions. Yet one could also say that the system wound up in state *Y because* it decayed to intermediate state *N* rather than decaying to *X*. Whatever the terminology, the causal chain from the initial state to the final state *Y* passes through the token event *c*. This example derives from one given by Salmon (1980, p. 65).

## 9.3 Other Difficulties and Refinements

A second problem with probabilistic theories such as Cartwright's and Eells's, which John Dupré pointed out, is that they imply that the very claims the theories were designed to explicate are false. If some people have a physiological condition that causes lung cancer more strongly if one does not smoke, then it will be false that smoking makes lung cancer more probable in every test condition and, according to **CCC**, smoking is not a cause of lung cancer. Dupré argues that this is unacceptable. According to Dupré, one must instead take the surgeon general as describing an average effect or a statistical correlation over a "fair sample."[10] The problem is general, since there are background contexts in which smoking lowers one's probability of getting lung cancer, drinking makes driving safer, and not wearing seat-belts makes one less likely to suffer injury or death. A theory of probabilistic causality ought not to imply that such causal generalizations are false.

Dupré's own revision is not satisfactory as a theory of probabilistic causation, because it seems to reduce causation to mere co-occurrence. On Dupré's account, the causal relevance of **a** to **b** depends on the actual frequencies of contexts, rather than exclusively on lawlike relations among properties. So the causal relevance of **a** to **b** can change merely with a change in the frequency of other types. To make the causal relevance of **a** to **b** depend on the frequencies of other factors fails to distinguish the question, "What causes what?" from the question "How often would such and such cause so and so?" (Eells 1987, p. 110). "Average effect is a sorry excuse for a causal concept" (Eells 1987, p. 113).

Dupré's difficulties do not resolve Eells's problems. Eells's theory still implies that seat belts don't save lives, drinking while driving doesn't cause accidents, and smoking is not a cause of lung cancer. Something has gone awry. Theories of probabilistic causality are supposed to explicate claims such as "Smoking causes lung cancer," not to reject them.

A third possible problem is that theories of probabilistic causation such as Cartwright's imply that probabilistic causation is not transitive. Most theorists of probabilistic causation accept this implication, but I have defended transitivity.[11] Suppose that **b** depends on intermediate factors

[10] 1984, pp. 172–3; 1993, pp. 201–4. See also Cartwright (1989, p. 100). Hitchcock (1997) has given a precise mathematical characterization of Dupré's suggestion. The basic idea is that **a** should increase the probability of **b** when one holds fixed the frequency of factors distinct from **a** that are not causally dependent on **a** upon which effect **b** depends. This is more plausible as a characterization of claims such as the surgeon general's than as an account of probabilistic causation. See also David Papineau's notion of population causation (1986, p. 123).

[11] See p. 107. I am indebted to Nancy Cartwright and Christopher Hitchcock here.

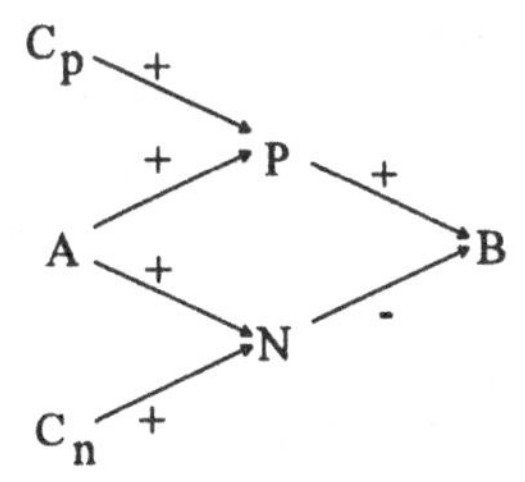


**Figure 9.2:** Transitivity without unanimity of intermediaries

which in turn depend on **a**. Eells and Sober prove that if the effect of *a*'s on *b*'s occurs only via the intermediaries,[12] then probabilistic causation will be transitive if there is only one intermediary or if the intermediaries are "unanimous." For example, probabilistic causation will be transitive if **a** increases the probability of intermediaries **d** and **f**, and **d** and **f** both increase the probability of **b**.

Eells and Sober establish only a sufficient condition for transitivity, and they deny that the condition is necessary. Suppose **a** increases the probability of two intermediaries, **p** and **n**, and **p** (the positive intermediary) increases the probability of **b**, while **n** (the negative intermediary) decreases Pr(B). If the influence via **p** and **n** does not cancel out, should one say that **b** or ~**b** depends on **a**, depending on whether the influence via **p** outweighs the influence via **n**? Eells and Sober believe so. They maintain that if the influence one way predominates over the influence the other way in all causally homogeneous circumstances, then probabilistic causation is still transitive.

John Dupré has argued that without unanimity of intermediaries, contextual unanimity will typically fail, too (1984, 1993, p. 198). With the help of the independence condition, his argument can be reconstructed as follows. Suppose (as in figure 9.2) that **b** depends on **p** positively and **n** negatively and that both **p** and **n** depend positively on **a**. Given $\mathbf{I_g}$ (the type-level version of the independence condition), **p** and **n** have other causes, $\mathbf{c_p}$ and $\mathbf{c_n}$, which are independent of one another and of **a**. If $Pr(P/A\&C_p)$ is much greater than $Pr(P/A\&{\sim}C_p)$ and $Pr(P/C_p\&{\sim}A)$, and similarly for $Pr(N/A\&C_n)$ compared to $Pr(N/A\&{\sim}C_n)$ and $Pr(N/C_n\&{\sim}A)$, then it will typically be the case that $Pr(B/A\&C_p\&{\sim}C_n\&K) > Pr(B/{\sim}A\&C_p\&{\sim}C_n\&K)$, while at the same time, $Pr(B/A\&{\sim}C_p\&C_n\&K) < Pr(B/{\sim}A\&{\sim}C_p\&C_n\&K)$. (Given $C_p\&{\sim}C_n$, **a** will increase the probability of **b**, while given ${\sim}C_p\&C_n$, **a** will decrease the

[12] This is the so-called Markov condition, that there can be no causal influence at a distance. It should not be confused with the causal Markov condition that will be discussed in chapters 10 and 12.

probability of **b**.) Since $\mathbf{c}_p$ and $\mathbf{c}_n$ are relevant factors determining the test conditions within which **a** is supposed to raise the probability of **b**, one sees that **a** does not always increase the probability of **b** and hence that it cannot cause **b**. When unanimity of intermediaries fails, contextual unanimity fails too, and **a** will be causally mixed for **b**.

$\mathbf{I}_g$ and the failure of unanimity of intermediaries do not, however, imply that there *must* be a failure of contextual unanimity. Suppose that the presence or absence of $\mathbf{c}_p$ and $\mathbf{c}_n$ makes a relatively small difference to the probability of **p** and **n** given both the presence and absence of **a** and that $C_p$ and $C_n$ are not necessary conditions for **n** or **p**. The positive impact of $A$'s instantiation to the probability of **b** via **p** may then always outweigh the negative impact via **n** – that is, **a** may increase the probability of **b** in all four cells of $C_p\&C_n$, $C_p\&{\sim}C_n$, ${\sim}C_p\&C_n$, and ${\sim}C_p\&{\sim}C_n$. The probability increase within each cell of the partition will be an average of an increase over the **p** path and a decrease over the **n** path, because, by assumption, both paths are open in each of the cells, although with differing probabilities. Whether **a** is a cause of **b** thus depends on the probabilities of **p** and **n**. If these were actual frequencies, which could vary when the causes of **p** and **n** are fixed, averaging over different intermediaries would be as objectionable as Dupré's "average effect." One would be conflating the question of whether **b** depends on **a** with the question of how often it depends on **a**. Averaging over intermediaries would provide one with "a sorry excuse for a causal concept." Averaging over causal intermediaries could be justifiable, however, if Pr(P) and Pr(N) are not just frequencies, but are *determined* by the causes of **p** and **n**. If one believes, as Eells and Sober do, that **a** is a probabilistic cause of **b** even though the intermediaries between **a** and **b** are not unanimous, and one holds that whether **a** causes **b** should not depend on actual frequencies, then one is committed to the view that probabilities are determined.[13] Theories such as Eells's assume that probabilities of effects will be determined by the probabilistic causes.

These complications concerning transitivity reflect semantic choices. What is at issue is the transitivity of a single *kind* of causal relevance, not the transitivity of causal relevance itself, and I would argue that causal relevance is the more important notion for a theory of causation.[14] Suppose

[13] And in that case, given Hitchcock's interpretation of Dupré discussed in footnote 10, Dupré's account will not necessarily be "a sorry excuse for a causal concept" either. For further discussion of Eells's views on transitivity, see Eells (1991, ch. 4).

[14] See p. 107. In a treatment of the pragmatics of explanation, in contrast, the kind of relevance may be of the utmost importance. "The general lesson is that within the probabilistic realm, . . . things almost always happen because of some circumstances and despite other circumstances. Moreover, in complex causal nets positive and negative causal influences may mix in countless ways. If we want to take account of all this, we have to attend to causal relevance *simpliciter*"

that one says that **a** is causally neutral with respect to **b** in circumstances *K* if and only if $Pr(B/A\&K\&S_j) = Pr(B/{\sim}A\&K\&S_j)$ for all *j*, and that **a** is causally relevant to **b** if it is not causally neutral with respect to **b**. The independence condition implies that causal relevance should be transitive. If there is only one intermediary or if intermediaries are unanimous (and if the Markov condition holds), then by Eells and Sober's sufficient condition for transitivity, causal relevance is transitive. Consider a case in which **a** is relevant for **b** along some paths in one way and either neutral for **b** or relevant for **b** in another way along other paths. **a** will be causally neutral for **b** if and only if the influence over the various paths cancels out given *every* test situation. Given $\mathbf{I_g}$, the intermediaries between **a** and **b** will each have independent causes that define the test situations in which one examines the probabilistic relevance of *A* to *B*. It will consequently be impossible for the influence of **a** on **b** over the different paths to cancel out in every test situation.

An alternative way of dealing with the issue of transitivity, which resembles a proposal of Richard Otte's (1985, pp. 123–4), would be to adapt Lewis's device of distinguishing between causal dependence and causation (1973; Hitchcock 1995a, p. 276). **b** is causally dependent on **a** in *K* if for all j, $Pr(B/A\&K\&S_j) > Pr(B/{\sim}A\&K\&S_j)$. **a** is a (type-level) cause of **b** (in *K*) if **b** causally depends on **a** in *K* or there is in *K* a chain of causal dependence running from **a** to **b**. When unanimity of intermediaries fails, **a** may turn out to be both a cause of **b** (along one path) and a preventative of **b** (along another) (as also in Cartwright 1989, p. 101; Spohn 1990, p. 138). One can still calculate its average effect in each background context, and the average effect might be of practical importance, but average effect would appear to be of little theoretical interest.

## 9.4 What Should One Hold Fixed?

The root idea of probabilistic theories of causality is that causes increase the probability of their effects in every test situation – that is, "in every situation which is otherwise causally homogeneous with respect to" the effects (Cartwright 1979, p. 25). Clauses 1–4 of Cartwright's **CCC** specify the membership of $\{\mathbf{c}_i\}$, the set of positive or negative causal factors that must be held fixed. A test situation is a set of values of those factors. **d** is in $\{\mathbf{c}_i\}$ if and only if it is a cause of **b** or ~**b**, other than **a**, and it is not caused by **a**.

The conditions of **CCC** have been controversial. One should presumably insist that everything in $\{\mathbf{c}_i\}$ be distinct from **a** rather than simply excluding

(Spohn 1994, p. 234). For another argument for distinguishing the determination of causal relevance from its classification, see Hitchcock (1993, 1996a, 1996c).

**a** itself. If one excludes factors that are caused by **a**, then one must exclude factors that are prevented by **a** as well. One cannot consistently require that {$\mathbf{c}_i$} include all causes of **b** except for **a** and also exclude everything caused by **a** (Ray 1992, pp. 220–1; see also Irzik and Meyer 1987). But the difficulties here are not just matters of detail. Here are some of the major questions that arise:

1. Should one hold fixed factors that are caused by **a**, but not in the circumstances that actually obtain on some particular occasion (Cartwright 1989, pp. 95–104, Eells 1991, pp. 173–97)? Let **a** be your calling me, **d** be my phone ringing, and **b** my picking up the phone. According to Eells (and Cartwright's original condition 4), one should not hold **d** fixed. According to Cartwright's true position, one should hold fixed all events of my phone ringing except those that are caused by the particular instantiation of *A*. It might appear that Eells's theory falsely implies that your calling me causes me to pick up my phone even when my phone is already ringing with another call and that Cartwright's proposal is the correct one. But if one holds fixed whether anyone else is calling, as one should, then Eells's theory has no mistaken implication.[15] Furthermore, Eells argues that if my phone occasionally and spontaneously becomes insensitive to incoming calls and rings by itself and this happens by chance on all occasions on which you call me, then Cartwright's proposal implies that **a** is causally irrelevant to **b**, when in fact it is causally mixed (1991, p. 177).
2. Does partitioning too finely lead to mistakes (Cartwright 1979; Eells and Sober 1983)? I touched on this question briefly in footnote 6. Given an interpretation of probabilities as propensities or hypothetical limits of frequencies, it seems as if partitioning on irrelevant factors can do no harm.
3. Does one need to include in {$\mathbf{c}_i$} factors that are neither causes nor preventatives of **b** (Eells 1991, ch. 3)? Eells argues compellingly that{$\mathbf{c}_i$} should consist of all factors distinct from **a** and not subsequent to **a** with which **a** interacts with respect to **b** (see also Spohn 1990, p. 132). **d** is subsequent to **a** if it comes after **a** and **a** is causally relevant to it (positive, neutral, or mixed for it) or would be if causation were transitive. **a** interacts with a factor **d** with respect to **b** if and only if $\Pr(B/A\&D) \neq \Pr(B/A\&{\sim}D)$ or $\Pr(B/{\sim}A\&D) \neq \Pr(B/{\sim}A\&{\sim}D)$.[16] So {$\mathbf{c}_i$} must include factors that are causally mixed for **b** and even some factors that are causally irrelevant to **b**.

By the definition of {$S_j$}, there is no finer partition of any cell that would change the conditional probability of **b** given **a** in that cell. That probability

[15] Ray (1992, pp. 227–8). Holding fixed such factors also provides a way of dealing with Richard Otte's objection (1985, p. 121) that if the positive direct causal influence of *a* on *b* cancels out the negative indirect influence via *v*, then if we cannot hold fixed **v**, we would mistakenly conclude that *a* is not a cause of *b*. If, however, *v* has other causes that are independent of *a* (as by $\mathbf{I}_g$ it must), then one can uncover the separate causal influences by holding fixed those other causes.

[16] One cannot proceed in a piecemeal fashion. The claims in the text can be correct only if one relativizes these claims to contexts that are otherwise causally homogeneous, and vicious circularity then threatens in the attempt to carry out this relativization. One must consider the complete partition of all the factors with which **a** interacts with respect to **b**. Eells is offering an account of what probabilistic causation is, not an account of how one draws causal conclusions from knowledge of probabilities.

will be strictly between 0 and 1 only if the circumstances are genuinely indeterministic. This means that the theory of probabilistic causation is relevant to the surgeon general's claim only if some of the mechanisms by which smoking contributes to lung cancer are indeterministic. But explicating the meaning of claims such as the surgeon general's is precisely one of the main goals of formulating a theory of probabilistic causation! Our appreciation of its relevance to this task should not wait until we have discovered whether the mechanisms responsible for lung cancer are deterministic or not.

### 9.5 Are Undetermined Events Caused?

Suppose, with respect to a partition that contains everything distinct from **a** and not subsequent to **a** to which Pr(B/A) is sensitive, Pr(B/A&$S_j$) = $p_j$ . Suppose that for some particular $j$, $p_j = p$, where $p$ is not 0 or 1. Given an instantiation of $A$ and $S_j$, the probability that $B$ will be instantiated is $p$. $B$ is then instantiated or it is not. If it is, one can cite in the explanation the factors responsible for Pr(B) = $p$, and one can contrast that probability and the factors in the situation that determine it to another situation where the probability of $B$ being instantiated is lower. If $B$ is not instantiated, one can cite in explanation the factors responsible for Pr(~B) = 1 - $p$, and one can contrast the situation to others in which Pr(B) is higher. One may focus on different conjuncts in the set that determines Pr(B) and Pr(~B) when one is concerned to contrast the situation to others in which $B$ was less likely to be instantiated than when one is concerned to contrast it to situations in which $B$ was more likely to be instantiated. But the complete set of factors responsible for Pr(B) is exactly the same as the set of factors responsible for Pr(~B). There is nothing more to be said in explanation of $B$'s instantiation or its noninstantiation. The causes only determined $p$. Whether $B$ was then instantiated cannot be explained by the causes, for they explain just as well why $B$ was not instantiated. Whether $B$ was instantiated cannot be explained at all. It just happened (Stegmueller 1973; cited by Humphreys 1989, p. 117).

Humphreys writes: "[O]nce the contribution of $A$ to the propensity has been cited, there is nothing more to say that is genuinely causal – it is simply that the higher the value of the propensity, the more likely $B$ is to occur" (1989, p. 36 [my relettering]). "After citing the causal factor. . . , the event $B$ either just happens or just does not happen" (1989, p. 35 [my relettering]). Nevertheless, Humphreys defends the view that specific outcomes *can* be explained. $B$ is instantiated *because of* the factors $F^+$ that increase its probability, *despite* the factors $F^-$ that decrease its probability, while $\sim B$ is instantiated because of the factors $F^-$ that decrease the probability that *B* will be instantiated, despite the factors $F^+$ that increase the probability that *B* will be instantiated (1989, p. 118). Such an account can be given only if one can offer a nonarbitrary account of contrast cases. Humphreys's account permits one to say that some of the factors that determine the probability of an event that occurs in part by chance explain the event itself, and this might be considered an advantage. In my view, such an account blurs the line between a metaphysical theory of causation and a pragmatic account of explanatory practices.

Consider the following case. One puts a lucky cat in a device that will feed it *foie gras* (**b**) if a geiger counter measures a certain level of radiation.[17] A shield blocks a radioactive source. One can then open the shield (**a**) or leave it closed (~**a**). Opening the shield makes it more likely that the cat will get fed, but there is some chance that it will not get fed if the shield is open and some chance that it will get fed if the shield is closed: $1 > \Pr(B/A) > \Pr(B/{\sim}A) > 0$. Suppose the shield is left closed and the cat nevertheless gets fed. If Pr(B/~A) is very low, most people would say that it was a lucky chance. If Pr(B/A) was much higher, people would say that the cat got fed *despite* the fact that the shield remained closed. Although they would concede that the cat wouldn't have gotten fed if there were no background radiation, they might deny that the background radiation *explains* why the cat got fed. Humphreys's account seems to fit this usage. If one takes the zero level of radiation to be the level of background radiation, Humphreys's account implies that there is no explanation for the cat getting fed, given ~**a**. Given the set-up, it just happened by chance.

If on the other hand Pr(B/~A) were high, then Humphreys's account would not match our explanatory practices nearly as well. His account would still deny that background radiation explains why the cat gets fed, but many people would now cite the level of background radiation as an explanation. One might account for this conflict between the implications of Humphreys's theory and our practices by claiming that there is some vacillation concerning whether one should regard the zero level of radiation as zero radiation or as the background level. Problems remain. If the shield is opened, then on Humphreys's account one can explain why the cat gets fed (if it does), but one cannot explain why the cat does not get fed (if it doesn't), regardless of what the level of Pr(B/A) is, just so long as Pr(B/A) > Pr(B/~A).

I think one should separate questions about explanatory practices from metaphysical questions concerning causation and explanation. In a case like the one just described, the boundaries between what is explanatory and what is not depend on factors that are not relevant to the theory of causation. On

[17] This is a humane elaboration of the case proposed by Dretske and Snyder (1972).

the view I endorse, all that gets caused are the conditional probabilities. There is no explanation for why the cat gets fed or why it does not (compare Lauwers 1978). In either case the same factors are responsible for the probabilities. If the shield remains closed and the cat does not get fed, then, depending on the context, one may focus on some of the causal factors that would have led to a different probability that the cat would get the *foie gras*. If there is a big difference between Pr(~B/A) and Pr(~B/~A), one would cite the fact that the shield remained closed to explain why the cat did not get fed, while if the difference is small, one would not cite it. The causal structure is, however, just the same. Our explanations are also influenced by irrelevant comparisons. If another cat in a differently designed apparatus had a higher probability of getting fed and did in fact get the *foie gras*, then one would point to the difference in the two set-ups to explain why this cat did not get fed. If the other cat did not get fed, then one would not point to the difference in the set-ups to explain why this cat did not get fed. But none of this bears on the causal structure. The causal facts are exhausted once one specifies the probabilities. The fact that factors responsible for the probabilities are often regarded as also responsible for the events themselves is of no metaphysical significance. The theory of causation should be concerned with whatever affects the probabilities. Questions about what people count as explanatory should be left for an interpretation of human practices.

## 9.6 Probabilistic Causation or Deterministic Causation of Probabilities?

Rather than regarding the cat getting fed (if it does) as probabilistically caused, one should regard the probability that it gets fed as deterministically caused (compare Papineau 1989, p. 320; Woodward, 1989, §3). Rather than theorizing about probabilistic causation, one should theorize about deterministic causation of probabilistic states. Alternatively, one can say that this is what probabilistic causation is: the deterministic causation of probabilistic states that then issue by chance into one outcome or another. Just as one can represent the valuc of one variable as a function of the values of other variables, so one can represent the probability distribution of a variable as such a function. Mathematical dependence is not, of course, causal dependence. According to **CPg**, the value of $y$ causally depends on the values of $x_1$, . . ., $x_n$ just in case for each $x_i$, $y$ and $x_i$ are nomically connected and everything nomically connected to $x_i$ and distinct from $y$ is connected to $y$. Increases in the value of $x_i$ may lead to increases or decreases in the value of $y$, and both the quantity and the direction of the change in the value of $y$ consequent to an increase in the value of $x_i$ may depend on the values of the other variables. Similarly, the probability density of $y$, $f(y)$ depends on $x_i$ just in case $f(y)$ and $x_i$ are connected and everything connected to $x_i$ and distinct from $f(y)$ is connected to $f(y)$. Indeed Eells himself toys with the idea of sometimes taking probability distributions to be causal relata (1991, p. 236). Increases in the value of $x_i$ may increase or decrease the probability $y$ that will take particular values, and both the quantity and the direction of the change in the probability of $y$ taking on particular values consequent on an increase in the value of $x_i$ may depend on the values of the other variables.

### 9.6.1 Three Objections

One might object that taking probabilistic causation to be deterministic causation of probabilities is inconsistent with the transitivity of causation. Suppose, for example, that *a* is a probabilistic cause of *b*, which is a deterministic cause of *c*. Since *a* is only a cause of a probability of b and not a cause of *b* itself, there is no causal chain between *a* and *c*. Even if Pr(C/A&K) > Pr(C/A&~K), so that *a* counts as a probabilistic cause of *c*, it will not cause *c via b*. But this difficulty can be repaired. One can guarantee transitivity by defining **a** is a probabilistic cause of **b** if and only if **a** is a deterministic cause of Pr(B) (where 0 < Pr(B) < 1) or **a** is a probabilistic cause of $\mathbf{a}_1$, which is a probabilistic cause of $\mathbf{a}_2$, which is . . . a probabilistic cause of $\mathbf{a}_n$, which is a probabilistic cause of **b**.[18] Although this speaks of probabilistic causation, it still defines it in terms of deterministic causation of probability.

Second, as Christopher Hitchcock pointed out to me, speaking of causing probabilities commits one to a certain metaphysics of probability. If features of probability distributions can be caused, then they must be the sort of entity that can be caused. If, for example, the occurrence of a token of **a** causes the chance that a token of **b** will occur to increase, then an increase in the chance of a token of **b**'s occurring at this place and time must count as an event or trope. But what kind of an event is this? In response, it seems to me that there is no real difficulty here, provided that one is willing to countenance single-case probabilities, such as propensities. If one is willing to regard the located value of a variable as an event, why shouldn't one countenance the located value of a propensity?

Finally, taking probabilistic causation to be deterministic causation of probabilities has intuitively implausible consequences (Mellor 1995, p. 53). For example, one must deny that assembling a critical mass of a fissile

[18] This proposal raises a problem that will be addressed in chapter 12: I have not yet discussed how one can draw a distinction between direct and indirect causes. Note that claims such as the operationalizing assumption will need restating: The link should be between probabilistic dependency and (direct) causal dependence, not between probabilistic dependency and causation. I am indebted here to David Papineau.

material like uranium 235 causes it to explode. One only causes the probability of an explosion to be *extremely* high. The explosion is due to chance. Committed as I am to the theory presented in this book, I can live with this implausibility. It stems, I believe, from our inability to take small chances seriously. Readers may not share my complacency. For those who don't, there is a cheap cure: simply define "**a** probabilistically causes **b** in circumstances *K*" as "Pr(B) depends on **a** in circumstances *K*." This notion of probabilistic causation corresponds to what Eells would call probabilistic causal relevance – on this definition probabilistic causes include factors that Eells regards as preventatives or causally mixed as well as factors that Eells would count as probabilistic causes. This account of probabilistic causality differs only terminologically from the position defended in this chapter, which denies that there is any such thing as probabilistic causation. It may nevertheless be more congenial to many readers.

### 9.6.2 Defense of Taking Probabilistic Causation to Be Deterministic Causation of Probabilities

Denying the existence of probabilistic causation or taking it to be deterministic causation of probabilities is preferable in six regards to taking causation itself to be probabilistic. First, what good reason is there to regard the outcome itself as *caused* when its occurrence or nonoccurrence is a matter of chance? If it is time that we changed our conceptual apparatus and ceased to find the idea of genuinely probabilistic causation paradoxical, then strong arguments ought to be given. But once one distinguishes questions concerning the metaphysics of causation from questions concerning the pragmatics of explanation, it is hard to find any argument for countenancing genuinely probabilistic causation.

Second, to take probabilistic causation to be deterministic causation of a probability permits causation to be univocal. It is a deterministic relation between located values of variables. One need only add that the values of variables can sometimes be features of probability distributions. This is preferable to saying that some causal relations are deterministic and some are probabilistic. Furthermore, if probabilistic causation is deterministic causation of a probability, then claims about causation at the type level can be construed as generalizations concerning token causation.

Third, this construal (or rejection) of probabilistic causation permits a more natural description of cases where causes appear to lower the probability of their "effects." Among the factors that determine the value of a variable $y$, some raise the value of $y$ (given the value of the others) and some lower it. All are equally causes of the value of $y$. Exactly the same is true if $y$ is the probability that some property $P$ will be instantiated. The probabilistic causal relevance of $x$ to $y$ is the function relating the probability distribution of $y$ to values of $x$.[19] It is then a purely pragmatic issue whether one takes the causally relevant factors to cause or prevent *P*'s instantiation (Hitchcock 1993, 1996a, 1996c). Furthermore, if anything is a token "cause" (i.e., causally relevant factor), then it is, in just those circumstances, always a type-level cause as well.

Fourth, this account provides a superior perspective on the difficulties concerning transitivity canvassed above. When there are no indeterministic links in a causal chain, then causation is deterministic and will be transitive. When there is an indeterministic link between **b** and its immediate causes, causation will still be deterministic in the sense that Pr(B) will be a function of its distal causes $x_1, \ldots, x_n$. If there are multiple intermediaries, then whether a particular value of $x_i$ makes Pr(B) larger or smaller than some other value of $x_i$, given a specification of the values of all other relevant variables, will depend on the sum of its effects across different intermediaries, but this is just the same as the case where a nonprobabilistic quantitative variable $y$ depends on distal causes. There's no problem concerning the transitivity of *causation*. The issues instead concern the calculation of quantitative effects across multiple intermediaries.

Fifth, as I argued before, theories of probabilistic causation such as Eells's assume that Pr(B) $= p$ is calculable from knowledge of the nomological relations and the given values of causal variables. If, given a specification of the values of all relevant variables, Pr(B) is not determined, then calculations of probabilities along causal chains will depend on actual frequencies, and the claim to have a theory of probabilistic *causation* will be questionable. Since the view that probabilistic causation implies determination of probabilities is implicit in theories of probabilistic causation, an account of probabilistic causation that takes it simply to *be* the deterministic causation of probabilities seems as credible as those theories.

### 9.6.3 Generalizations Across Inhomogeneous Contexts

Last, construing probabilistic causation as deterministic causation of probabilities clarifies the problem of accounting for generalizations such as the surgeon general's. If there's nothing indeterministic involved in the development of lung cancer, then the basic causal truths about the relations between smoking and lung cancer consist of deterministic generalizations concerning when smoking does or does not give rise to cancer in particular circumstances. If there are indeterministic elements, then the basic truths

[19] This draws on Hitchcock (1993, esp. pp. 349f). Hitchcock is concerned with the probabilistic causal relevance of values of $x$ to a particular event, which is probabilistically caused, rather than to a probability distribution, which is deterministically caused.

consist of generalizations about the probability of acquiring lung cancer in various circumstances.

The surgeon general is not enunciating any such basic truths. A generalization such as "smoking causes lung cancer" implies instead that there is a broad range of contexts in which some aspect of smoking is an INUS condition of lung cancer or of a significant probability of lung cancer, and there are few circumstances in which smoking deterministically prevents lung cancer or lowers its probability (except by way of causing prior death by other causes). This proposal might be refined along the lines of Hitchcock's reformulation of Dupré's suggestion (see footnote 10). Such an interpretation of the surgeon general's claim makes better sense of the importance of experiments and clinical trials than taking the surgeon general to be claiming that smoking raises the probability of lung cancer in all test situations. Through experimentation one seeks to learn the physiological mechanisms and thus the precise circumstances in which smoking is an INUS condition for lung cancer or for some probability of lung cancer.

Theorists such as Patrick Suppes thought that claims such as the surgeon general's called for a theory of probabilistic causation. I have argued that they were mistaken. Such claims require instead an analysis of generalizations about the causal impact of factors across *inhomogeneous* contexts. Causal generalizations like the surgeon general's raise pragmatic rather than ontological questions, and studies of the nature of causation need not say more about them.

Just the same thing could be said about causal generalizations such as the surgeon general's by proponents of theories of genuinely probabilistic causation. But they would then be conceding that the theory is irrelevant to the central tasks it was designed to fulfill. The fact that life is full of generalizations such as the surgeon general's gives one no reason to develop theories of probabilistic causation, because nothing in the analysis of these generalizations is incompatible with a deterministic view of causation. The only real motivation for constructing a theory of probabilistic causation is indeterminism within causally homogeneous contexts, and we have yet to see any reason why what happens there should be regarded as probabilistic causation of outcomes rather than deterministic causation of probabilities of outcomes.

## 9.7 Conclusion: Causation and Determination

The existence of indeterministic relations in quantum mechanics does not necessitate either abandoning or revising a traditional deterministic view of causation. Permitting located values of probabilities to be causal relata accommodates the most obvious difficulties posed by indeterminism, though I make no claim to resolve all the causal puzzles of quantum mechanics. I see no compelling reason for abandoning a traditional view of causation. For the reasons just given, the best alternative is to take causation in an indeterministic context to be deterministic causation of probabilities rather than probabilistic causation of nonprobabilistic outcomes. Not only is this a conservative revision, which enables one to hold on to the conclusions established in the other chapters, but it copes more adequately with the problems that motivated theories of probabilistic causation in the first place.

It remains true, as argued at the beginning of the chapter, that the claim that causation is always deterministic is not mandated by the evidence and less obviously in accord with everyday experience than is a view like Elizabeth Anscombe's, whereby causes are "enough" to make something happen without being sufficient or necessary (1971, p. 91; Rosen 1982, pp. 111f). But a view like Anscombe's casts an equal shadow over the hypothesis that there are probabilistic regularities. In any case, everyday experience is not decisive, or else we would still endorse Aristotle's view that objects stop unless they are subject to a force, instead of accepting Newton's first law of motion. The analogy is unfair, since Newton's first law of motion, unlike the view that causation is deterministic, has impressive indirect empirical vindication. The reason for holding on to the view that causation is deterministic is rather that no alternative that better fits everyday experience permits one to develop a significant theory of causation.

# 10

# Causation and Conditional Probabilities

In chapter 4 I made the operationalizing assumption that there are causal connections among events if and only if there are probabilistic dependencies in the background circumstances among the relevant properties or variables. If this assumption is a reasonable approximation, then the claims about causal asymmetry made in chapters 4–7 have implications concerning probability distributions, which deserve further scrutiny. This chapter begins this scrutiny by exploring David Papineau's theory relating causation to *conditional* probabilistic dependencies.

## 10.1 The Fork Asymmetry

Suppose event types **a** and **b** are probabilistically dependent on one another:

1. Pr(AB) > Pr(A)·Pr(B).

Suppose further that

2. **a** and **b** are not related as cause and effect.

Then, according to Hans Reichenbach, there exists an event type **c** satisfying the following conditions:

3. Pr(AC) > Pr(A)·Pr(C)
4. Pr(BC) > Pr(B)·Pr(C)
5. Pr(AB/C) = Pr(A/C)·Pr(B/C)
6. Pr(AB/~C) = Pr(A/~C)·Pr(B/~C)
7. Tokens of kind **c** *precede* tokens of kinds **a** and **b**.
8. Tokens of kind **c** are *common causes* of tokens of kinds **a** and **b**.

The probabilistic dependence between **a** and **b** is explained by the existence of the common cause **c**. Reichenbach proves that if none of the probabilities is one or zero, then 3–6 entail 1. This proof is given in chapter 4*, pp. 76–7.

7 or 8 is called in the literature "the fork asymmetry": the causal arrows point away from the event type that does the screening-off and toward the future. 7 might be called the temporal fork asymmetry, while 8 might be called the causal fork asymmetry. There can be events of kind **c** satisfying

3–6 that are common effects of *a* and *b* and that succeed them in time, but only when there are also common (preceding) causes. What is called in the literature, the "principle of the common cause" is either (a) the claim that if **a** and **b** are probabilistically dependent and not related as cause and effect, then they are related as effects of a screening-off common cause, or (more modestly) (b) the claim that the best explanation of a probabilistic dependency between **a** and **b**, when they are not related as cause and effect is that they are effects of a screening-off common cause (see Forster 1988; Sober 1988; Arntzenius 1993).

The fork asymmetry apparently does not permit one to reduce causation to facts about probabilistic dependencies, because the thesis that there will be screening-off common causes presupposes that one already knows that *a* does not cause *b* and *b* does not cause *a*.

## 10.2 An Asymmetry of Screening-Off?

David Papineau has attempted to *reduce* causal asymmetry to what is essentially the fork asymmetry.

> (S-O) Take any event **c**. Then among the events which are correlated with **c** will be some that are correlated with each other in such a way that their correlation is screened off by **c** – these are **c**'s effects; and among the events which are correlated with **c** will also be some that are not correlated with each other – these will be **c**'s causes.[1] (1993, pp. 239–40)

On Papineau's view, if $e_1$ and $e_2$ are of types that are correlated with each other and with **c** and the correlation is screened off by **c**, then $e_1$'s and $e_2$'s are effects of *c*'s. The asymmetry Papineau is concerned with is thus an asymmetry of screening-off, not an asymmetry of independence and connectedness. (But, as we shall see shortly, the two are intimately connected.) Unlike the independence theory of chapter 4 and Reichenbach's and Salmon's accounts of the fork asymmetry, Papineau's theory is reductive. It purports to analyze causal asymmetry in terms of screening-off.

As Papineau knows, (S-O) is at best a first approximation. Causal intermediaries screen off just as well as do common causes. Papineau responds,

> Still, this complication is not necessarily fatal to the proposed analysis of the direction of causation. For it still remains true that the absence of the screening-off structure distinguishes the case where *c* is a common *effect* of two causes *a* and *b*, from those where *c* is either a common cause or an intermediate cause, . . . . (1990, p. 84 [lowercase italics substituted for uppercase italics in the original])

[1] Papineau wrote "C" where I wrote "c." I have taken Papineau to be talking about event types, because one event token cannot be correlated with another.

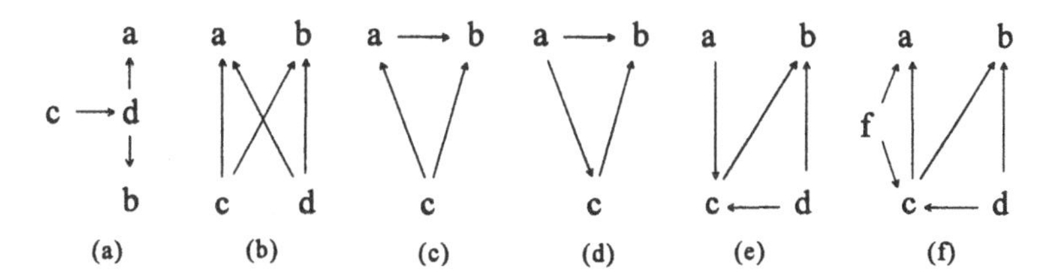


**Figure 10.1:** Common causes and causal intermediaries that don't screen off

Common causes and causal intermediaries do not, however, always screen off. Consider the six graphs displayed in figure 10.1. These are "normal" graphs in which edges represent causal relations that are direct relative to the variables under consideration. So figure 10.1e, for example, represents a structure in which *d* is both a direct and an indirect cause of *b*. In figure 10.1a–10.1c, *c* is a cause of both *a* and *b* yet **c** does not screen off **a** and **b**. In figure 10.1a, this is because *c* is a distal cause and, as Sober proved (see pp. 77–8), if proximate causes screen off, distal causes typically do not. In figure 10.1b, **c** fails to screen off because *a* and *b* have another common cause *d* that may be responsible for a probabilistic dependency between *a*'s and *b*'s even conditional on **c**. In figure 10.1c, **c** fails to screen off **a** and **b**, even though it is a single common cause, because there is another causal connection between *a* and *b*. I have shown this connection as direct, but it could be indirect. **c** fails to screen off in figure 10.1d for the same reason, but in figure 10.1d, *c* is a causal intermediary.

The most interesting cases are shown in figures 10.1e and 10.1f, in which **c** fails to screen off **a** and **b** even though *c* is the *only* causal intermediary or the *only* common cause. To see why *c* does not screen off, let us embody these causal structures in arrays of salt basins (like those in figures 6.2 and 6.3). These systems are shown in figure 10.2a and 10.2b. Suppose as before that the concentration in each basin is the average of the concentration of the basins above and the concentration of basin 4, $x_4$, is fixed at some value $k$. $x_4 = .5(x_1 + x_2)$ and so if $x_4 = k$, $x_2 = 2k - x_1$ and $x_5 = .5(k + 2k - x_1)$. In figure 10.2a, $x_5$ corresponds to *b* and $x_1$ corresponds to *a*. Since $x_5$ depends on $x_1$, *a* and *b* are not independent. In figure 10.2b, $x_6$, which is $.5(x_4 + x_1)$, corresponds to *a*, while $x_5$ still corresponds to *b*. Since both $x_5$ and $x_6$ depend on $x_1$, *a* and *b* are not independent. In figure 10.1e, the fact that *d* causes *b* via two paths – both directly and via *c* – prevents *c* from screening-off its causes from its effects. In figure 10.1f, the fact that there is an additional causal pathway between *d* and *b* and an additional causal pathway between *f* and *a* prevents *c* from screening-off its effects from one another.

Even if these problems could be solved and the absence of screening-off were to demonstrate that *c* is neither a common cause nor a causal interme-

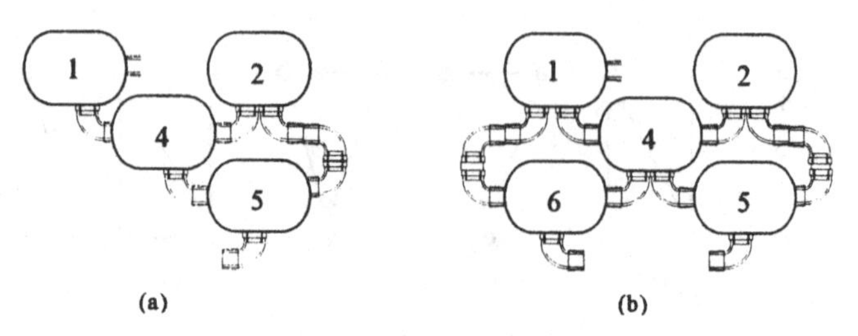


**Figure 10.2:** Multiple connections again

diary, Papineau still needs some way to distinguish common causes and causal intermediaries.[2] He suggests that probabilistic facts about more complicated structures will demarcate common causes from causal intermediaries. Suppose that **c** screens off **a** and **b** and that one concludes that *c* is either a common cause of *a* and *b* or is a causal intermediary. The three possibilities are represented in figure 10.3. If *c* has some other independent cause, *d*, then the three cases can be distinguished. When *a* causes *c* and *c* causes *b* as in figure 10.3a, **d** and **a** will be probabilistically independent. When the causal chain runs the opposite direction as in 10.3b, **d** and **b** will be independent. When *c* is a common cause of *a* and *b* as in 10.3c, **d** will not be independent of either **a** or **b**, though it will be screened off from them by **c**. As the discussion of figure 10.1 makes clear, qualifications are needed. In particular, there must be no other causal connection between **a** and **b**, and **d** must not have any causal connections to **a** or **b** apart from those that follow from **d** causing **c**.

Although Papineau attempts to analyze the asymmetry of causation in terms of an asymmetry of screening-off, he does not take the asymmetry of screening-off as a primitive fact about the world. On the contrary, he argues that if causation is (as he believes) deterministic, and if the separate causes of a given effect are (unconditionally) independent of one another, then the fact that causes screen off their effects follows. Papineau writes:

> I want to claim that the asymmetry of causation derives from the fact that the background conditions together with which causes determine their effects are independent of each other, whereas the same does not hold of the background conditions together with which effects "determine" their causes (1985a, p. 280).

If the asymmetry of screening-off is, as proved in §10.1*, a consequence of independence assumptions that are stronger than **I**, and if existence of independent causes must be assumed in Papineau's purported reduction,

[2] Papineau flirts with the possibility of using spatiotemporal notions (1989, pp. 332–3). In his emphasis on probabilistic connections between **c** and background conditions, there is a hint (that is never developed explicitly) that simultaneity among effects of a common cause might mark a difference between the two cases of screening-off (1989, p. 336).

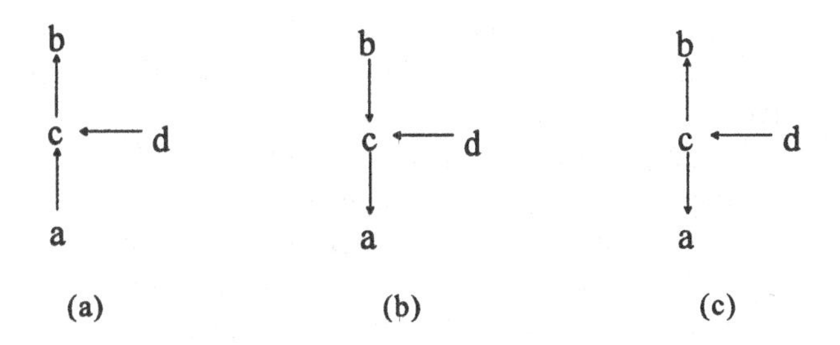


**Figure 10.3:** Distinguishing common causes and causal intermediaries

why not then account for the asymmetry of causation directly in terms of independence in the way suggested in chapter 4? I think the answer is that Papineau does not see how to give a probabilistic reduction of causal asymmetry without invoking screening-off. But there is a way, which I shall explore in chapter 12: To effect the reduction, one can use the operationalizing assumption **OA** and claim that *a* and *b* (or *x* and *y*) are causally connected if and only if **a** and **b** (or *x* and *y*) are unconditionally probabilistically dependent in the circumstances.

## 10.3 Causal Graphs and Probability Distributions

In more recent work, Papineau has turned to the work of Spirtes, Glymour, and Scheines (1993) concerning the relations between causal graphs and conditional independence conditions to reduce causation to facts about probabilistic dependencies. Although Spirtes, Glymour, and Scheines (SGS from now on) do not join in Papineau's reductionist program, their work is relevant to it, and some comments on their exciting project are in order. At the same time I shall comment on related work of Judea Pearl and his co-researchers (Verma and Pearl 1990; Verma 1992; Pearl 1993, 1995; Pearl and Verma 1994), even though Papineau does not directly rely on it. Many of the results of SGS that are most relevant to Papineau's project derive from the work of Pearl and his collaborators. Pearl and SGS are concerned with methods to infer causal relations from data concerning probability distributions, not with the metaphysics of causation. They prove striking theorems establishing relations between directed graphs and probability distributions (see also Glymour et al. 1987).

The fundamental axioms upon which this work depends are:

**CM** (*Causal Markov condition*) If *x* and *y* are probabilistically dependent conditional on all the direct causes of *x*, then *x* causes *y*.

**F** (*Faithfulness condition*) If *x* causes *y*, then *x* and *y* are probabilistically dependent conditional on all the direct causes of *x*.

The notion of direct causation is relative to the set of variables under consideration. Given a set of variables $V$ and a normal graph that represents the causal relations among those variables, $y$ directly depends on $x$ if and only if there is an edge from the vertex representing $x$ to the vertex representing $y$. Suppose one accepts the rough equivalence between causal connection and probabilistic dependence stated by the operationalizing assumption (**OA**). Then, given the connection principle, there is a probabilistic dependence between $x$ and $y$ if and only if $x$ depends on $y$, $y$ depends on $x$, or $x$ and $y$ both depend on some third variable. If $x$ depends on $y$, and the set of *all* the direct causes of $x$ screens $x$ off from its indirect causes, then conditional on the set of all the direct causes of $x$, there should be no probabilistic dependence between $x$ and $y$. If $x$ and $y$ are effects of a common cause, then the set of *all* the direct causes of $x$ should screen off $x$ and $y$. And if $x$ and $y$ have no causal connection, they should have no probabilistic dependence conditional on all the causes of $x$. So if $x$ and $y$ are probabilistically dependent conditional on the set of all the direct causes of $x$, $y$ must depend on $x$. Conversely, if $y$ depends on $x$, $x$ and $y$ should not be probabilistically independent conditional on all the causes of $x$ (unless there are multiple paths between $x$ and $y$ and the coefficients cancel out). Later I shall offer a proof that the causal Markov and faithfulness conditions follow from the connection principle, strong independence, and stronger operationalizing assumptions (theorem 12.2, p. 258).

The fundamental axioms give a truth condition for "$x$ causes $y$" in terms of conditional probabilities *and* causation itself. So they do not provide the reduction of causation to probabilities that Papineau seeks. But sophisticated formal work of Pearl and Verma (1994) shows that when the causal Markov and faithfulness conditions are satisfied, then one can offer a reductionist truth condition for direct causal dependence. They show that variables are adjacent (one is directly dependent on the other) if and only if they are probabilistically dependent conditional on *every* set of variables that does not include them, and they show that in every triple in which $x$ and $y$ and $y$ and $z$ are adjacent, but $x$ and $z$ are not adjacent, $x$ and $z$ are direct causes of $y$ if and only if $x$ and $z$ are probabilistically dependent conditional on every set of variables that includes $y$ but not $x$ or $z$.

I shall have more to say about this theorem in §10.2, where I will sketch a proof. The following remarks may help motivate the result. When variables are adjacent and there are no failures of faithfulness, it should not be possible to screen them off. If $x$ and $y$ and $y$ and $z$ are adjacent, there are four possibilities:

1. $x \rightarrow y \rightarrow z$,
2. $x \leftarrow y \leftarrow z$,
3. $x \leftarrow y \rightarrow z$, and
4. $x \rightarrow y \leftarrow z$.

In the first three cases, it cannot be the case that *x* and *z* are dependent conditional on every set of variables including *y*, for if that set includes all the causes of *z* in cases 1 and 3 or all the causes of *x* in case 2, *x* and *z* must, by the fundamental axioms be probabilistically independent. So (4) is the only possibility. The necessary condition – if *x* and *z* are both direct causes of *y*, then they are probabilistically dependent conditional on every set containing *y* and not containing *x* and *z* – is less obvious. Consider again the example involving the two switches controlling my upstairs hall light (pp. 83–4): Their positions are unconditionally independent, but probabilistically dependent given that the light is on.

Pearl and SGS apparently provide Papineau with the reductive truth condition he seeks. The truth condition they provide does not, however, say or imply that common causes or causal intermediaries screen off. As we saw in the discussion of figure 10.1, that claim is false. It says only that when *y* is adjacent to two variables, *x* and *z*, which are not adjacent to one another, then *x* and *z* are direct causes of *y* whenever *x* and *z* are dependent conditional on every set of variables including *y* but not including *x* or *z*. Even when *x* and *z* are adjacent to *y* and not to each other, it is false to maintain that *y* will screen them off whenever values of *y* are common causes or causal intermediaries. Of the six cases depicted in figure 10.1, adjacency rules out only the case depicted in figure 10.1a. The decisive asymmetry SGS and Pearl capture might be described as an asymmetry of conditional dependence rather than as an asymmetry of screening-off. Whatever one calls it, it might appear that something like the reduction that Papineau has sought has been secured.

### 10.4 Problems with this Account

This reductive truth condition for "*x* causes *y*" derives from the fundamental axioms, and both this truth condition and the axioms appear to be subject to many counterexamples. Consider the causal Markov condition first. It appears to be obviously false. There may be a correlation between *x* and *y* conditional on all the direct causes of *x in the set of variables V*, even though *y* does not depend on *x*, because *x* and *y* have a common cause that is left out of *V*. One remedy is to assume that if *x* and *y* depend on some variable *z*, then either *z* or a descendent of *z* that *x* and *y* depend on is in *V*.[3] This makes the truth condition harder to use in practice and less suitable as a basis for a reduction of causal asymmetry to facts about probabilities. In speaking of probabilities conditional on all subsets of variables *V* that

[3] Probabilistic relations among measured variables can also provide evidence for the existence of latent variables that are common causes. See SGS (1993, esp. ch. 10) and Pearl and Verma (1994).

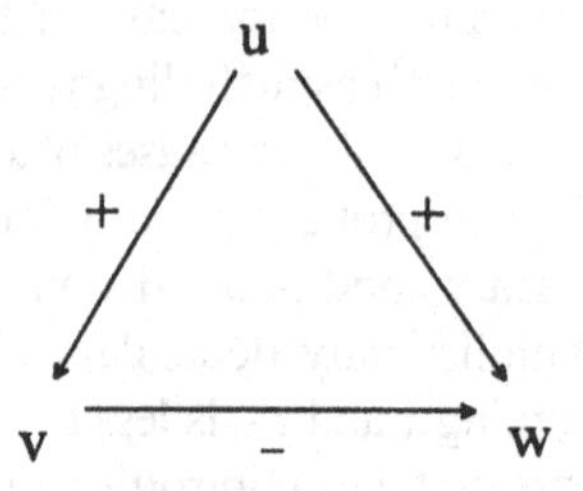


**Figure 10.4:** Papineau's example

satisfy various conditions, one is implicitly relying on causal information in defining *V*.

With this assumption, the causal Markov condition is more plausible, and indeed it is, as we will see in chapter 12, an implication of the connection principle, the independence condition, and operationalizing assumptions. Variables are connected in virtue of being dependent on one another or dependent on some common variable.[4] When one examines the probabilistic dependence between *x* and *y* conditional on some fixed values for all the direct causes of *x*, we are, as it were, considering a possible state of affairs in which *x* has no causes apart from its own independent source of variation. In that possible state of affairs the only causal connection between *x* and *y* would be if *x* caused *y*, and given the link between causal connections and probabilistic dependencies, the conditional probabilistic dependence shows that *x* causes *y*. This argument for the causal Markov condition is, of course, no stronger than is the claim that probabilistic dependencies are sufficient for causal connections, and in chapter 4, we already saw reasons for doubt. I will return to this issue in §12.1.

The faithfulness condition is also implausible. Why shouldn't conditional independence sometimes result from parameter values rather than causal structure? Consider figure 10.4. Suppose, to use Papineau's example,

> that drinking cola (*u*) both stimulates people to exercise more (*v*), but also causes them to put on weight (*w*). And suppose further that exercise *v* independently has a negative influence on weight increase *w*, to just the extent required to cancel out the direct positive influence of *u*, and leave us with an overall zero correlation between cola *u* and weight increase *w*. (1993, p. 244 [*u*, *v*, and *w* written as C, E, and W in the original])

Such a case violates the faithfulness condition (and also shows that *unconditional* probabilistic dependence is not necessary for causal connection).

[4] The work of Pearl and of SGS is committed to the strong independence condition as well as to the connection principle. In most of the models it is explicitly assumed that each variable *x* has its own independent source of variation – that is, that each has a cause that is not represented in the causal graph that is causally independent of everything that is not causally dependent on *x* and that is screened off from all the other variables by *x*.

Even though cola consumption ($u$) causes weight increase ($w$), $u$ and $w$ are probabilistically independent. The independence here results from the values of the parameters rather than from the causal structure.

If such cases are rare (as one has good reason to believe), such possibilities are no serious objection in principle to the use of conditional dependency information in inferring causal relations from probabilistic evidence.[5] But such cases are a serious problem for any attempted *reduction* of causation to probabilities. Papineau suggests two ways to deal with failures of faithfulness. One (1993, p. 244) is to suggest a different reduction of causation to probabilities based on SGS's minimality condition (SGS 1993, p. 55). Minimality says that adding a directed edge to a causal graph always eliminates some conditional independence relation. There is, however, no theorem in SGS that will serve reductionist purposes and that relies only on minimality. Papineau's second proposal is that one embed the system in which faithfulness failed into a larger system (1993, p. 245). Examination of this proposal is postponed to §10.3*.

There are other problems for Papineau's reductionist program (Irzik 1996). If the probability of $x$ given all its direct causes is 1,[6] then the probability of $y$ conditional on the direct causes of $x$ will be the same as the probability of $y$ conditional on $x$ and all of the direct causes of $x$, and there will be no conditional probabilistic dependency between $x$ and $y$ even if $y$ causally depends on $x$. Papineau maintains that all causation is deterministic, so if all variables were included, there would be many spurious conditional independence relations. Thus Papineau cannot allow the set of variables $V$ to include all the relevant variables. The conditional probabilities he is concerned with are "mixed." They take intermediate values only because one does not condition on all the relevant variables. But at the same time, nothing that $x$ directly depends on that $y$ also depends on can be omitted if dependence conditional on all subsets of $V$ is to establish causal adjacency.

Papineau must leave out some of the relevant variables, but he must not leave out common causes. There is no guarantee that these two demands can be satisfied simultaneously. Furthermore, even when including all the common causes is consistent with keeping the probabilities intermediate, one must rely on knowledge of the direction of causation to know *which* variables to include. The problems are perfectly general. Given determinism, probabilistic facts will tell one the direction of causation only if some

[5] There may be practical problems, though, because low sample covariances that one mistakes for zero population covariances are not rare.

[6] To assume that each variable has its own independent source of variation has the additional advantage of ruling out this possibility.

of the relevant variables are omitted, but none of the common causes is. One must know the direction of causation to know what variables must be part of the set considered and so one cannot reduce causation to facts about probabilities.

These are not the last words to be said concerning Papineau's reductionist program, because I have not yet considered the full range of things that might be said about whether probabilistic dependencies are sufficient or necessary for causal relations. This task is undertaken in chapter 12. But we already have grounds for pessimism.

# 10*

# Causal Graphs and Conditional Probabilistic Dependencies

## 10.1* Independence Implies Screening-Off

If causes are INUS conditions, the following two theorems (which are sketched by Papineau (1985b, p. 63 and 1985a, p. 279)) establish that effects are probabilistically dependent on their causes and that common causes screen off their effects.

**Theorem 10.1:** If $(x)(Ax \leftrightarrow Cx\&Sx \text{ or } Zx)$, $C$ is probabilistically independent of $S$ and $Z$, and Pr(S or Z) > Pr(Z), then Pr(A/C) > Pr(A/~C).

Proof: Pr(A/C) = Pr(S or Z) and Pr(A/~C) = Pr(Z). So Pr(A/C) > Pr(A/~C).

**Theorem 10.2:** If (1) $(x)(Ax \leftrightarrow Cx\&Sx \text{ or } Zx)$, (2) $(x)(Bx \leftrightarrow Cx\&Yx \text{ or } Wx)$, (3) $C$ is probabilistically independent of $S$, $Z$, $Y$, and $W$, (4) $S$, $Z$, $Y$, and $W$ are probabilistically independent of one another, and (5) all probabilities are intermediate, then $C$ and $\sim C$ screen off $A$ and $B$.

Proof: Given premises 1–3 and 5, Pr(AB/C) = Pr[(S or Z) & (Y or W)]. Given premise 4, Pr[(S or Z) & (Y or W)] = Pr(S or Z)·Pr(Y or W), which by 1, 2, 3, and 5 equals Pr(A/C)·Pr(B/C). Since $Z$ and $W$ are independent, Pr(Z&W) = Pr(Z)·Pr(W), and Pr(AB/~C) = Pr(A/~C)·Pr(B/~C).

In Papineau's view, screening-off is explained by the independence among different deterministic causal facts.

## 10.2* Causal Graphs and Probability Distributions – Some Formal Results

In this section I shall sketch the proof of the striking theorem discussed in §10.3.

**Theorem 10.3 CM** and **F** imply

1. (*Direct causal connection*) For all $x$ and $y$ in $V$, $x$ and $y$ are adjacent if and only if they are probabilistically dependent conditional on *every* subset of $V$ that does not include them,[1] and

[1] Compare this to Spohn's account of direct causation (1983a, p. 388; 1983b, p. 82; 1990, p. 128), in which $a$ is a direct cause of $b$ if it precedes $b$ and Pr(B/A&Z) > Pr(B/A), where $Z$ consists of the whole past of $b$ apart from $a$.

2. (*Direct causation*) For all $x$, $y$ and $z$ in $V$, if $x$ and $y$ and $y$ and $z$ are adjacent, and $x$ and $z$ are not adjacent, then $y$ causally depends on $x$ and $z$ if and only if $x$ and $z$ are probabilistically dependent conditional on every subset of $V$ that contains $y$ but not $x$ or $z$.

Theorem 10.3 is a restatement of theorem 3.4 in SGS (1993), and it is proven in Verma and Pearl (1990). The null set counts as a subset of $V$ that does not include $x$ and $y$. The theorem is restricted to acyclic graphs, and so causal dependence must be asymmetric. The first implication, which I called "direct causal connection," gives a necessary and sufficient condition for adjacency, while the second, "direct causation," gives a sufficient condition for $y$ to directly depend on $x$ and on $z$. The second does not provide a necessary condition, because there may be no variable $z$ adjacent to $y$ and not to $x$. If one assumes that one can always expand the variables to find such a $z$ (e.g., by an intervention), then information about probabilities provide truth conditions for "$x$ causes $y$," and the reduction of causal asymmetry to probabilistic facts will apparently be complete.

Proof sketch: Consider first the conclusion that dependence conditional on every set that does not contain $x$ and $y$ is necessary and sufficient for adjacency. Suppose that $x$ and $y$ are probabilistically dependent conditional on every subset of $V$ that does not contain $x$ and $y$. Then if $y$ is not a direct cause of $x$, $x$ and $y$ are probabilistically dependent conditional on the subset of $V$ consisting of all the direct causes of $x$, and if $x$ is not a direct cause of $y$, then $x$ and $y$ are probabilistically dependent conditional on the subset of $V$ consisting of all the direct causes of $y$. So by **CM** and **F**, if $x$ and $y$ are not adjacent, $x$ causes $y$ and $y$ causes $x$. Since, by assumption, causation is asymmetric, $x$ and $y$ are adjacent. Suppose conversely that $x$ and $y$ are adjacent. Then given **CM** and **F**, no subset of $V$ not containing $x$ or $y$ screens off the probabilistic dependency between them.[2]

To prove the second part of the theorem, suppose first that $x$ and $y$ and $y$ and $z$ are adjacent and $x$ and $z$ are not adjacent and that $x$ and $z$ are probabilistically dependent conditional on every subset of $V$ that contains $y$ and that does not contain $x$ or $z$. By the argument given on page 213, three of the four possible configurations of the edges between $x$ and $y$ and $y$ and $z$ imply that $x$ and $z$ are independent conditional on some sets of variables including $y$ but not $x$ or $z$. So $x$ and $z$ cause $y$.

Conversely, suppose that $x$ and $y$ and $y$ and $z$ are adjacent and $x$ causes $y$ and $z$ causes $y$. The task is to prove that $x$ and $z$ are dependent conditional on all subsets of $V$ containing $y$ but not $x$ or $z$. I shall sketch the proof only for the case of dichotomous variables that take the values 1 and 0. In what follows "Pr($x$)," etc. should be read as $Pr(x = 1)$, and "Pr(~$x$)," etc. should be read as $Pr(x = 0)$.

Since $x$ and $z$ are not adjacent, they are probabilistically independent conditional on some set that does not include $x$ or $z$. Conditioning on the set of all the direct causes of $x$ or the set of all the direct causes of $z$ will render them independent, though they might also be unconditionally independent. We may suppose, without loss of generality that $x$ and $z$ are independent conditional on $P_x$, the set of all the "parents"

[2] This part of the proof is not very satisfactory. For a better proof of this part of the theorem from somewhat different premises, see theorem 12.3, p. 258.

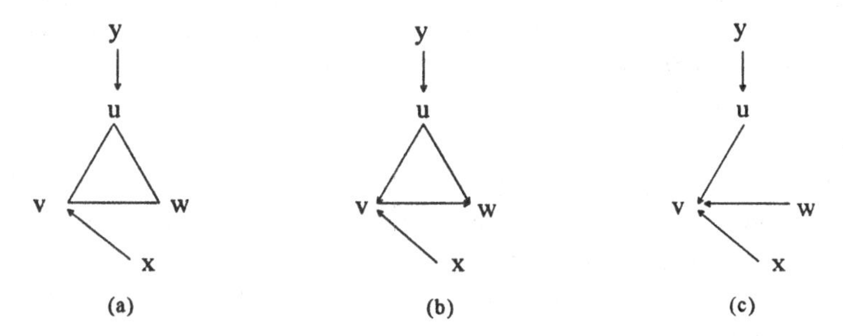


**Figure 10.5:** Dealing with failures of faithfulness

– that is, direct causes – of $x$. To keep the notation simple, this condition will be left implicit. In what follows, "$\Pr(xz) = \Pr(x) \cdot \Pr(z)$," for example, should be read as the claim that $\Pr(x{=}1)$ and $\Pr(z{=}1)$ are independent conditional on $P_x$. We thus have the following premises:

1. $\Pr(y/x) \neq \Pr(y/{\sim}x)$
2. $\Pr(y/z) \neq \Pr(y/{\sim}z)$
3. $\Pr(x\&z) = \Pr(x) \cdot \Pr(z)$
4. None of the probabilities is zero or one (from the assumption that each variable has its own source of variation.

To prove: $\Pr(xz/y) \neq \Pr(x/y) \cdot \Pr(z/y)$.

By the algebra in steps 4–8 on pp. 76–7, one can prove that

5. $\Pr(x/y) \cdot \Pr(z/y) P(y) + \Pr(x/{\sim}y) \cdot \Pr(z/{\sim}y) \cdot P({\sim}y) \neq$
   $[\Pr(x/y) \cdot \Pr(y) + \Pr(x/{\sim}y) \cdot \Pr({\sim}y)] \cdot [\Pr(z/y) \cdot \Pr(y) + \Pr(z/{\sim}y) \cdot \Pr({\sim}y)]$
6. the r.h.s. of 5 is $\Pr(x) \cdot \Pr(z)$ (probability theorem)
7. $\Pr(xz/y) = \Pr(x/y) \cdot \Pr(z/y)$ (hypothesis)
8. $\Pr(xz/{\sim}y) = \Pr(x/{\sim}y) \cdot \Pr(z/{\sim}y)$ (from 3, 7)
9. the l.h.s. of 5 = $\Pr(xz/y) \cdot \Pr(y) + \Pr(xz/{\sim}y) \cdot \Pr({\sim}y)$ (from 7, 8)
10. the l.h.s. of 5 = $\Pr(xz)$ (from 9)
11. $\Pr(xz/y) \neq \Pr(x/y) \cdot \Pr(z/y)$ (from 3, 5, 6, 10)

For dichotomous variables, this proves that $x$ and $z$ are probabilistically dependent conditional on some set containing $y$ and not containing $x$ and $z$. It does not prove the result for all kinds of variables and for all sets containing $y$ and not $x$ and $z$.

## 10.3* Papineau on Failures of Faithfulness

Papineau's second proposal, mentioned at the end of §10.4, is that one embed the system in which faithfulness failed into a larger system (1993, p. 245). In his example, one has probabilistic dependencies between the values of $u$ (cola consumption) and $v$ (exercise) and between $v$ and $w$ (weight gain), which cannot be screened off. So one would mistakenly conclude that $v$ depends on $u$ and $w$. Papineau writes:

> Note, however, that the misleading conclusion. . . could be overturned if we embedded these variables in a larger structure of variables. . . . Thus, in the

diagram below, [reproduced with changes in notation in figure 10.5a], a correlation between $x$ and $w$ would undermine the former claim [that $v$ depends on $w$], and a correlation between $y$ and $w$ would undermine the latter [that $w$ does not depend on $u$]. (1993, p. 245)

Figure 10.5a reproduces Papineau's figure. Figure 10.5b shows what is (by assumption) the true causal situation. Figure 10.5c shows the structure that is suggested by the original joint probability distribution. Suppose that there are no other failures of faithfulness apart from the misleading independence between $u$ and $w$ and suppose that one knows that the only causal relation between $x$ and $y$ and the other variables is that $v$ depends on $x$ and $u$ depends on $y$. (This last supposition is of course much more than a reductive analysis is entitled to.) The structure in Figure 10.5b then implies the following independence relations:

$x \perp y$ $\quad$ $x \perp u$ $\quad$ $y \perp x/u$ $\quad$ $y \perp v/u$ $\quad$ $y \perp w/u$ $\quad$ $x \perp w/u, v$

where "⊥" denotes probabilistic independence conditional on what follows the "/." The structure in Figure 10.5c implies all these independence relations except for the last one, and in addition it implies $y \perp w$, $u \perp w$, and $x \perp w$. Given the original failure of faithfulness, one should observe the independence relations implied by figure 10.5b *plus* $u \perp w$ and $y \perp w$. The original failure of faithfulness guarantees the spurious independence between $y$ and $w$. If there is a failure of faithfulness with respect to the subgraph consisting of just $u$, $v$, and $w$, then there will still be a failure of faithfulness in the larger graph.[3]

What one observes follows from *neither* the structure in figure 10.5b nor the one in figure 10.5c. One of the probabilistic independence relations implied by the structure in figure 10.5c is missing (since $x$ is not independent of $w$), and there is an additional conditional independence relation between $x$ and $w$ given $v$ and $u$. On the other hand, there are two probabilistic independence relations (between $u$ and $w$ and between $y$ and $w$) that are not implied by figure 10.5b. The only explanation of what is observed that does not require postulating an *additional* failure of faithfulness is to conclude that figure 10.5b represents the true structure and that the independence between $u$ and $w$ and between $y$ and $w$ results from a failure of

[3] An easy way to see that this is true is to note the connection between causal graphs and systems of linear equations. If there are linear relations among the variables, then, corresponding to the subgraph is a system of equations such as:

1. $u = \mu_1$
2. $v = au + \mu_2$
3. $w = bu - cv + \mu_3$

where $\mu_1$, $\mu_2$, and $\mu_3$ are error terms that are normally distributed and probabilistically independent of one another. Given the failure of faithfulness described – that is, that $b = ac$ – the value of $w$ does not depend on the value of $u$ or on any variable whose only causal connection to $w$ is via $u$.

faithfulness. Postulations of additional failures of faithfulness are less parsimonious, and they will have false implications concerning probabilistic dependencies in larger graphs, unless there are further failures of faithfulness. Embedding the variables in a larger structure does not enable one to see through the failure of faithfulness. The independence never disappears. To reveal the causal structure decisively, one needs not a larger graph, but a finer-grained one, which includes some causal intermediary between $u$ and $w$ that does not depend directly on $v$. Even a finer-grained graph will not eliminate all failures of faithfulness, though. See the example on p. 244.

So considering larger structures or – more effectively – finer-grained structures, permits one to discover failures of faithfulness. But the proposed reduction of causal asymmetry to facts about probability distributions is still in trouble, because Papineau has not told us how to transform claims about enlarged or finer-grained structures into a reductive analysis of causal asymmetry. There is also no reason why one might not have further spurious independence relations in larger or finer-grained systems, no matter how many times one iterates. At this point, Papineau digs in his heels: "But I do not accept that such failures of faithfulness can be universal. A world in which no probabilistic dependencies at all manifested some supposed causal structure would be a world in which that causal structure did not exist" (1993, p. 246). Those who do not already accept this reduction may not find this response persuasive.

# 11

# Intervention, Robustness, and Probabilistic Dependence

This chapter will be concerned with yet another asymmetry of causation. Recently, several authors have argued that there is an asymmetry of *robustness* or *invariance*. In particular, this chapter will consider three related purported asymmetries:

> (*Asymmetry of robustness*) $x$ causes $y$ if and only if the relationship between $x$ and $y$ and other variables is invariant with respect to the value of $x$ but not with respect to the value of $y$ (Arntzenius 1990, pp. 90–1; Hoover 1990, 1991, 1993; Hoover and Sheffrin 1992; Hoover and Perez 1994).
>
> (*Asymmetry of parameter independence*) Suppose that $S$ is a complete and correct specification of a causal system involving a set of variables including $x$ and $y$. $x$ causes $y$ if and only if one can calculate from $S$ the correct value (or the best estimate) of $y$ but not of $x$, when one substitutes a changed parameter value into $S$ (Simon 1953, p. 69; Hoover 1990, pp. 211–12; Woodward 1995, 1998).
>
> **MI** (*Asymmetry of modal invariance*) Suppose that $S$ is a complete and correct specification of a causal system involving a set of variables including $x$ and $y$. $x$ causes $y$ if and only if the value of $y$ one calculates when one substitutes a new value of $x$ into $S$ is a correct prediction or a best estimate of what the value of $y$ would be if the value of $x$ were set to its new value via intervention (Cartwright 1989, ch. 4; Hoover 1991, 1994; Woodward 1995, 1998; Forster 1996b).

The phrase "the relationship between $x$ and $y$" is intentionally vague, because different theorists have focused on different relationships. It covers the conditional probability distribution of $y$ given $x$, a structural equation system relating $x$ and $y$, and any function relating $x$ and $y$. We have come across the notion of a causal system before. Although vague, the notion is indispensable. Among other things, one considers only alternative values of variables and parameters that are consistent with the system maintaining its integrity. A specification of a causal system is complete if it captures all the mechanisms that are relevant to the values of the variables in which one is interested.

What these three claims mean and why they might seem plausible is not obvious from these brief statements. I shall argue that the asymmetry of robustness breaks down and that the asymmetry of parameter independence is somewhat obscure. Rather than grapple with these obscurities, one should turn to the asymmetry of modal invariance, which is plausible and in a revised form offers the possibility within a limited domain of a reduction of causation to probabilities.

### 11.1 The Asymmetry of Robustness

The asymmetry of robustness is something of a straw man, although it is suggested by remarks Frank Arntzenius (1990) and Kevin Hoover (1990, 1991, 1993) make. I shall use their remarks to put flesh on the position that probabilities conditional on causes, unlike probabilities conditional on effects, are robust, but I do not mean to attribute this position to Arntzenius or Hoover. Arntzenius gives the example that Pr(S/C), the probability of smokers among those who had lung cancer, was extremely low before smoking became common, but if **c** indeed depends on **s**, Pr(C/S) was just the same then as Pr(C/S) is now for people in the same circumstances with regard to other causes of lung cancer.[1] Similarly, Kevin Hoover writes, "If *money* [*M*] *causes price* [*P*] and if the intervention was in the money-determination process, . . . one would not expect either D(M/P) or D(M) to remain stable; but one would expect D(P/M) to be stable, although not D(P)" (1991, p. 384). D(·) here is the probability distribution. If prices depend on the money supply, then changing the probability distribution of *M* changes the probability distributions of *P* and *M* given *P*, but it does not affect the conditional probability distribution of *P* given *M*.[2]

These remarks suggest what I called "the asymmetry of robustness." As I understand the asymmetry of robustness, it refers to the actual joint probability distribution of *x*, *y*, and other variables, and it claims that the conditional distribution of *y* given *x* is independent of the value of *x*, while the conditional distribution of *x* given *y* is not independent of the value of *y* (Woodward 1998, p. 17). But the contrast between the actual probability distribution and other probability distributions that might, for example, arise from interventions is unclear. See p. 230. One must read claims about robustness as specifying explicitly or implicitly limits to the extent of variation in the values of variables. In the sort of apparatus one might find in a high school physics lab, the relation between current and voltage is robust

[1] 1990, p. 90. This example is misleading, because Arntzenius is concerned only with the robustness of the relationship between an event and the set of *all* its causes.

[2] Michael Redhead has written a good deal about robustness, but he has not invoked robustness to account for causal *asymmetry*. Redhead argues that a necessary condition for a causal relation – regardless of its direction – is that "sufficiently small disturbances of either relata do not affect the causal relation" (1987, 102–3; see also Elby 1992).

– but not if the voltage is boosted from 6 volts to 1000 volts and the wire melts.

Is there an asymmetry of robustness, and how does this purported asymmetry relate to the other asymmetries we have discussed? Since the function relating $y$ to $x$ and to other variables is independent of the value of $x$, the value of $y$ given any particular value of $x$ must be independent of the values of those variables that caused $x$ to have that value. So the asymmetry of robustness turns out to involve an asymmetry of screening-off: $x$ causes $y$ if and only if for all causes $z$ of $x$ $\Pr[y=y^*/(x=x^* \& z=z^*)] = \Pr(y=y^*/x=x^*)$. But as we saw in chapters 6 and 10, this claim is false (see also Healey 1992, p. 286). When there are multiple connections, the conditional distribution will not be insensitive to the value of $x$, because differences in the value of $x$ may represent differences in a variable that both $x$ and $y$ depend on.

One might reformulate the claim connecting causation and robustness as: When $x$ causes $y$ then the conditional distribution of $y$ given $x$ is robust with respect to all the causes of $x$ that do not have any independent causal influence on $y$. This implies that when $y$ depends on $x$, then $x$ screens off $y$ from everything it depends on that does not have any independent connection to $y$. As the discussion of figures 10.1e and 10.1f on page 209 shows, this claim is false, too. Even though the only connection between the concentrations in saltwater basins 2 and 5 in figure 6.3 is via the concentration in basin 4, $x_4$ does not screen off $x_2$ and $x_5$. $x_1$ and $x_2$ will co-vary when one controls for $x_4$, and so will $x_2$ and $x_5$. Only the complete set of direct causes screens off remote causes.

Although the asymmetry of robustness breaks down, claims about robustness remain of interest. Significant multiple connections between causes and their effects are not ubiquitous, and robustness may be a good practical test for the direction of causation. Just as there is an approximate asymmetry of counterfactual dependence, so there is an approximate asymmetry of robustness.

## 11.2 The Asymmetry of Parameter Independence

Consider the two following systems of linear equations:

(System 1)
$$x = \alpha_x + \mu_x$$
$$y = \alpha_{xy}x + \mu_y$$
$$z = \alpha_{xz}x + \alpha_{yz}y + \mu_z$$

(System 2)
$$x = \alpha_x + \mu_x$$
$$y = \alpha_{xy}x + \mu_y$$
$$z = \alpha^*_{xz}x + \mu^*_z.$$

If $\alpha^*_{xz} = \alpha_{xz} + \alpha_{yz}\alpha_{xy}$ and $\mu^*_z = \mu_z + \alpha_{yz}\mu_y$, then these two sets of equations are empirically equivalent. Data on values of $x$, $y$, and $z$ are compatible with system 1 if and only if they are compatible with system 2. But if variables on the right-hand side of an equation are taken as causes of variables on the left-hand side, the two systems differ in what they say about the causal structure. In particular, they disagree about whether *z* depends on *y*. If the two systems are empirically equivalent, how could this disagreement matter? Herbert Simon's answer, which has recently been revived and endorsed by Kevin Hoover and James Woodward, is

> We can picture the situation, perhaps somewhat metaphorically, as follows. We suppose a group of persons whom we shall call "experimenters." If we like, we may consider "nature" to be a member of the group. The experimenters, severally or separately, are able to choose the nonzero elements of the coefficient matrix of a linear structure, but they may not replace zero elements by nonzero elements or vice versa. . . . Once the matrix is specified, the values of the *n* variables in the *n* linear equations of the structure are uniquely determined. Hence, the experimenters *control indirectly* the values of these variables. The causal ordering specifies which variables will be affected by an intervention at a particular point. . . . (1953, p. 69).

The $\alpha$'s (including $\alpha^*$) in systems 1 and 2 are the "nonzero elements of the coefficient matrix" that the experimenters can choose. When the experimenters choose a new value for $\alpha_{xy}$ – suppose it is twice the old value – then if the error terms have zero mean and their distribution does not change, system 1 implies that the expected value of *y* will double and the expected value of *z* will increase by $2\alpha_{yz}$, while system 2 in contrast implies that the expected value of *z* will not change. If the expected value of *z* does not in fact change, then system 2 is correct. If it increases by $2\alpha_{yz}$, then system 1 is correct.

This argument depends on two crucial presuppositions. The first was already mentioned. In a correct causal system, the distribution of the error terms must not change when the experimenters set the values of the parameters (coefficients). The second is that if a system of equations correctly specifies the causal relationships, then experimenters can set the values of parameters *independently*. Suppose system 2 specifies the causal structure correctly, and investigators mistakenly accept system 1. Then, when the experimenters double the value of $\alpha_{xy}$, substitute that new value into the equations, and calculate the values of *y* and *z*, the calculated value for *z* will be mistaken. In setting the value of $\alpha_{xy}$, it appears to experimenters who mistakenly accept system 1 that they have also changed one of the parameters in the third equation in system 1. If the investigators correctly believe that system 2 specifies the causal structure, they can make correct predictions or best estimates of the values of *y* and *z* when they substitute the new value of $\alpha_{xy}$ without making any other adjustments to the equations. If, on the other hand, system 1 captures the causal structure, and the experimenters believe this, then when they double the value of $\alpha_{xy}$, they can calculate the value of *z* without supposing that the intervention that doubled $\alpha_{xy}$ somehow changed some other coefficient. If they mistakenly employ system 2 for their calculations, it will appear that the intervention that set $\alpha_{xy}$ also changed $\alpha^*_{xz}$ (or the zero value of the missing $\alpha_{yz}$) in system 2. There is an asymmetry of parameter independence.

Both Kevin Hoover and James Woodward talk of *invariance* rather than *independence*.[3] Hoover writes, "The fact that variables show a particular functional relationship or pattern of covariance does not capture the essence of causality. The important thing is that those relations remain stable in the face of interventions of control (actual or hypothetical)" (1990, p. 212). Woodward writes, ". . . it should be possible to intervene to change each of the coefficients in these equations separately without changing any of the other coefficients. Call this condition coefficient invariance" (1998, p. 12). But what is at issue is independence, *not* invariance. Of course one expects parameters not to change, but unless every possible mechanism is modeled, it may happen that relationships will break down and that parameters will change. Such a change in one parameter may occur at the same time as an intervention that changes the value of another. If the causal structure is correct, this must be a coincidence. The change in one parameter must not causally depend on the intervention that changes the value of another.

It may seem plausible to maintain that parameter independence will hold when and only when equations in complete systems of linear equations each have effects on their left-hand side and causes of those effects on their right-hand side. This claim is, however, problematic, because it is not generally clear what it means to change the value of a coefficient without so altering the system that the consequences are meaningless. Although parameter independence implies the independent variability and insensitivity conditions of chapter 8, independent variability and insensitivity refer only to interventions that change the values specified for variables. Since (or so I will shortly argue) modal invariance captures the intuition behind parameter independence without requiring that one explicate the notion of a coefficient-changing but in some sense system-preserving intervention, there is little reason to pursue the asymmetry of parameter independence.

It is important to distinguish parameter independence from another invariance notion, which Trygve Haavelmo called "autonomy." The relationship between the pressure on a gas pedal and the speed of a car, unlike the relationship between the pressure, temperature, and volume of a gas, has very little autonomy (1944, pp. 27–8). The relationship between the

[3] Hoover writes that the "true" parameters "may be chosen *independently* of each other" (1990, p. 211 [my emphasis]). Indeed, it is this feature of parameters that, in Hoover's view, distinguishes them from coefficients, some of which may not be parameters. It may be that in Hoover's account the independent manipulability of parameters is primary and that invariance is a derivative indicator of causal order.

pressure on the gas pedal and the speed is liable to break down. However useful it might be in a particular context, one cannot count on it to hold in a different context. Nonautonomous relations depend on details of a specific context and cannot be relied on in different contexts.

In his exploration of the relevance of autonomy to *explanation*, Woodward argues for an intimate connection between autonomy and parameter independence (which, recall, he calls "coefficient invariance"):

> Suppose it is possible to intervene in such a way as to separately change each of the coefficients in (2.4) [a system of simultaneous linear equations]. Then if the relationships described in (2.4) are autonomous, each of the coefficients should be invariant under interventions that produce changes in any of the other coefficients. (1995, pp. 12–13)

Woodward then goes on to argue that "relationships that are nonautonomous are noncausal" (1995, p. 13). Woodward's thought seems to be that if the equations in the system are autonomous, then the coefficients should remain invariant in different contexts, including the different contexts created by interventions that change the values of some of the coefficients. Absolutely autonomous relations would, of course, have all the invariance one could want, and absolutely nonautonomous relations are arguably noncausal. But autonomy is best understood as a comparative notion. Relations are more or less autonomous, and a significant degree of autonomy is neither sufficient nor necessary for parameter independence. It is not sufficient, because anything short of absolute autonomy is consistent with one parameter changing in response to an intervention that changes the value of another. It is not necessary, because, as Haavelmo's own example illustrates, relations with little autonomy can still be causal.

Other things being equal, relationships with little autonomy have little utility, because their scope of applicability is so narrow; and they are hard to test, because few data are relevant. Scientists seek autonomous relations, and people prefer knowledge of more autonomous relations to knowledge of less autonomous relations. Explanations that rely on less autonomous relations are shallower than explanations that rely on more autonomous relations. Furthermore it is plausible to maintain that $x$ has a causal capacity to bring about $y$ only if the relationship between $x$ and $y$ has a good deal of autonomy. Although the notion of autonomy is important in all these ways, I believe it has little to contribute to an account of the metaphysics of causation.[4] Equations and equation systems that are not autonomous have a small scope of application and are of relatively slight explanatory value, but the relations they identify can be causal all the same.

[4] Nancy Cartwright (1989, pp. 142–58) and James Woodward emphasize different aspects of autonomy, and I am indebted to them here. See also Engle et al. (1983) and Cartwright (1995, pp. 54–6). For related criticism of linking causality and autonomy, see (Hoover 1995, pp. 79–80).

## 11.3 The Asymmetry of Modal Invariance

Hoover and Woodward both come to their concern with invariance out of a recognition of the *practical* importance of causal relations. Even though neither attempts to analyze causation in terms of manipulation or agency, both insist on the importance of the links between causation and manipulability. In his forthcoming book Woodward writes:

> Although I have denied that causation can be reductively defined in terms of agency, my view is nonetheless that considerations having to do with human agency and with our practical interests as agents in manipulating the world can play a central role in illuminating our notion of causality. Most fundamentally, it is these interests that explain or help to largely explain why we have a notion of causality at all, and why it takes the form or features that it does. It is our interest in manipulation that explains why we have (or provides the underlying motivation for our having) a notion of causality that is distinct from the notion of correlation.[5]

One can control the value of $y$ by manipulating the value of $x$ only if the relationship between $x$ and $y$ does not break down when one intervenes to manipulate the value of $x$. Manipulability thus apparently requires one kind of invariance: If $R$ is a quantitative specification of how $x$ causes $y$ and one intervenes and sets the value of $x$, then, ceteris paribus, one should be able correctly to calculate the value of $y$ by substituting the new value of $x$ into $R$. "[T]he conditional analysis of causality is partly the assertion that causal relations are invariant to attempts to use them to control the effects" (Hoover 1994, p. 66). In this way, one arrives at one version of what I called the asymmetry of modal invariance: $x$ causes $y$ if and only if one can calculate correctly (or make the best estimate of) what the value of $y$ would be if the value of $x$ were set via intervention. This invariance, like parameter independence, can be possessed by relations that have very little autonomy.

One needs to formulate modal invariance carefully. Let $z$ be an intervention variable that sets the value of $x$ and suppose one interprets modal invariance as the claim that $y$ depends on $x$ in background circumstances $K$ if and only if $y = f(x, K)$ whether or not the value of $x$ is set by intervention or as $\Pr(y{=}y^*/x{=}x^*\&K\&z{=}z^*) = \Pr(y{=}y^*/x{=}x^*\&K)$. These claims are false. When there are multiple connections, it matters to the value of $y$ whether the value of $x$ varies as the result of intervention or as a result of causes that have independent influences on the value of $y$. Consider, for example, the saltwater basins again and suppose that all the variation in $x_4$, the concentration in basin 4, is due to variations in the concentration in basin 1. It will then be the case that $x_5 = r.x_4$. This relationship will break down when one intervenes and changes the value of $x_4$, even though $x_5$ causally depends on $x_4$. I argued in chapter 6 that when there are multiple connections, as in the

[5]Woodward unpublished, ch. 5 (draft of March 26, 1996), pp. 44-5.

case of the basins, effects are not counterfactually dependent on their causes. The value of $x_5$ is not counterfactually dependent on the value of $x_4$, but when the value of $x_4$ is set by intervention, then the value of $x_5$ is counterfactually dependent on it. Thus, a simplistic version of modal invariance breaks down. Multiple connection cases like those involving the basins, which presented difficulties for Lewis's counterfactual theory, provide counterexamples to a simplified formulation of the asymmetry of modal invariance as well.

What one needs to say is more complicated: Only if an equation system or its qualitative analogue gets the causal structure right and is a complete account of the relevant variables in the system will coefficients or analogous conditional probabilities be invariant. The coefficient in $x_5 = .5(x_3 + x_4)$, in contrast to the "$r$" in the equation above, is invariant to interventions that set the value of $x_4$ (but not to interventions that cut the pipes!). Similarly, in the equation systems 1 and 2 above, if $z$ truly depends on $y$ as specified in system 1, then, ceteris paribus, $\alpha_{yz}$ will not change when one intervenes and changes the value of $y$, while $\alpha^*_{xz}$ or the (zero) value of $\alpha_{yz}$ is not invariant in system 2. Multiple connections create no difficulty for a proper formulation of modal invariance. Modal invariance implies the independent variability and insensitivity conditions of chapter 8.

Modal invariance sounds like a version of an agency theory that is, like all versions of agency theory, subject to objections on grounds of circularity, scope, and anthropomorphism. Because this account is not intended as an analysis, the circularity objection is blunted. Since Woodward emphasizes that he is thinking of "abstract" interventions, which need not be human actions, the scope and anthropomorphism objections are mooted. But then one also wonders whether, except as a heuristic, the reference to intervention is needed.

It is not. In an important unpublished essay, my colleague Malcolm Forster argues that – within the context of path models – $x$ causes $y$ if and only if the partial regression coefficient $\alpha_{xy}$ remains the same in every possible joint probability distribution of the variables in the system that is consistent with the integrity of the system (1996a). A system, such as a dimmer switch connected to a light bulb, constrains what joint probability distributions of switch settings and degrees of illumination are possible. In the actual probability distribution it might be that the switch is rarely turned more than halfway up. This limitation is not a property of the system. If it is only happenstance, it might be argued that the actual *probability distribution* is not in fact truncated in this way. Probabilities are not the same thing as frequencies and the actual probability distribution is not the same thing as the actual frequencies. But perhaps the fact that the dimmer switch is rarely turned more than halfway reflects the preferences of the home owners. Since these preferences are (by assumption) not a part of the light-switch–light-bulb system, the fact that the dimmer switch is rarely turned more than halfway is a feature of the actual probability distribution that is not a feature of all possible probability distributions that are consistent with the integrity of the given system. That the switch can be rotated only 150 degrees, in contrast, might be a property of the system itself and hence true of all probability distributions consistent with the integrity of the system. (By brute force one might crack the plastic and rotate the switch further, but then the system would be changed.) The distinction between features of the given distribution and features of all distributions consistent with the given system is not easy to draw. Is there, for example, a single joint distribution of lung cancer and smoking governing the last few millenia, or is the probability distribution now, like the frequencies, very different from what it used to be?[6]

Forster shows that the rules for constructing a set of regression equations from a path diagram have an inverse, and he thereby achieves a reduction of causation to properties of sets of probability distributions. Since this work is not yet published, it would be inappropriate for me to work through the details. Forster's reduction of causation to probabilities is only partial: Path diagrams involve linear and hence additive relations among causes and their effects. It is not yet clear whether the idea extends to nonlinear relationships. Furthermore (as he is well aware), Forster does not provide the sort of reduction that a naive empiricist seeks. It is no easier to know the set of probability distributions consistent with a given system than to know what the causal relations are. But the fact that Forster reduces causation to modal claims concerning sets of probability distributions rather than to easily observable regularities does not render it uninteresting. In particular, one might argue that Forster's framework subsumes the formulation of modal invariance in terms of abstract interventions. The talk of interventions is relegated to a heuristic role in specifying the set of probability distributions consistent with a particular system and in showing how modal invariance underwrites the connection between causation and manipulation.

## 11.4 Modal Invariance and Independence

I argued before that what Woodward called "coefficient invariance" ought to be regarded as an independence condition rather than as an invariance condition. What about modal invariance? When $y$ causes $z$ as specified in system 1, should one require that a coefficient such as $\alpha_{yz}$ be *invariant* to

[6] I speculate that the set of probability distributions consistent with a given system can be identified with a single probability distribution defined over a larger system. See chapter 11*.

interventions that change the value or probability distribution of *y*, or should one require that the features of the system whose influence the coefficient summarizes be causally independent of interventions that change the value or probability distribution of *y*? Independence is of course implied by invariance. Should one accept the stronger invariance claim or only the weaker independence claim?

The choice is misleading, because there is an ambiguity concerning the sort of invariance claimed. Is one making the counterfactual claim that if one *were* to intervene and change the value of *x*, the coefficients *would* remain unchanged, or is one making the factual assertion that when one intervenes and changes the value of *x* the coefficients do not change? The counterfactual claim plus the prediction condition of chapter 6 (p. 120) implies that if one has no other knowledge, the best prediction of the values of the variables that depend on *x* is based on the assumption that the coefficients are invariant. But the counterfactual claim is not refuted if, owing to some *independent* factor, some coefficient changes, and the factual claim is false. *The modal invariance claim is an independence condition.*

**MI**, modal invariance, is deducible from the definition of an intervention and the assumption that interventions are possible (which imply the strong-independence condition) and the connection principle. Assume first that *x* causes *y*. A genuine intervention with respect to *x* has no causal connection to anything except via causing *x*. So the relationship between *x* and *y* must be undisturbed, and one can calculate what the value of *y* would be. Similarly, an intervention with respect to *y* will leave the value of *x* unaffected, and since the specification is complete, it must specify a relationship between *x* and *y* that will be disrupted. Conversely, suppose that an intervention with respect to *x* leaves the relationship between *x* and *y* undisturbed. Then *y* cannot cause *x* and *x* and *y* cannot be effects of a common cause, and so the nomic relationship between *x* and *y* must be *x* causing *y*.[7]

### 11.5 Conclusion: Causation and Modularity

Although I took similar logical relations in the case of the agency theories discussed in chapters 5 and 7 as supporting the independence theory of

[7] See also theorems 8.4 and 8.5. Theorem 11.1, p. 235 states that if the probability distribution over the variables in a graph satisfies the causal Markov condition, then the distribution is factorizable: The joint probability density function equals the product of the density functions for each variable conditional on its direct causes. If an intervention changes only the value of a variable *x*, it will leave unchanged the density functions of the direct effects of *x* conditional on *x* and of all other conditional densities not involving *x*. In this way **MI** follows from the causal Markov condition. This line of thought is developed further in SGS's manipulation theorem, which is discussed in chapter 11*. Since the causal Markov condition in turn can be deduced from **CC** and $\mathbf{I}_s$ (theorem 12.2), one can construct another derivation of **MI** from **CC** and $\mathbf{I}_s$.

causal priority (**CP**), I think that different conclusions are called for in the case of the work of Hoover, Woodward, and Forster. Since their work is not subject to the scope or anthropomorphism objections and (particularly in Forster's case) avoids objections on grounds of circularity, there is little reason to prefer **CP**$_g$ to **MI** in the domain where **MI** is applicable.

Until Forster's work is in print, it is impossible to formulate **MI** precisely or to scrutinize it with care. I believe that this work complements rather than competes with **CP**$_g$. In particular, both share the same underlying intuition: What characterizes causation and causal explanation is a certain *modularity*,[8] which permits us to factor out influences and gives us reason to pick out some nomological relations, to dub them "causal," and to use them asymmetrically in explanations. This factoring does not take a simple form. The influence of *x* on *y* can depend on the values of other variables. Nor can we always reach conclusions about what the value of *y* would be if the value of *x* had been different (though we typically can). What we can do, if we know the system and the relevant laws well enough, is calculate what the value of *y* would be if the value of *x* were *set* at some other value (within the range of values that leave the given system intact). Human beings are concerned about the possibilities of intervening in the course of nature, and consequently they want such knowledge, and they take it to explain – that is, to show why things "had to be" as they are.

[8] I borrowed this suggestive terminology from James Woodward.

# 11*

# Interventions and Conditional Probabilities

Conditioning on a variable is not the same thing as intervening to fix the value of a variable, and the conditional probability of *y* given some particular value of *x* is not the same thing as the probability of *y* conditional on an intervention that "sets" this value of *x*. When one conditions, one takes as given the probability distribution. When one intervenes, one changes the probability distribution (but see pp. 229–30). These observations suggest a problem. The difference in the probability of **b** with and without an intervention that brings about **a** seems a surer guide to causal relations than does a comparison of Pr(B/A) and Pr(B/~A). Do attempts to reduce causation to facts about probability distributions or merely to operationalize them in those terms rest on a confusion of conditional probabilities and probabilities given interventions? Do they collapse when this distinction is drawn? For example, suppose (as Ronald Fisher hypothesized) that tokens of smoking, **s**, and lung cancer, **c**, are related as effects of some common cause and not as cause and effect. In that case, even though Pr(C/S) > Pr(C/~S), the probability of lung cancer is unaffected by interventions that bring about or prevent smoking. Following Pearl (1993, p. 267; 1995, p. 673), let us indicate the conditional probability that *B* will be instantiated given an intervention that brings about a token of kind **a** as Pr(B/set-A) and the conditional probability that $y = y^*$ given an intervention that causes the value of $x$ to be $x^*$ as Pr($y$=$y^*$/set-$x$=$x^*$) or less precisely as Pr($y$/set-$x$). In this notation, Fisher's hypothesis implies that Pr(C/set-S) = Pr(C/set-~S). What is of interest to a theory of causation are these probabilities, not Pr(C/S) and Pr(C/~S).

The relations between Pr(C/set-S) and Pr(C/set-~S) depend on comparing *different* probability distributions rather than on facts about any single probability distribution. It thus seems that facts about any individual probability distribution could not serve as truth conditions for causal claims and that any reduction or operationalization of causal asymmetry in terms of facts about a single probability distribution must fail (Woodward 1997, pp. 297–300). Here is the objection in James Woodward's words:

> [T]he distinctive feature of a causal relationship is *not* the presence of a correlation per se or even of a correlation that persists when we statistically control for various possible confounding variables, but rather the presence of a correlation that is stable or invariant under some class of changes. This is a modal or subjunctive claim that (or so I would argue) cannot be cashed out just in terms of facts about covariances, partial correlation coefficients, conditional probabilities and the like. (1995, p. 24)

Woodward may be right that modal claims about what would be the case if one intervened cannot be reduced to claims about actual probability distributions (it depends on how much modal content one builds into the given probability distribution). Even if he is right, these modal claims about what would be the case if one intervened can still be reduced to other (modal) claims concerning a single hypothetical probability distribution. One can translate claims about interventions and their consequences for probability distributions into claims about features of a single probability distribution defined over an expanded set of variables that includes the intervention variables.

Suppose one accepts $\mathbf{DI_g}$, the definition of an (abstract) intervention. In the directed (normal) graph that represents the causal relations, there is an edge from the vertex $z$ representing the intervention variable to the vertex representing the manipulated variable and no other edges into or out of $z$.[1] Suppose one also accepts **SPI** (p. 153) and assumes that situations with and without an intervention differ with respect to the value of the intervention variable, not with respect to causal structure. This is a substantial and controversial assumption, since it seems plausible to maintain that an intervention that sets the value of a variable *breaks* the links between that variable and its causes and thereby changes the causal structure. On the other hand, I have already argued for **SPI**, and apart from some intuitive discomfort, there seems to be no objection here to regarding the intervention variable as always there, though sometimes with an off value.

The fact that $\Pr(y=y^*/\text{set-}x=x^*) \neq \Pr(y=y^*/x=x^*)$ is perfectly compatible with defining $\Pr(y=y^*/\text{set-}x=x^*)$ in terms of a hypothetical probability distribution over an expanded set of variables. In particular, if one considers a probability distribution $\boldsymbol{Q}$ that is defined over the larger graph containing the intervention variable, one can then define $\Pr(y=y^*/\text{set-}x=x^*)$ as $Q(y=y^*/x=x^* \,\&\, z=z^*)$, where $z^*$ is some "on" value of the intervention variable. Except for $x$, each variable will have the same parents, and since all the causal relations in the expanded system are the same as those in the actual manipulated and unmanipulated systems, it should be possible to calculate the new probability distribution from the old one and from knowing how $x$ depends on $z$.

[1] I am using the same symbol to refer to both a vertex and the variable it represents.

Doing this requires a theorem that SGS mention and attribute to Kiiveri and Speed (1982).

**Theorem 11.1:** Let $X = \{x_1, \ldots, x_n\}$ be the finite set of $n$ variables represented by the $n$ vertices of some directed acyclic graph $G$ and let $\boldsymbol{Q}$ be a probability distribution over $X$ that satisfies the causal Markov condition – for all $x$, $y$ in $X$, if there is no path in $G$ from $x$ to $y$, then $x$ and $y$ are probabilistically independent conditional on the set of all the parents of $x$, $Px$. Then $Q(x_1, \ldots, x_n) = Q(x_1/Px_1)\cdot Q(x_2/Px_2)\cdot \ldots \cdot Q(x_n/Px_n)$.

Proof: $X$ is the union of the disjoint subsets $X_0, \ldots, X_m$ where for all $x_i$ in $X$, $x_i$ is in the subset $X_k$ if and only if the longest path in $G$ from $x_i$ to a terminal node (a vertex with no edges out of it) has $k$ edges. $X_0$ is, of course, the set of variables represented by the terminal nodes of $G$. For notational convenience, number the $n$ variables in $X$ so that for all $k$, every variable in $X_k$ has a smaller subscript than any variable in $X_{k+1}$.

Suppose there are $j$ variables in $X_0$. Since $\boldsymbol{Q}$ satisfies the causal Markov condition and there are no edges out of vertices representing variables in $X_0$, every variable in $X_0$, conditional on its parents, is probabilistically independent of all the other variables in $X$. So

$Q(X) = Q(x_1/Px_1)\cdot \ldots \cdot Q(x_j/Px_j)\cdot Q(x_{j+}, \ldots, x_n)$.

Consider the subgraph $G_1$ of $G$ consisting only of the vertices representing variables in $\{x_{j+1}, \ldots, x_n\}$. Every vertex in $G_1$ has exactly the same parents as it has in $G$, and so $\boldsymbol{Q}$ restricted to the variables represented in $G_1$ satisfies the causal Markov condition with respect to $G_1$. In $G_1$ the variables in the subset $X_1$ of $X$ are all represented by terminal nodes. Suppose there are $r$ vertices in $X_1$ and that $j+r = s$. Then, given the causal Markov condition, $Q(X) = Q(x_1/Px_1)\cdot \ldots \cdot Q(x_s/Px_s)\cdot Q(x_{s+1}, \ldots, x_n)$.

Repeating the same argument with respect to the subgraph $G_2$ containing the vertices representing variables in $\{x_{s+1}, \ldots, x_n\}$ and so on, it follows that $Q(X) = Q(x_1/Px_1)\cdot \ldots \cdot Q(x_n/Px_n)$.

Given the factorization made possible by the causal Markov condition, the probabilities of all variables conditional on their immediate causal antecedents must be the same as those in the graph without the intervention variable, while the probabilities of the manipulated variables conditional on their antecedents are determined by the properties of the intervention. These facts should tell one enough about the probability distribution over the larger structure to enable one to deduce the consequences of the intervention.

This line of thought is made rigorous in a theorem Spirtes, Glymour, and Scheines prove concerning interventions – their Manipulation Theorem (theorem 3.6; 1993, p. 79). The manipulation theorem is complicated to state and prove, and I shall approach it via a modification of the basins example. Suppose each basin also has a device that can inject additional salt or can remove some of the salt from the solution. The operation of these devices is reflected in random variables, $u_1 - u_5$ (one for each basin), each with a mean zero effect. So knowledge of the values of $x_3$ and $x_4$, for

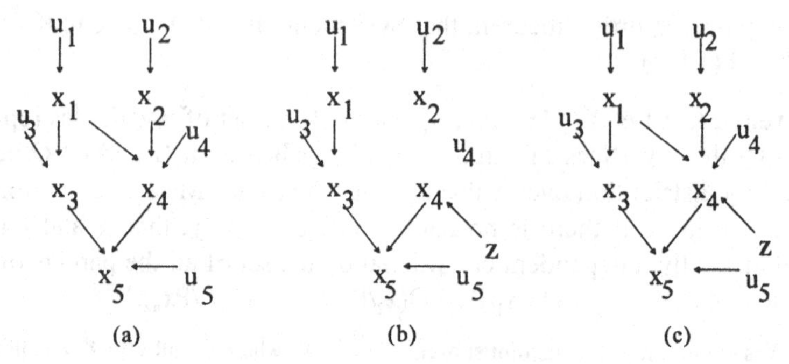


**Figure 11.1:** Manipulations

example, enables one only to deduce the expected value of $x_5$, not its actual value.

Figure 11.1a represents what SGS call an unmanipulated situation. Suppose that an experimenter intervenes and fixes the salt concentration in basin 4 at $x_4$**. Such an intervention apparently eliminates the edges between $x_1$ and $x_4$ and between $x_2$ and $x_4$, and one then has the graph shown in figure 11.1b, which represents the manipulated situation. The first probability distribution does not say what the second will be. One needs causal knowledge to draw conclusions concerning the second distribution. Hence it seems that there is more to the causal relations than facts about probability distributions. This is the argument against any proposed reduction.

Given **SPI** (the structural presupposition of intervention, p. 153), both the unmanipulated and manipulated situations should be represented by a single larger structure that includes the intervention variable. Like Forster's notion of the set of probability distributions consistent with a given system, features of the system (and $\mathbf{DI_g}$) define that larger structure. This larger structure – the combined graph – is shown in figure 11.1c. When the intervention variable $z$ has an off value, the relations between $x_1$, $x_2$ and $x_4$ are as in figure 11.1a. When the policy variable has an on value, then it counteracts the effect of salt concentrations in basins 1 and 2, and the probabilistic independence relations match those of figure 11.1b.

With the help of these three graphs, one can state a version of SGS's manipulation theorem. Let ***P*** be a probability distribution over the combined graph that satisfies **CM** – that is, every variable is probabilistically independent of everything except its descendants (effects) conditional on its parents (direct causes). By theorem 11.1, a conditional probability distribution over ($z$, $x_1$, $x_2$, $x_3$, $x_4$, and $x_5$) faithful to the combined graph, given $z$ = off, is equal to the product of the conditional probability distributions of every variable given its immediate parents in the unmanipulated graph. The probability distribution over $(z, x_1 - x_5)$ given $z$ = off, is equal to $P(x_5/(x_3 \& x_4 \& [z = \text{off}]))\cdot P(x_4/(x_1 \& x_2 \& [z = \text{off}]))\cdot P(x_3/(x_1 \& [z = \text{off}]))\cdot P(x_2/[z = \text{off}])\cdot P(x_1/[z = \text{off}])$. According to the manipulation theorem, the probability distribution given that $z = z^*$ (that is, when the value of $x_4$ is influenced or fixed by intervention) is equal to the product of the distributions of all variables in the combined graph immediately dependent on $z$, conditional on their immediate causes and $z = z^*$ – that is, $P(x_4/[x_1 \& x_2 \& (z = z^*)])$, times the product of the distribution of all other variables, conditional on their immediate causes and $z$ = off. The distribution given that $z = z^*$ is thus a product of factors of the same magnitudes as is the distribution given that $z$ = off, except for the factor $P(x_4/(x_1 \& x_2 \& [z = z^*]))$, which replaces the factor $P(x_4/(x_1 \& x_2 \& [z = \text{off}]))$. An intervention need not break all connection between $x_4$ and $x_1$ and $x_2$, but if it does, then $P(x_4/(x_1 \& x_2 \& [z = z^*])) = P(x_4/[z = z^*])$.

It must be emphasized that the manipulation theorem makes claims about a *single* hypothetical probability distribution ***P*** over the combined graph. Before carrying out the experiment, all one knows is the original (unmanipulated ) probability distribution $\boldsymbol{P}^u$ over the five variables. To make inferences concerning ***P***, one has to assume that (1) $P(x_5/(x_3 \& x_4 \& [z = \text{off}])) = P^u(x_5/x_3 \& x_4)$, (2) $P(x_4/(x_1 \& x_2 \& [z = \text{off}])) = P^u(x_4/x_1 \& x_2)$, (3) $P(x_3/x_1 \&[ z = \text{off}]) = P^u(x_3/x_1)$, (4) $P(x_2/[z = \text{off}]) = P^u(x_2)$, and (5) $P(x_1/[z = \text{off}]) = P^u(x_1)$. If the unmanipulated situation is identical to a situation in which the policy variable has an off value, these equations are unobjectionable. To make inferences concerning the results of manipulations, one has to assume that (6) $P(x_4/[z = z^*]) = P^m(x_4)$, where $\boldsymbol{P}^m$ is a probability distribution over the variables that is faithful to the manipulated graph. Finally one must assume that the probability of all other variables conditional on their parents is the same in the manipulated and unmanipulated graphs. This follows from the definition of an intervention. The manipulation only affects the value or distribution of $x_4$.

The probabilities of variables conditional on their parents change only for the directly manipulated variables. The other factors in the probability distribution over the combined graph remain unchanged. $\boldsymbol{P}^u$, the probability distribution in the unmanipulated graph, will be a product of factors that are identical to those in ***P***, conditional on $[z = \text{off}]$. The probability distribution over the manipulated graph, $\boldsymbol{P}^m$, will be a product of factors that are identical to those in ***P***, conditional on $[z = z^*]$. The factors multiplied together in ***P*** conditional on $[z = \text{off}]$ and $[z = z^*]$ differ only with respect to those stating the conditional probabilities of the directly manipulated variables. $\boldsymbol{P}^m$ will thus be the product of the probabilities given by $\boldsymbol{P}^u$ for the unmanipulated variables conditional on their parents times the probabilities of the directly manipulated variables conditional on their parents as determined by the intervention. So even though one cannot *reduce* the notion of an intervention to facts about a probability distribution over the original variables, one can relate interventions to facts about hypothetical probability distributions and operationalize causal claims concerning the consequences of interventions.

To draw the combined graph and to talk about probability distributions over its variables requires assumptions concerning the relations between probability distributions in different circumstances. These further assumptions are implicit in the definition of an intervention. To link probabilities in $\boldsymbol{P}$ to probabilities in $\boldsymbol{P}^u$, one needs to assume that the causal structure in the initial situation is the same as the causal structure in the combined graph when [$z$ = off]. An intervention is equivalent to the state of affairs in which an exogenous variable in the combined graph has another value. To say that interventions are equivalent to states of affairs in which exogenous intervention variables in combined graphs have specific values reiterates **SPI**.

# 12

# Operationalizing and Revising the Independence Theory

The operationalizing assumption (**OA**) linking causal connections and probabilistic dependencies appears to be false – probabilistic dependencies seem to be neither sufficient nor necessary for nomic connections. In chapter 4, I urged the reader instead to regard **OA** as a useful approximation. But perhaps appearances are misleading. Is it truly impossible to reduce claims about causation to claims about probabilistic dependencies? Chapter 11 suggested that a reduction may be possible within the context of path models. Could **OA** and the independence theory provide a reductive account of causation? And if they cannot, just how closely linked are lawful connections and probabilistic dependencies? To what extent can the independence theory of causation be made operational?

## 12.1 Do Probabilistic Dependencies Imply Causal Connections?

There seem to be cases in which *a* is not causally connected to *b* – that is, that there is no nomological link between *a* and *b* – even though *a* and *b* are distinct events and in the background circumstances the kinds **a** and **b** are probabilistically dependent. This section will be concerned with three difficulties for the claim that probabilistic dependencies are *sufficient* for causal connections.

The first concerns "spurious" probabilistic dependencies, like those between water levels in Venice and bread prices in England. Papineau has argued that "correlations between the stages of *different* time series are not to bc counted as causally significant unless they display co-variation beyond that due to co-variation within *each* time series" (1993, p. 243). Suppose one can make no better prediction of the sea level in Venice at some date from knowledge of wheat prices than one can from knowing the trend within the time series of water levels, and one can make no better prediction of wheat prices from knowledge of water levels than from knowledge of the autocorrelation within the time series of wheat prices. Then the probabilistic dependence is not sufficient for causal connection.

The most one could claim is that correlation beyond what follows from the autocorrelations is sufficient for causal connection; it is not necessary.[1] Suppose a small explosion sets two toy cars moving in opposite directions. The measurements of their velocities, which are performed every few seconds, constitute two time series. The predictions one can make from studying just one of these time series are, we may suppose, not improved by studying the other. If one maintained that it is necessary for causal connections between time series that knowledge of one improve one's predictions of the other, then one would have to maintain (mistakenly) that the toy-car velocities are not lawfully connected.

Does this response resolve the difficulties? Surely one can find coincidental deviations from a trend in time series data or other coincidences. For example, whether the National League wins the World Series in an election year is correlated with whether the Republicans win the presidency.[2] In response, Papineau would deny that these are genuine probabilistic dependencies. Probabilities are not sample frequencies. If a dependency is accidental – if one cannot assert the counterfactual, "if the series were to go on, the correlations would persist" – then there is no real probabilistic dependency (see §2.6.2).[3] To respond this way is apparently to say that true probabilistic dependencies imply lawlike connections, because a probabilistic dependency is not a true one if it does not. In that case one has arguably given nomic or causal conditions for the existence of probabilistic dependencies rather than probabilistic conditions for the existence of causal connections.

Like most other accounts of causation, this one relies on the notion of a law, and all the difficulties involved in distinguishing laws from accidental generalizations arise in distinguishing lawful covariation from accidental covariation. It may be that there is no way to say what is a genuine probability without relying on the notion of a nomological connection or – worse – on the notion of causation, in which case any sufficient probabilistic condition for the existence of a causal connection would be uninteresting. The adequacy of claims about the relations between causal connections and probabilistic dependencies awaits a satisfactory independent account of lawfulness, which I shall not attempt to supply.

[1] To say that it is necessary and sufficient would be much stronger than Granger's definition of causality, "We say that $Y_t$ is causing $X_t$ if we are better able to predict $X_t$ using all available information than if the information apart from $Y_t$ has been used" (1969, p. 428).

[2] I borrow this example from Woodward (forthcoming a), who attributes it to Jonathan Katz.

[3] In an unpublished talk, David Papineau has also argued that these should not be regarded as probabilistic dependencies at all, on the grounds that the values of the variables in neither series is generated independently. Instead of finding covariation between many values of two variables, one is instead reacting to a common feature of two individual trends.

The claim that probabilistic dependencies are sufficient for causal connections might also appear vulnerable to problems that arise when populations are mixed (Yule 1903; SGS 1993, pp. 57–64). Suppose, for example, that one finds the following results from a study of the effectiveness of a new treatment:[4]

| | *Recovers* | *Does not recover* |
|---|---|---|
| *Treated* | 240 | 140 |
| *Untreated* | 260 | 350 |

Treatment and recovery are strongly correlated. Yet it could be that when one analyzes the data further, one finds the following:

| *Women* | *Recovers* | *Does not recover* |
|---|---|---|
| *Treated* | 200 | 60 |
| *Untreated* | 100 | 30 |

| *Men* | *Recovers* | *Does not recover* |
|---|---|---|
| *Treated* | 40 | 80 |
| *Untreated* | 160 | 320 |

Among men and among women there is no correlation between treatment and recovery. The correlation in the aggregate data results from the facts that women are more likely to recover and that a larger proportion of the women than the men in the sample are treated. One might then infer that treatment and recovery are causally connected after all as effects of a common cause (gender). If this explanation is ruled out (for example by evidence that individuals were randomly assigned to treatment or nontreatment groups without regard to gender), one can maintain that there is no genuine probabilistic dependency. One has only a flukey sample in which a lower percentage of men received treatment. In either case there is no counterexample to the claim that probabilistic dependency is sufficient for causal connection.

A further problem with taking probabilistic dependencies as sufficient for causal connection was touched on earlier in §4.1*. Just as claims about

[4] I borrowed this format from Meek and Glymour (1994, p. 1012).

causal connections are relativized to the causal field, so the probabilistic dependencies upon which they depend must be relativized to some set of background circumstances. But in relativizing one has to be careful not to include any common effects of *a* and *b*, because, conditional on a common effect, independent causes *a* and *b* will be probabilistically dependent. In practice one relativizes only to a very general background of prior conditions. Distinguishing what counts as background introduces an element of circularity and may not be admissible in a genuine *reduction* of causal connection to probabilistic dependencies.

As we shall see in §12.6, quantum physics also raises serious difficulties for the claim that probabilistic dependencies are sufficient for causal connections.

## 12.2 Do Causal Connections Imply Probabilistic Dependencies?

Intuitively, a causal connection is a nonaccidental, nomological, or necessary linkage. Such linkages should typically issue in lawful *covariation*, and indeed it is arguable that humans construct the notion of such linkages from their typical consequences, but it seems that there can be lawful connections without variation or covariation. It is false that if *a* and *b* are nomically connected relative to some causal field, then events of these kinds are always probabilistically dependent in the background circumstances. When *a* causes *b* by two different pathways, the causal influences along the two pathways may cancel out, and consequently there may be no probabilistic dependence in the background circumstances between events of kind *a* and events of kind *b*. Such cases may be rare, and it may also be possible to find evidence for the existence of such canceling out when one considers larger sets of variables. But it is still the case that *a* and *b* may be causally connected even though events of kinds **a** and **b** are not probabilistically dependent in the background circumstances.

This canceling out cannot arise when there are no multiple connections (see pp. 127–8), or, equivalently, if none of the proximate causes are causally connected. Let us say in such cases that *a* and *b* are *singly* connected or that there is a single connection between them. One might conjecture that if *a* and *b* are singly connected, then in the background circumstances the event types **a** and **b** will be probabilistically dependent on one another.

This conjecture has some plausibility, but faces problems. Consider a case in which nomological connections without probabilistic dependencies arise from mixing results from different populations. Suppose that we have the following data concerning the effect of a new treatment on recovery rates:

| | *Recovers* | *Does not recover* |
|---|---|---|
| *Treated* | 400 | 400 |
| *Untreated* | 400 | 400 |

Treatment and recovery are uncorrelated, and if probabilistic dependency is a necessary condition for causal connection, one would conclude that treatment and recovery are not causally connected. But suppose that the results for men and women, separately considered, paint a different picture.[5]

| *Women* | *Recovers* | *Does not recover* |
|---|---|---|
| *Treated* | 120 | 160 |
| *Untreated* | 40 | 80 |

| *Men* | *Recovers* | *Does not recover* |
|---|---|---|
| *Treated* | 80 | 40 |
| *Untreated* | 160 | 120 |

Among both men and women, one is more likely to recover if one is treated. The fact that there is no probabilistic dependency in the combined figures is due to two circumstances: Women are less likely to recover whether or not treated, and the percentage of women treated is larger than the percentage of men treated.

How should one respond? One might maintain that treatment and recovery are nevertheless probabilistically dependent in the *population*. It is only because one has an unrepresentative sample that one fails to find this probabilistic dependence. Alternatively, one could argue that in this case one does not have only a single connection between treatment and recovery. On the contrary, both are also connected as effects of a common cause (gender). Restricting one's attention to single connections might resolve this objection to the original necessary condition.

Other difficulties remain. (1) Suppose that *a* and *b* are causally connected but that the background circumstances themselves are sufficient for the

[5] As SGS note (1993, pp. 64–7), the possibility that there may be independence in a combined sample even though there is dependence in samples that are combined was first noted by Kendall in 1948. Simpson in 1951 first noted that there may be positive dependence in all subsamples even though there is independence in the full sample. The fact that probabilistic relations in combined samples may differ completely from probabilities in the samples that are combined has been much discussed in the literature under Nancy Cartwright's label of "Simpson's Paradox."

occurrence of an event of kind **b**. In those background circumstances, there will be no probabilistic dependence between events of kinds **a** and **b**. (2) Suppose one has an "exclusive or" connection like the connection between the two switches that control my upstairs hall light. $x$, $y$, and $z$ are dichotomous variables. $z = 1$ if $x = 1$ and $y = 0$ or if $x = 0$ and $y = 1$; otherwise $z = 0$.[6] $z$ causally depends on $x$ and $y$, but $z$ and $x$ are probabilistically independent, and so are $z$ and $y$. Or consider a case where there is a nomological connection between $x$ and $y$ but neither vary. There is surely a nomological connection between the mass of the earth and the gravitational acceleration near its surface even if neither varies and there is thus no covariation. There is no way to sustain the claim that unconditional probabilistic dependencies are necessary for causal connections.

## 12.3 The Quasi-Reduction of Causality to Probabilistic Dependence

The discussion in the first two sections of this chapter shows that there is no way to reduce the notion of a causal or nomic connection to probabilistic dependency. Yet that discussion nevertheless enhances the theoretical respectability of the notion. It is significant that one can come close to giving a sufficient condition, even if it is one that relies on the problematic notion of a law. On reflection, this irreducibility, even to a modal notion such as a probabilistic dependency, should have been expected. Probabilistic dependencies depend on the existence of variation, while nomic connections do not. Other circumstances in which the relations between probabilistic dependencies and nomic connections break down involve either probabilities of one or zero or the same features of the world that overturn claims about the asymmetries of counterfactual dependence.[7]

The notion of a nomological connection is a theoretical relation; it is that relation that typically obtains among events when they covary.[8] It permits one to derive exact conclusions concerning causation in just the same way that idealizations in physics enable one to derive exact functional relations. One can only dispense with the notion of a necessary, lawful, or causal connection in favor of the notion of a probabilistic dependency if one is willing to recognize and work with the *approximate* relations between causation and probabilistic dependencies. Causal connection among tokens

[6] Judea Pearl gave me this example.

[7] As we saw in chapter 11, multiple connection also overturns simple claims relating causation and invariance.

[8] On such definitions of theoretical entities and relations see Lewis (1972) and, for an application to the theory of causation, Menzies (1996).

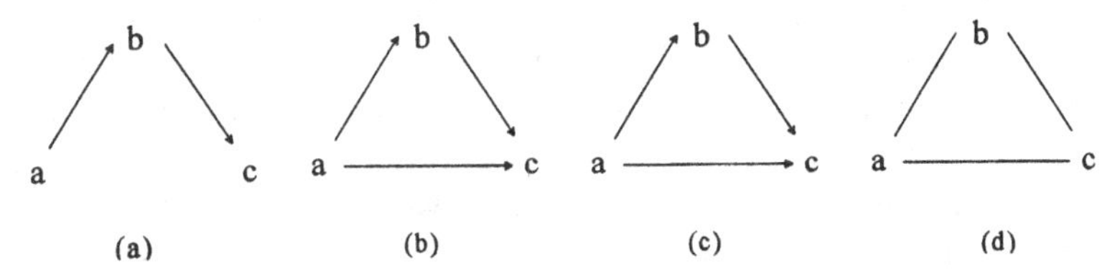


**Figure 12.1:** The indistinguishability of direct and indirect causes

is not the same thing as lawful probabilistic dependence in the background circumstances among the relevant types, but one will seldom go wrong by pretending that it is.

Since the link between causal connections and probabilistic dependencies almost provides a reduction of the notion of a causal connection, the independence theory of causal priority (**CP**) almost provides a reduction of causation itself – provided, of course, that transitivity, the connection principle, and the independence condition hold. Even though these principles mention causation, relying on them creates no circularity in the quasi-reduction. The quasi-reduction depends on their truth, not on our ability to determine whether they are true. **CP** plus the link between causal connections and probabilistic dependencies provides a quasi-reduction of causality to probabilities. In particular, the independence theory (**CP**) and the operationalizing assumption (**OA**) imply that the following three conditions are almost individually necessary and jointly sufficient for "*a* causes *b*": (1) *a* and *b* are distinct, (2) the types **a** and **b** are probabilistically dependent in the background circumstances, and (3) everything dependent on **a** in the background circumstances is probabilistically dependent on **b**.

## 12.4 Why One Needs to Know More than Causal Connections and How to Know It

One often wants to know not merely *whether* two variables are related as cause and effect, but *how*. Suppose we know that *a* is an indirect cause of *b* via *c* and want to know whether it is also a direct cause. We want to know whether the correct normal graph is as shown in figure 12.1a or as shown in figure 12.1b. (This is just the same problem as discriminating between systems 1 and 2 on p. 224 of chapter 11.) The problem is that the causal relations shown in figures 12.1a and 12.1b have exactly the same path and connection graphs (which are shown in figure 12.1c and d). To know whether a causal relation is both direct and indirect or only indirect can be of great importance if, for example, one wants to block the influence of *a* on *b*. How can one, in principle, distinguish direct from indirect causes?

One possibility is to say that *a* is a direct cause of *b* if and only if *a* causes *b* and *a* and *b* are contiguous. But this suggestion conflates two notions of directness. Vertices that are adjacent in a causal graph need not represent contiguous tropes or events. In addition, I've already registered my objections to *stipulating* whether there could be causal action at a distance. A better alternative is to consider conditional as well as unconditional probabilities. If *x* is only an indirect cause of *y*, *x* and *y* should not covary when there are no variations in any of the (direct or indirect) causes of *y* apart from *x*. Since one already has a quasi-reduction of the notion of a direct-or-indirect cause, there is no circularity here. Were it not for the difficulties already mentioned in the linkages between causal connection and probabilistic dependence, this would in principle enable one to draw a unique normal graph on the basis of information about conditional and unconditional probabilistic dependencies. In this way one can establish the empirical credentials of claims about both causation and direct causation. But information concerning *unconditional* probabilistic dependencies does not suffice to do so.

## 12.5 Operationalizing CP

If one takes *x* and *y* to be causally connected if and only if they are distinct and unconditionally probabilistically dependent, then (given **CP**), causal claims have definite consequences for probability distributions. These consequences are of limited use in making causal inferences because of the amount and precision of the information they require, and because they do not permit one to distinguish cases in which there is both direct and indirect causation from cases in which there is only indirect causation. To make such distinctions and to improve the efficiency of inference generally, one needs to consider conditional probability distributions.

In particular, I suggest that one interpret conditioning on a variable or on some particular value of a variable as operationalizing counterfactuals with antecedents such as "if *x* were not to vary at all" or "if *x* were to have value *x** and were not to vary at all." In particular, one might offer the following (again inexact) operationalizing assumption:

> **COA** (*Counterfactual operationalizing assumption*) If the variables in some set *S* were never to vary, *x* and *y* would be causally connected if and only if *x* and *y* are distinct and in the background circumstances probabilistically dependent conditional on the fixed values of the variables in set *S*.

This counterfactual operationalizing assumption implies the simpler operationalizing assumption presented in chapter 4, which links causal connections and probabilistic dependencies. Conversely, if one assumes that the probabilistic dependency between *x* and *y* conditional on *z* is the same as the probabilistic dependency between $x$ and $y$ in a possible world in which $z$ never varied, then one can derive **COA** from **OA**. See theorem 12.1, p. 258.

According to the counterfactual operationalizing assumption, if $y$ and $z$ are causally connected only as effects of a common cause $x$, then they would not be causally connected if the value of $x$ never varied. This seems correct. But consider now the probabilistic dependence of causes conditional on their effects. Suppose that $u$ and $v$ are causes of $x$ that are causally independent and unconditionally probabilistically independent of one another. Then $u$ and $v$ are probabilistically *dependent* on one another conditional on $x$. This fact lies at the heart of the algorithms SGS and Pearl employ in their computer programs for discovering causal structure, and many have found it odd. Indeed, at first glance this fact might lead one to question whether **COA** is unacceptable. For **COA** implies that if $x$ were never to vary, its independent causes, $u$ and $v$, would be causally connected.

Suppose that $x = \alpha u + \beta v + \mu$, where $\mu$ is an independent normally distributed error term with mean zero that is independent of $u$ and $v$. Given **COA**, controlling for $x$ operationalizes the counterfactual supposition that the value of $x$ is unchanging. Then as the value of $u$ increases, either (1) the expected value of $v$ must decrease so as to cancel out the influence the increase in $u$ would otherwise have on the expected value of $x$ or (2) the relationship between $x$, $u$, and $v$ must break down. In *conditioning* on $x$, one is considering worlds where the value of $x$ remains unchanged, not worlds where one intervenes to keep it unchanged. Since one is conditioning on $x$, rather than intervening to *fix* its value, worlds where the relationship among $x$, $u$, and $v$ continues to hold will be more similar (by clause 4 of my account of similarity among possible worlds, **SIM**) than are worlds where the relationship breaks down. So the value of $v$ must vary so as to cancel out any effect on the value of $x$ that would be caused by variation in the value of $u$. If the operationalizing assumption **OA** is true in the most similar possible worlds in which $x$ never varies, there would be causal connections in those worlds between $u$ and $v$. So **COA** makes sense of the puzzling result that conditioning on a common effect induces probabilistic dependencies.

The work of Pearl and SGS develops the statistical implications of the independence view of causal asymmetry. My metaphysics interprets their graph-theoretic results and provides independent arguments for their axioms and theorems. It could even help prove theorems concerning their systems.

### 12.6 Lessons from Quantum Mechanics

There is one further and more interesting difficulty about the link between

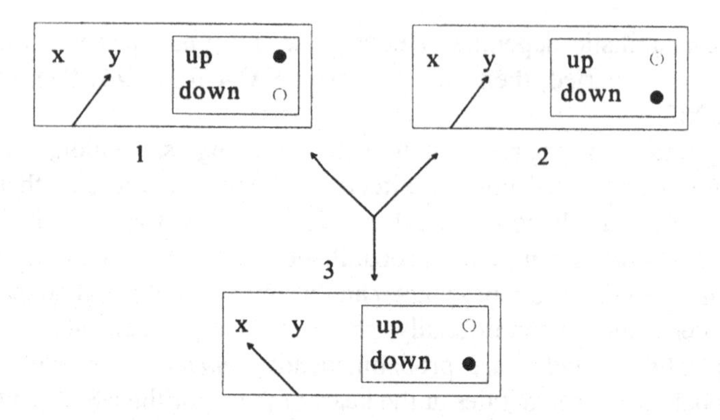


**Figure 12.2:** The GHZ correlations

nomological connections and probabilistic dependencies. Quantum mechanics, especially the so-called EPR phenomena,[9] apparently provides a striking counterexample to the claim that probabilistic dependencies derive from causal relations. Consider the following results predicted by quantum theory. In figure 12.2, three spin-1/2 particles (i.e., electrons) fly apart toward three spin-measuring devices. These devices are far enough apart that no causal influence (unless it travels faster than the speed of light) could link the events measuring the spins of the three particles. In each device Stern-Gerlach magnets can be oriented along one of the two axes perpendicular to the motion of the electron. Call the three measuring devices 1, 2, and 3 and the two axes $x$ and $y$. Each device indicates whether the spin of a particle that interacts with it is up or down along the axis measured. The first predicted result is that when the magnets in exactly two of the devices are oriented in the same direction (suppose it is the $y$ direction), one finds either one or three spin-ups. One never finds zero or two spin-ups.

*If* one assumes that the measurements obtained by each detector depend only on the orientation of its magnet and the state of the electron it interacts with, then these results imply that the state and orientation jointly *determine* the measurement outcome. If there were an element of chance in the measurements, and the measurements depended only on the orientation and the state of the individual electron, then one would sometimes observe no spin-ups or two spin-ups.

Let the particles moving toward detectors 1, 2, and 3 also be numbered 1, 2, and 3. Let us denote "X1 = +1" $=_{df}$ "electron 1 is in that state that leads detector 1 to indicate spin-up when its magnet is oriented along $x$ axis." "X1

[9] So called because a famous thought experiment by Einstein, Podalsky, and Rosen (1935) called attention to it. For discussion of the Einstein, Podalsky, and Rosen paper, see Fine 1987.

$= -1$" $=_{df}$ "electron 1 is in that state that causes detector 1 to indicate spin-down when its magnet is oriented along the $x$ axis." "$Y1 = +1$" $=_{df}$ "electron 1 is in that the state that causes the first detector to register spin-up when its magnet is oriented along the $y$ axis." +1 and −1 values for X2, Y2, X3, and Y3 are similarly defined.

There are three ways in which two of the magnets can be oriented along the $y$ axis, and however they are oriented one or three spin-ups will be measured. If the magnets in detectors 1 and 2 are oriented along the $y$ axis, there must be zero or two −1 values among Y1, Y2, and X3, and hence the product of Y1, Y2, and X3 is +1: $Y1 \cdot Y2 \cdot X3 = +1$. Similarly, $Y1 \cdot X2 \cdot Y3 = +1$ and $X1 \cdot Y2 \cdot Y3 = +1$.

The product of the three left-hand sides, $Y1 \cdot Y2 \cdot X3 \cdot Y1 \cdot X2 \cdot Y3 \cdot X1 \cdot Y2 \cdot Y3$ is equal to the product of the three right-hand sides, or +1, and so $X1 \cdot X2 \cdot X3 \cdot Y1^2 \cdot Y2^2 \cdot Y3^2 = +1$. Since Y1, Y2, and Y3 are either +1 or −1, $Y1^2 = Y2^2 = Y3^2 = +1$. Hence $X1 \cdot X2 \cdot X3 = +1$. A further fact predicted by quantum theory leads to a contradiction: When all the magnets are oriented along the same axis there are zero or two spin-ups measured. This implies that $X1 \cdot X2 \cdot X3 = -1$.

The contradiction is unavoidable if one assumes (1) that the values of the state variables (X1, Y1, X2, Y2, X3, and Y3) are probabilistically independent of the orientation of the magnets in any of the detectors and (2) that the orientation of the magnets in the detectors are probabilistically independent of one another.[10] The probabilistic dependencies among the measurements cannot be explained in terms of one measurement causing another or in terms of their being effects of some common cause. One has an apparent counterexample to the claim that probabilistic dependency arises from causal relations.

Notice that one does not have a counterexample to the claim that probabilistic dependencies are sufficient for nomological connections, since the measurement outcomes are lawfully linked. But one cannot accept *both* the operationalizing assumption and the connection principle (which says that distinct events are nomologically linked if and only if they are related as cause and effect or as effects of a common cause). It looks as if nomological connections and causal connections are coming apart.

There are four possible responses to the EPR phenomena. First, one might maintain that probabilistic dependencies are sufficient for causal connections, except for quantum mechanical phenomena. This is ad hoc. What is it about quantum mechanical phenomena which makes it the case that there can be genuine probabilistic dependencies among events that are

[10] I have taken his formulation from Forster (1996a) and Mermin (1990). For general discussions of these phenomena see Bell (1964, 1966), Cushing and McMullin (1989), and Price (1996, ch. 8, 9).

not related as cause and effect or as effects of a common cause?

Second, one might deny that the interactions of the electrons at the different measuring devices are causally connected on the grounds that they are not distinct. One has instead a single wave state extended over a large distance. I shall not pursue this line, partly because of the amount of physics required to do it justice and partly because I believe that the difficulties concerning causality go deeper.

A third possibility is to maintain that the states of the particles are causally dependent on the settings of the magnets (Price 1996). The interaction of electron 1 with the magnetic field of detector 1 with its particular orientation exerts a causal influence *backward* in time on whether $X1 = +1$ or $-1$ or on whether $Y1 = +1$ or $-1$. The orientations of the magnets in the other two detectors similarly exert causal influences backward in time on the states of the other electrons. Those causal influences are carried backward in time to the moment when the three electrons fly apart and there determine the states of the three electrons. Their states determined, the electrons fly apart, and the detectors measure the states. The measurement outcomes depend on the orientation of the magnets and the states of the particles. The states of the electrons depend on the orientation of the magnets and the physical laws that constrain what sets of states are possible.

Notice that in these circumstances "bilking" (p. 72) is impossible. There is no way *after* the effect has occurred (i.e., after the time when the particles separate) to intervene and to alter the cause (the orientation of a magnet) so as to destroy the supposed backward causal relation, because one cannot learn about the backward "effect" until the measurements are made. The fact that intervention is empty does not entail that the independence condition is not satisfied, since the independence condition is logically weaker than is the assumption that an abstract intervention is possible. But it should put us on our guard. It should also ring some alarm bells for Huw Price, who ties causation to agency.

A cartoon may be useful: Electron 1 arrives at detector one, which has its magnet oriented along the $x$ axis. The electron flips a miniature coin to determine whether its spin along the $x$ axis is up or down and conveys the result (suppose it is "up") back to an earlier conference among the three electrons that took place as they were separating. If the other electrons do not interact with measuring devices, then the result of the coin flip holds, and the only determinate spin state of the three-electron system is that the spin of electron 1 in the $x$ direction is "up." If all the electrons interact with the measuring devices, then the results they send back concerning their coin flips may not be compatible. For example, the magnets in all three detectors might be oriented along the $x$ axis and all three coins might have come up heads. In that case the three electrons at the conference draw straws and the one with the short straw must change its tentative orientation. As this heuristic picture illustrates, an account of the phenomenon that relies on backwards causation need not maintain that the backwards causation results in determinate spin states for all the electrons along both axes. Price's account does not require hidden variables, although it permits them.

This account relies on probabilistic dependencies among states that obtain in virtue of future interactions and thus abandons the claim that things cannot be correlated in virtue of future interactions. Nevertheless, the probabilistic dependencies still have a causal explanation. The states of the electrons interacting with the three detectors are correlated in virtue of the backward causal influence of their interactions with the detectors on the initial state from which they separate. There is no counterexample to the connection principle, since the particle's state is (backwards) causally dependent on the orientations (as well as forward causally dependent on the interactions among the separating electrons).

It is hard to say whether it is plausible to take the lawful dependencies between the measured properties as involving backwards causation, and in any case plausibility is a rare commodity among interpretations of quantum phenomena. For a forceful defense of this view, see Price (1996). I am inclined to doubt whether the story about backward causation adds anything to a description of the phenomena that simply denies that they can be causally explained. As the discussion in Chapter 8 showed, an apparent advantage of explanations that identify causes of what is to be explained is that they provide the "handles" as it were with which to "grab on" to the explanandum and to do something about it. Identifying causes should lead to further testable implications concerning what will happen when these handles are pulled. But this is not the case in the mixed forward-backward causal explanation of the GHZ correlations. The explanations add no testable implications (Dowe 1996 implausibly suggests otherwise). This fact is reflected in the failure of the independence condition. Although the measurements of the three devices all depend on multiple independent causal factors, they all depend on just the same factors, and so, given the connection principle, the independence condition must be violated.

This line of thought leads to a fourth reaction to these phenomena, which I am inclined to think is the right one: One should deny that the probabilistic dependencies among these measurement results have a causal explanation.[11] These phenomena show that probabilistic dependencies are not sufficient for specifically *causal* connections. They show that causal explanation and causal analysis are out of place, not that there is any mistake in relating

[11] I am here agreeing with van Fraassen (1982), though I draw no anti-realist conclusions. See also Skyrms (1984, esp. p. 254) and Papineau (1991). My argument owes a great deal to Leslie Graves, Malcolm Forster, and David Papineau.

causation to patterns of causal connection. Indeed one might turn around and invoke the independence condition to criticize Price's effort to offer a causal explanation involving backwards causation. These quantum phenomena reveal the limits of causation and causal explanation rather than problems within theories of causation (Teller 1989). The cardinal feature of causal explanation, which is the identification of separable and independent factors that are in principle instruments for controlling the explanandum, is irretrievably missing.

But if the GHZ correlations are not causal, then causal connections are not, as I have been pretending, the same thing as nomological or nonaccidental connections. Since the GHZ correlations are clearly nomological, there are nomological connections among things that are not related as cause and effect or only as effects of a common cause. Either the connection principle, **CC**, must be revised, or the interpretation of causal connection as any sort of nonaccidental link must be dropped. Pursuing these difficulties demands a revision of the independence theory of causal priority.

One might wonder whether such a revision is possible. If one grants that there is some species of causal connection among the measurements, it would appear that everything causally connected with any one of the measurements would have a causal connection to the other measurements and, by **CP**, the measurement events would all be causes of one another. This new kind of causal connection would collapse into causation. This result is completely unsatisfactory both because of the physical difficulties it raises concerning nonlocality and because the relations among the measurement events lack the independence that gives causal language its point.

## 12.7 Revising CP

The discussion of quantum mechanical phenomena shows that one cannot consistently hold **CC** and maintain that probabilistic dependency operationalizes the notion of a causal connection. In the EPR phenomena either the rough link between probabilistic dependency and causal connection is broken, or the connection principle, **CC** must be revised.

The latter path is the right one. The connection principle needs to be revised as follows:

> **CC′** (*Revised connection principle*) *a* is nomically connected to *b* if and only if *a* and *b* are distinct and either *a* causes *b*, *b* causes *a*, *a* and *b* are effects of a common cause, or *a* and *b* are mutually dependent.

To avoid misleading terminology, I now speak of "nomic connections" rather than "causal connections." The intuitive idea of a nomic connection is still that of necessary or lawful connection. I called the "other" category "mutual dependence." (iv) is meant to be a residual category capturing nomic connections that are not specifically causal.

The independence condition needs no revision, though its references to causal connections should be read as references to which I am now calling "nomic connections." Restricted to circumstances in which there is no mutual dependence, **CC′** and **I** enable one to prove all the exact same theorems, and the revised theory presented here is simply a generalization of the original version.

With the help of a new axiom governing mutual dependence, this generalization of the independence theory permits one to prove variants of all the theorems proven in this book (see §12.4*).

> **SSMD** (*Strong symmetry of mutual dependence*) Mutual dependence is symmetrical, and if *a* and *b* are mutually dependent, then everything nomically connected to *a* and distinct from *b* is nomically connected to *b*.

Strong symmetry implies that if *a* and *b* are mutually dependent, then they will have the same causes and effects. This seems justifiable in the case of the GHZ phenomena. There are causes of the measurement events – for example, the setting of the magnets, but these are causes of all the events. It might be questioned whether the measurement events all have the same effects. Suppose that the pointer on the first measuring device is connected to a switch in a circuit including a light bulb, so that when the first device measures spin-up, the circuit is closed and the bulb goes on. It seems plausible to say that the orientation of the first electron caused the bulb to go on and that the orientation of the other electrons did not. However plausible this claim might appear, I don't think it can be sustained. None of the theories of causation canvassed in this book except Hume's supports this claim.

With the revised connection principle, one does not prove exactly the same theorems. Some further changes are demanded. The most important of these is that the independence theory, **CP** itself, must be revised. The fact that *a* is nomically connected to *b* and that everything nomically connected to *a* and distinct from *b* is nomically connected to *b* is no longer sufficient for *a* causes *b* – *a* and *b* could be mutually dependent. And it is a good thing that this condition is no longer sufficient, or else one would have the disastrous implication (mentioned at the end of the last section) that the measurement events in the GHZ phenomena would be causes of one another. A correct formulation of the independence theory is:

> **CP′** (*Revised independence theory of causal priority*) *a* causes *b* if and only if *a* is nomically connected to *b*, everything nomically connected to *a* and distinct from

*b* is nomically connected to *b*, and something nomically connected to *b* is independent of *a*.

**CP′** (like **I** and **CC** or **I** and **CC′**) guarantees the asymmetry of causation.

With these revisions of the independence theory, I can in good conscience make this response to the difficulties raised by quantum mechanics. The separate measurements in the GHZ thought experiment are nomically connected, but they are not related as cause and effect or as effects of a common cause. Causal explanation is out of place because the separate measurements do not have independent causes. The modularity that makes causal explanation appropriate does not obtain.

## 12.8 Theories of Causation and Theories of Causal Inference

The claims in this chapter and this book are not intended to provide the outlines of a practical theory of causal inference. If one does not measure all the variables (and one never does!), the independence theory can be an unreliable basis for inferences concerning causation. The correlations implied by distant nomic connections will be too small to detect in manageable samples. Algorithms based on these connections between causality and probabilistic dependencies would be inefficient. Nor should the work of SGS or Pearl and his collaborators be regarded as providing an alternative metaphysical theory of causation. In my criticisms of Papineau's proposals, I have already shown the problems that arise if one attempts to reduce causation to claims about conditional probabilistic dependencies.

The independence theory of causal asymmetry complements current work on causal inference. This book explores the fundamental relations between causation, nomic connections, and probabilistic dependencies. The work on causal inference shows how the relations between causation and probabilistic dependence might be made practical. This book is about the metaphysics of causation, not about statistical methods of inferring causal structure from knowledge of probabilistic dependencies.

# 12*

# Probability Distributions and Causation

## 12.1* Causation and Causal Connection: A Graphical Exposition

The relations between causation and causal connection (and ultimately probabilistic dependencies) can be clarified with the help of the three kinds of graphs. Figure 12.3a, 12.3b, and 12.3c show respectively the normal, path, and connection graphs that correspond to the causal relations among the salt concentrations of the five salt basins in figure 6.3, which is reproduced in figure 12.3d. To each normal graph, there corresponds a single path graph, but a path graph may correspond to more than one normal graph. The connection principle[1] implies that for each path graph, there is a unique connection graph: For all vertices *u* and *v* in a connection graph $G_c$, there is an undirected edge between *u* and *v* if and only if in a corresponding path graph $G_p$ with the same vertices there is an edge from *u* to *v* or from *v* to *u* or there is an edge from some vertex to both *u* and *v*. If one substitutes the word "path" for the word edge, this same claim defines the unique connection graph corresponding to every normal graph.[2]

If **I** (or $\mathbf{I_g}$) is true, then there is a unique path graph corresponding to each connection graph that correctly represents the causal relations: For all vertices *u* and *v* in a path graph $G_p$, there is a directed edge from *u* to *v* if and only if in the corresponding connection graph $G_c$ with the same vertices there is an undirected edge between *u* and *v* and for all vertices *w* other than *v* if there is an edge in $G_c$ between *w* and *u*, then there is an edge between *w* and *v*. Notice that the causal relations depicted in figure 12.3 do not satisfy $\mathbf{I_g}$, though if one filled in some of the further causal factors upon which $x_3$ and $x_5$ depend, one would satisfy $\mathbf{I_g}$. Let $u_3$ and $u_5$ represent independent causal factors upon which $x_3$ and $x_5$ depend. One would then have the connection graph shown in figure 12.4a.

[1] I shall continue to talk about **CC** and **CP** rather than the revised versions **CC′** and **CP′** until §12.4*.

[2] I am as before using "*v*" to refer to a vertex as well as to what the vertex represents.

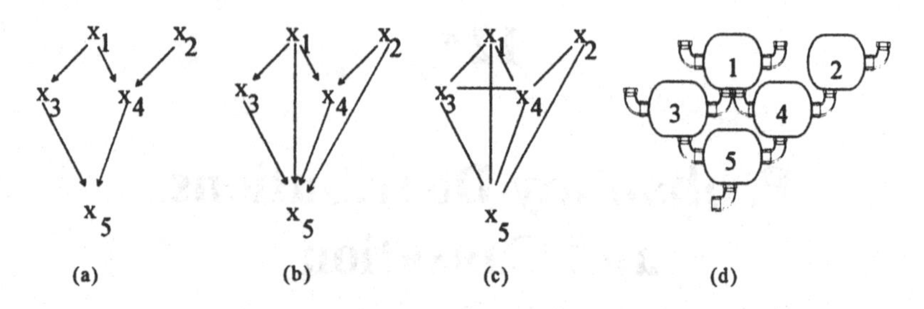


**Figure 12.3:** Normal, path, and connection graphs

In the corresponding path graph, shown in figure 12.4b, there is no directed edge between $x_1$ and $x_2$ because there is no unconnected edge between them in $G_c$. One can conclude that $x_1$ causes $x_3$ and not vice versa, because there is an edge between $x_3$ and some vertex ($u_3$) that does not have an edge to $x_1$, while everything with an edge to $x_1$ has an edge to $x_3$. There is no directed edge in $G_p$ from $x_3$ to $x_4$ because in $G_c$ there is an undirected edge between $u_3$ and $x_3$ and no edge between $u_3$ and $x_4$. Similarly, there is no directed edge in $G_p$ from $x_4$ to $x_3$ because there is an undirected edge between $x_2$ and $x_4$ and no undirected edge between $x_2$ and $x_3$. Since the question of whether there is a directed edge in $G_p$ between any ordered pair of vertices is determined by the pattern of undirected edges in a connection graph with the same vertices, there is a one-to-one mapping between connection graphs and path graphs. This is exactly what **CP′** or **CP′$_g$** says.

## 12.2* Conditional Probabilities and Probabilistic Dependencies in Possible Worlds

Compare with the counterfactual operationalizing assumption, **COA**, with the operationalizing assumption, **OA**, reformulated in terms of variables:

> **OA$_g$** (*Operationalizing assumption*) Variables *x* and *y* are connected if and only if they are distinct and probabilistically dependent in the background circumstances.

> **COA** (*Counterfactual operationalizing assumption*) If the values of the variables in some set *S* were never to vary, then *x* and *y* would be nomically connected if and only if *x* and *y* are distinct and in the background circumstances probabilistically dependent conditional on the fixed values of the variables in set *S*.

If one takes **OA$_g$** to be true in all possible worlds similar to the actual world, then **OA$_g$** and the following assumption concerning probabilities and possible worlds imply **COA**:

> **PDPW** (*Probability distributions and possible worlds*) The actual probability distribution of *y* conditional on $x = x^*$ is equal to the probability distribution *y* would have if the value of *x* never varied from $x^*$.

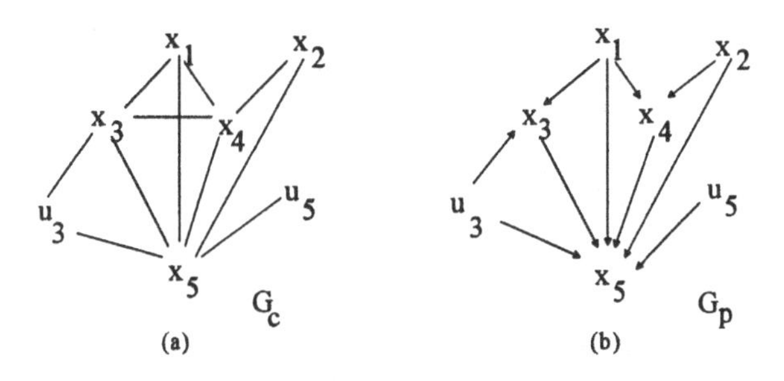


**Figure 12.4:** A connection graph satisfying independence

**PDPW** relates conditional probabilities to what unconditional probabilities would be. It says, for example, that if Pr(C/A) = *p*, then if *A* were always instantiated, Pr(C) would be *p*. It does not define the probability of the conditional: "if *A* were always instantiated then *C* would be instantiated," and it is thus not subject to David Lewis's trivialization results (1976, pp. 79–83). Lewis argues as follows: Let Pr(**A**→**C**) be the probability of the conditional, "If **A** were the case, **C** would be the case."[3] Suppose that one equates the probability of conditionals with conditional probabilities.

1. Pr(**A** →**C**) = Pr(**C**/**A**).

If Pr(**A**&**C**) and Pr(**A**&~**C**) are both positive, then

2. Pr(**A**→**C**/**C**) = Pr(**C**/**AC**) = 1
3. Pr(**A**→**C**/~**C**) = Pr(**C**/**A**&~**C**) = 0.

For any **D** the calculus of probabilities says

4. Pr(**D**) = Pr(**D**/**C**)·Pr(**C**) + Pr(**D**/~**C**)·Pr(~**C**).

If **D** is the proposition **A**→**C**, then

5. Pr(**A**→**C**) = Pr(**A**→**C**/**C**)·Pr(**C**) + Pr(**A**→**C**/~**C**)·Pr(~**C**).

If one substitutes for the conditionals using 1,2, and 3, one can deduce

6. Pr(**C**/**A**) = Pr(**C**).

Thus one has deduced that if Pr(**A**&**C**) and Pr(**A**&~**C**) are both positive, then **A** and **C** are probabilistically independent. Since this is absurd, it cannot be the case that Pr(**A**→**C**) = Pr(**C**/**A**).[4]

**PDPW** does not permit one to derive any such absurdity, because it does

[3] Lewis is assigning probabilities to propositions, rather than to properties or events types.

[4] The problem that Lewis points out has nothing special to do with counterfactual conditionals. It arises in just the same way if one equates conditional probabilities with probabilities of any proposition whatsoever.

not equate conditional probabilities to probabilities of any propositions. If *C* were always instantiated, Pr(C/A&C) would of course be 1, Pr(C/A&~C) would be zero, and Pr(C/A) would be Pr(C). But there is no paradox here.

**Theorem 12.1: PDPW** and the assumption that $\mathbf{OA_g}$ is true of all similar possible worlds imply **COA**.

Proof: Suppose that if the value of *x* were not to vary, then the values of *y* and *z* would be connected. Given $\mathbf{OA_g}$, *y* and *z* would be probabilistically dependent. Given **PDPW**, *y* and *z* are probabilistically dependent conditional on *x*.

To prove the converse, suppose *y* and *z* are probabilistically dependent conditional on *x*. By **PDPW**, *y* and *z* would be unconditionally probabilistically dependent if the value of *x* were not to vary. By $\mathbf{OA_g}$, the values of *y* and *z* would be connected.

I do not know how to evaluate **PDPW** directly, and I am unsure of how much weight to place on this derivation of **COA**. Other arguments in defense of **COA** are that its implications appear to be correct, that it gives counterfactuals (as interpreted in chapter 6) the right empirical content, that it links the theory articulated here to practical methods of causal inference such as those developed by SGS and Pearl, and that it enables one to interpret otherwise puzzling results that Pearl and SGS prove.

## 12.3* Independence Theory and Statistical Methods of Causal Inference

With the help of the sufficiency condition **SUF** defined in chapter 8 (p. 183), **CM** and **F** can be deduced from elements of the independence theory ($\mathbf{CC_g}$ and $\mathbf{I_{gs}}$) and the counterfactual operationalizing assumption (**COA**).

**Theorem 12.2: SUF, $\mathbf{CC_g}$, $\mathbf{I_{gs}}$,** and **COA** entail **CM** and **F**.

Proof: For any variable *x* in *V* suppose that all the direct causes of *x* in *V* had constant values. The only source of variation in *x* would be its own independent source, which (by $\mathbf{I_{gs}}$) exists, and, given **SUF**, *x* would not be an effect of any variable in *V*, nor would *x* and any variable in *V* be effects of a common cause. *x* would be exogenous. For any variable *y* $\mathbf{CC_g}$ implies that *x* would be connected to *y* if and only if *x* causes *y*. **COA** implies that *x* and *y* would be connected if the direct causes of *x* had constant values if and only if they are probabilistically dependent conditional on all the direct causes of *x*. So *x* causes *y* if and only if they are probabilistically dependent conditional on all the direct causes of *x*.

Conversely, the fundamental axioms of the causal modeling literature imply the independence condition.

**Theorem 12.3: CM, F, $\mathbf{CC_g}$, $\mathbf{T_g}$,** and **COA** imply $\mathbf{CP_g}$

Proof: Since $\mathbf{CC_g}$ and $\mathbf{T_g}$ imply the necessary condition in $\mathbf{CP_g}$ by themselves (theorem 4.1), all that needs to be proved is that if *x* and *y* are causally connected and everything causally connected to *x* and distinct from *y* is causally connected to *y*, then *x* causes *y*. If everything causally connected to *x* and distinct from *y* is causally connected to *y*, then all causes of *x* that are distinct from *y* are causally connected to *y*. So *x*'s independent source of variation must cause *y* and, conditional on all the direct causes of *x* (except its own independent source of variation), *x* and *y* are probabilistically dependent. Hence by **CM** and **F**, *x* causes *y*.

This last theorem is, however, scarcely worth proving, because in the causal models that employ **CM** and **F**, it is almost always assumed that each variable has its own independent source of variation, which directly implies $\mathbf{I}_s$ and hence **I**.

One can use **COA** and the propositions established in previous chapters to argue directly for conclusions that Pearl and SGS derive from **CM** and **F**. For example,

**Theorem 12.4: SUF, $\mathbf{CC}_g$, $\mathbf{I}_{gs}$,** and **COA** imply that *x* and *y* are adjacent if and only if they are probabilistically dependent conditional on every subset of the sufficient set of variables *V* that does not contain *x* or *y*.

Proof: First assume that *x* and *y* are probabilistically dependent on every subset of *V* that does not contain *x* or *y*. Suppose that nothing except *x* and *y* were to vary. By **COA**, *x* and *y* would be connected. By **SUF**, they would not be connected only as effects of a common cause, and (by $\mathbf{CC}_g$) they must be related as cause and effect. Since nothing varies in this possible world, the causal relation between *x* and *y* must be direct.

Conversely, assume that *x* and *y* are adjacent. If the values of variables in any subset of *V* that does not contain *x* and *y* were unchanging, *x* would still cause *y*, and *x* and *y* would still be connected. Given **COA**, *x* and *y* are probabilistically dependent conditional on any subset of *V* that does not contain *x* and *y*.

Given the previous theorem, there is nothing surprising about this one, but the theorem and its proof shows that elements of the independence theory coupled with the counterfactual operationalizing assumption might actually be useful in arriving at Pearl's and SGS's results.

Theorems concerning causation and actual or possible causal connections translate into theorems concerning the relations between causation and unconditional or conditional probability distributions. One can thus use the arguments of previous chapters to reinforce and extend Pearl's and SGS's results, and one can use their results to reinforce and to extend the arguments I have made concerning causation and nomological connections.

## 12.4* The Revised Independence Theory

This section proves that variants of the principal theorems of the previous chapters continue to hold when **CC** is replaced by **CC′** and one adds **SSMD**. In particular I shall prove variants of what one might regard as the six central theorems of this book: theorems 4.5, 4.6, 5.6, 6.5, and 11.2.

**Theorem 12.5: T, CC′, I, and SSMD** entail **CP′** (revising theorem 4.5, p. 84: **T, CC, and I** entail **CP**).

Proof: Suppose *a* causes *b*. I need to prove three things: (i) *a* is nomically connected to *b*, (ii) everything nomically connected to *a* and distinct from *b* is nomically connected to *b*, and (iii) something is nomically connected to *b* and independent of *a*. (i) is trivial from **CC′**. To prove (ii) use the proof of theorem 4.5 and note that there is one new way that some event *c* can be nomically connected to *a*, by being mutually dependent on *a* and by **SSMD**, *c* is nomically connected to *b*. Finally, by **I**, *b* has a cause that is independent of *a* and so there is something nomically connected to *b* and independent of *a*.

Conversely suppose (i), (ii), and (iii) hold. (ii) and **I** imply that *b* does not cause *a* and that *b* and *a* are not nomically connected only as effects of a common cause. (iii) and **SSMD** imply that *a* and *b* are not mutually dependent. So *a* causes *b*.

**Theorem 12.6: T, CC′, SSMD,** and **CP′** imply $\mathbf{I_w}$: If *a* causes *b* then *b* has a cause that is independent of *a* (revising theorem 4.7, p. 85: **T, CC, A,** and **CP** imply **I**).

Proof: Given **CP′**, if *a* causes *b*, then *a* and *b* are nomically connected and something *c* is nomically connected to *b* and not to *a*. *c* cannot be an effect of *b* or by **T** and **CC′** *c* would be nomically connected to *a*. *b* and *c* cannot be mutually dependent or else by **SSMD** and **CC′** *c* would be nomically connected to *a*. So *c* must be a cause of *b* or an effect of some other cause *d* of *b*. In the latter case *d* cannot be nomically connected to *a* for if it were *c* would either be an effect of *a*, an effect of a cause of *a*, or (by strong symmetry) *a* and *d* would be mutually dependent. So if *a* causes *b* then *b* has a cause that is independent of *a*.

Theorem 12.6 is weaker than the theorem it replaces. **T, CC′** and **CP′** do not imply **I**, because **T, CC′** and **CP′** do not exclude the possibility that effects of a common cause have all the same causes in common.

**Theorem 12.7:** Given $\mathbf{CC'_g}$, and $\mathbf{NIC_g}$, $\mathbf{I'_{gs}}$ is entailed by $\mathbf{DI'_g}$ and $\mathbf{PI_g}$ (revising theorem 5.6, p. 109: Given $\mathbf{CC_g}$, and $\mathbf{NIC_g}$, $\mathbf{I_{gs}}$ is entailed by $\mathbf{DI_g}$ and $\mathbf{PI_g}$).

I will not write out the obvious type-level restatement of **CC′**. The strong independence condition requires one additional clause:

$\mathbf{I'_{gs}}$ (*Revised strong independence condition*) In circumstances *K*, every nonexogenous variable *y* has some additional cause *z* such that for all variables *x*,

1. If *y* does not cause x and *x* and *y* are not mutually dependent and there is a causal path that does not go through *z* from *x* to *y* or from some cause of *x* to *y*, then *x* and *z* are causally independent.
2. If *y* causes *x*, then there is no path that does not pass through *y* from *z* to *x* or from any cause of *z* to *x*.
3. If *x* and *y* are mutually dependent, there is no directed or undirected path that does not pass through *y* between *z* and *x* or between any cause of *z* and *x*.

Similarly, the revised version of the definition of an intervention $\mathbf{DI'_g}$ adds a clause stating that the intervention variable is not mutually dependent on any of the variables. The proof of theorem 12.7 then follows the same lines as the proof of theorem 5.6.

Proof of theorem 12.7: Given $\mathbf{PI}_g$, for every *y*, there is some intervention variable *z*. Given $\mathbf{NIC}_g$, there must be some other variable *x* that causes *y*. Given $\mathbf{DI'}_g$ and $\mathbf{CC'}_g$, *z* is not nomically connected to any variable connected to *y* that is not an effect of *y* and is nomically connected to effects of *y* only via causing *y*. $\mathbf{I'}_{gs}$ thus follows.

Given theorems 12.5 and 12.7 one can easily prove that $\mathbf{DI'}_g$, $\mathbf{PI}_g$, $\mathbf{CC'}_g$, $\mathbf{NIC}_g$, and $\mathbf{T}_g$ entail $\mathbf{CP'}_g$. Here are two more conditions:

**CMD** (*Counterfactuals and mutual dependence*): If *a* and *b* are mutually dependent, then they are counterfactually dependent on one another.

**L′** (*Lewis's theory revised*): *a* causes *b* if and only if *b* counterfactually depends on *a* and *a* does not counterfactually depend on *b*.

**Theorem 12.8: CC′, I, DC, SIM, CMD,** and **CDCC** entail **L′** restricted to circumstances in which there are no multiple connections (revising theorem 6.5, p. 137: **CC, I, DC, SIM,** and **CDCC** entail **L** restricted to circumstances in which there are no multiple connections).

Proof: First I shall prove that these conditions imply the sufficient condition in **L′**, that if *b* counterfactually depends on *a* and *a* does not counterfactually depend on *b*, then *a* causes *b*. Suppose (i) *b* counterfactually depends on *a* and (ii) *a* does not counterfactually depend on *b*. Then by **CDCC** *a* and *b* are nomically connected. Given **CC′** there are four possible ways *a* and *b* can be nomically connected, and I shall prove that *a* causes *b* by eliminating the other three possibilities. **I** says that if *b* causes *a* or if *a* and *b* are nomically connected only as effects of a common cause, there will be some other cause *c* of *a* causally independent of *b*. Given **CDCC**, *b* and *c* will also be counterfactually independent. Clause 2 of **SIM** then implies that for any non-*a* world without *b*, there will be a non-*a* world without *c* that is at least as similar to the actual world. So if *b* causes *a* or if *a* and *b* are nomically connected only as effects of a common cause, *b* is not counterfactually dependent on *a*. So *b* does not cause *a*, and *a* and *b* are not causally connected only as effects of a common cause. **CMD** and the assumption that *a* is not counterfactually dependent on *b* establish that *a* and *b* are not mutually dependent. Hence *a* causes *b*.

To prove the necessary condition, suppose that *a* causes *b* and there is no preemption or overdetermination. Then, by **DC** *a* is necessary in the circumstances for *b*. Suppose (counterfactually) that *a* does not occur, and consider the possible worlds in which *b* occurs anyway. In some of these worlds *b* occurs miraculously, but these worlds will be very unlike the actual world. In other possible worlds *b* occurs because of some causes. Those causes cannot be preempted actual causes or causal overdeterminers, because by assumption there is no preemption or overdetermination. They cannot be actual causes of *a*, since there are no multiple connections. So whatever causes *b* must be some new occurrence or nonoccurrence or something that does not cause *a* or *b* in the actual world, but which, as a consequence of some new law, causes *b* in this possible world. By clauses 3 and 4 of **SIM**, possible worlds with *b* will be less similar to the actual world than are some possible worlds without *b*. So *b* is counterfactually dependent on *a*. *a* cannot be counterfactually dependent on *b*, because (by **I**) *b* must have another cause *c* that is independent of *a*, and given **CDCC**, *a* and *c* will also be counterfactually independent. Clause 2 of **SIM** then implies that for any non-*b* world without *a*, there will be a non-*b* world without *c* that is at least as similar to the actual world. So if *a* causes *b*, *a* is not counterfactually dependent on *b*.

**Theorem 12.9: SUF, $CC'_g$, $I'_{gs}$, and COA** entail that *x* causes *y* if and only if *x* and *y* are probabilistically dependent conditional on all the direct causes of *x* and *x* and *y* are probabilistically independent conditional on all the direct causes of *y* (revising theorem 12.2, p. 258: **SUF, $CC_g$, $I_{gs}$,** and **COA** entail **CM** and **F**).

Proof: For any variable *x* in *V* suppose that all the direct causes of *x* in *V* had constant values. The only source of variation in *x* would be its own independent source, which (by $I'_{gs}$) exists. Given **SUF**, *x* would not be an effect of any variable in *V*, nor would *x* and any variable in *V* be effects of a common cause. For any variable *y* in *V* $CC'_g$ implies that *x* would be nomically connected to *y* if and only if *x* causes *y* or *x* and *y* are mutually dependent. But *x* and *y* cannot be mutually dependent because they are probabilistically independent conditional on all the direct causes of *y*. So *x* causes *y*. Conversely, suppose that *x* causes *y*. In a possible world where all the causes of *x* had fixed values, *y* would still be nomically connected to *x*, but (given $I'_{gs}$) *y* would not be nomically connected to *x* in a possible world in which all the causes of *y* had fixed values. **COA** says *x* and *y* would be nomically connected if the direct causes of *x* had constant values and would not be nomically connected if the direct causes of *y* had constant values if and only if they are probabilistically dependent conditional on all the direct causes of *x* and probabilistically independent conditional on all the direct causes of *y*. So *x* causes *y* if and only if they are probabilistically dependent conditional on all the direct causes of *x* and probabilistically independent conditional on all the direct causes of *y*.

# 13

# Complications and Conclusions

In the discussion above I postponed addressing the problems of preemption and overdetermination as well as the difficulties raised by arbitrary event fusions and complex facts. In this chapter, I confront these problems and then end by drawing overall conclusions in which the pieces examined in the previous chapters will, I hope, find their places.

## 13.1 Overdetermination

Suppose an event *e* is causally overdetermined by two events, *a* and *c*. For example, suppose *a* and *c* are simultaneous. *a* involves pushing button one, which closes a circuit and causes a light to go on. *c* involves pushing button two, which closes the same circuit at exactly the same time and causes the same light to go on. *e* is the light going on. If *a* had not occurred, *c* alone would have caused *e*. If *c* had not occurred, *a* would have caused *e*. As things are, however, it is not the case that only one of *a* or *c* caused *e*. Either they both caused *e* or neither caused *e*. Each of them is a "causal overdeterminer" of *e*.

What is this relation of causal overdetermination? The most common approach in the literature, which I will endorse, denies that causal overdeterminers are causes. If *a* and *c* are causal overdeterminers of *e*, then neither *a* nor *c* is a cause of *e*. Instead one takes the instantiation of the disjunctive property *A* or *C* at the relevant place and time as a cause of *e*. Authors who have defended this approach have devoted considerable efforts to lessening what they take to be the counterintuitiveness of denying that causal overdeterminers are causes (for example, Mellor 1995, p. 102). Yet intuition is ambiguous. There also seems to be something right about the claim that what causes *e* is the occurrence of one-or-the-other of *a* or *c* rather than *a* or *c* or both (Berofsky 1977, p. 107).

Suppose the independence theory of causal priority (**CP** or the revised version **CP'**) is correct, and suppose that if *a* overdetermines *e*, then *a* is lawfully connected to *e*. (This supposition would require that the connection principle be revised again to admit yet another category of nomological connection.) Assume also that the relation "is a cause or a causal overdeterminer" is transitive and that if *a* causes or causally overdetermines *e* and *a* causes or causally overdetermines *d*, then *e* and *d* are lawfully connected. Then everything nomically connected to *a* and distinct from *e* would be nomically connected to *e*, and something would be nomically connected to *e* and independent of *a*, and it would follow from **CP** and from **CP′** that causal overdeterminers were causes.

One could accept the conclusion that overdeterminers are causes. One would then need no separate treatment of causal overdetermination, though there is, of course, nothing to prevent one from noticing when causal overdetermination takes place.[1] The conclusion is, however, intuitively uncomfortable, and it has the decisive disadvantage for me that it would require rewriting this whole book. Few of the claims explored in this book would remain true, without qualification or restatement. For example, the account of deterministic causation, **DC**, obviously no longer holds. The theorem that proves that effects are counterfactually dependent on their causes when there are no multiple connections no longer goes through, because an intervention with respect to a causal overdeterminer would make no difference to the overdetermined outcome. Explanations that cite only one of several overdeterminers will be flawed. In the circumstances, which include the presence of the other overdeterminers, there will be no probabilistic dependence between an overdeterminer and its effect. I had better maintain that causal overdeterminers are not causes and that causal overdetermination is not a species of nomic connection.

Let us then reject the supposition that if *c* is a causal overdeterminer of *e*, then there is a necessary connection between *c* and *e*. My intuitions are too corrupted to provide any guidance on whether this denial is (as it seems to me) plausible. In one sense of the term "necessary," there is certainly no necessary connection between an effect and each of its causal overdeterminers. Moreover, in the conditions (which include the presence of tokens of kind **c**), there is no lawful probabilistic dependence between **a** and **e**. If one denies that causal overdetermination is a species of nomic connection, then the conditions and theorems stated and proved above still stand – even claims such as **DC**, which state that causes must be necessary in the circumstances for their effects. For causal overdeterminers are not causes.

One still needs an account of the relation of causal overdetermination. In general, one should say if *a* causally overdetermines *e*, then some trope *d* explains *e*, where *d* is an instantiation of a disjunction of properties in which a property *A* instantiated by *a* is essential. This necessary condition is not sufficient: Event *d* can instantiate a disjunctive property without instantiations of the disjuncts being causal overdeterminers. One can also say that

[1] Spohn's account of direct causation permits direct overdeterminers to be causes (1991, p. 176).

the members of a set of $j$ events $c_1, \ldots, c_j$ are causally determiners of an event $e$ if the instantiation of $C_1$ or . . . or $C_j$ at the time and place where $c_1, \ldots, c_j$ causes $e$ and if any of the members of the set had occurred without any of the others occurring, it would have caused $e$. More needs to be said, including some account of how causal relations among complex events (like the instantiation of $C_1$ or . . . or $C_j$) are tied to the relations among components.

## 13.2 Preemption

Suppose that $c$ causes $e$ and that $b$ is a backup cause of $e$ that would have caused $e$ if $c$ had not. Everything nomically connected to $c$ is nomically connected to $e$, while something is nomically connected to $e$ and not to $c$, and so (according to **CP′** or **CP**), $c$ is a cause of $e$. It is not, however, the case that everything nomically connected to $b$ is nomically connected to $e$. Indeed unless $b$ bears some other relation to $e$ apart from being a backup cause, *everything* nomically connected to $b$ is independent of $e$. So obviously $b$ does not cause $e$.

This quick resolution of the problems of preemption seems a cheat: One first judges that $b$ does not cause $e$ and on that ground concludes that $b$ is not nomically connected to $e$. Then, having established that $b$ is not nomically connected to $e$, one concludes that $b$ is not a cause of $e$. If the basis for one's denial that $b$ is nomically connected to $e$ is one's judgment that $b$ does not cause $e$, then this account provides no justification for the conclusion that $b$ does not cause $e$ and no way to determine whether $b$ causes $e$. At most it shows that **CP** does not falsely imply that preempted causes are genuine causes.

Moreover, these apparent trivialities are questionable. I have emphasized the association between nomic connections and probabilistic dependencies. Since it appears that there are often probabilistic dependencies between the properties instantiated by backup causes and those instantiated by their preempted effects, there seems to be powerful evidence that there is after all a causal connection between backup causes and the effects they do not cause, but would cause if not preempted. Such nomic connections would permit one to generate counterexamples to the independence theory of causal asymmetry and to the connection principle.

In response, one might maintain that *in the background circumstances, in which the preempting cause is present*, there is no probabilistic dependence between the preempted cause and the effect, and hence there is no conflict between taking the existence of a probabilistic dependence as an almost sufficient condition for a nomic connection and denying that preempted causes are causally connected to the effects. There is, however, an obvious problem. On just the same grounds one can maintain that *in the circumstances* in which the preempted cause is present, there is no probabilistic dependency between events like the preempting cause and the effect either. If the existence of a probabilistic dependence were necessary for the existence of a nomic connection, this would be disastrous. But I have denied that correlations are necessary for nomic connections and so need not deny that the preempting cause really is a cause.

But this leaves us back where we started, for the basis for the claim that the preempting cause is really the cause, while the preempted "cause" is no cause at all, apparently does not lie in any facts about probabilistic dependencies, conditional or unconditional. What does it rest on?

I think that David Lewis gives almost the right answer: There is a causal chain leading from the real cause to the effect, while the causal chain from the preempted cause never gets to the effect. For Lewis the links in a causal chain are forged by counterfactual dependence. In the case of the two enemies, one of whom poisons the water in the reserve canteen and the other who drills a hole in it, the death does not counterfactually depend on either the poisoning or the drilling, since it would have occurred if the poisoning had not occurred or if the drilling had not occurred.[2] Yet the death is counterfactually dependent on the reserve can being empty, which is counterfactually dependent on drilling the holes, so the drilling but not the poisoning counts as a cause of the death.

Lewis's construal of the links in causal chains gets him into trouble, however, in cases of what he calls "late preemption." Suppose you and I are both hunting a deer and that your shot misses because my shot takes the deer down just before your bullet reaches it. Unless one insists that events are so fragile that they could not occur at different times, no matter how slight the difference, there is no event upon which the effect counterfactually depends that is causally between my shot and the effect. So Lewis's account mistakenly implies that my shooting the deer does not cause its death. A theory of causal preemption that construes the links in causal chains to be links of counterfactual dependence thus fails (Goosens 1979).

The right way to distinguish preempting from preempted causes is to examine other features of the causal process that links the preempting cause to the effect. In the case of the reserve canteen, the poison does not cause the death because the man never drinks the poison, and we know that such poisons do not cause death if they are not imbibed. There is a causal gap between the poisoning and the death. As we follow out chains of causes and

[2] The last claim is questionable. Could *that* death have been a death by poisoning rather than a death from thirst? See pp. 51–2 and Lewis (1986d, pp. 193–212). These issues are relevant here insofar as taking events to be *extremely* fragile permits one to escape the problems of late preemption. See Lewis (1986d, p. 204).

effects beginning with poisoning the water, we never get to the man's death. From the hole-drilling, on the other hand, we can follow a causal chain all the way to the death. If my shot penetrates the deer's body, and yours does not, then my shot, not yours, kills the deer.[3]

In distinguishing in this way between the preempting and the preempted cause, one is not necessarily assuming that causes are always contiguous to their immediate effects. All one assumes is that the poison doesn't kill the man if the man doesn't drink it, and that your bullet does not kill the deer if it never touches the deer.[4] To cope with the problems of preemption, one relies on context-specific knowledge concerning how particular causes operate. The independence theory does not need to be modified. Even though **CP** and **CP′** do not tell us how to distinguish the preempting cause from the preempted backup, they have no false implications. Nor should they be modified to require that direct causes be contiguous with their effects (though they could be). For the question of whether causes might sometimes act at a distance is one for science, not philosophy, to decide. The fact that we use our knowledge that *poison* does not act at a distance to judge that the poisoning did not cause the death does not commit us to the conclusion that no causes can act at a distance.

Without spatiotemporal contiguity, questions about which is the real cause and which is only the backup become much more delicate. Suppose that flipping a special switch in New York or flipping another special switch in Chicago each can cause a light to go on some time later in Moscow without any chain of causal intermediaries. With knowledge of the time delay, one may be able to tell which of the two flippings caused the light to go on. For example, if it takes four minutes for these special switches to have their effect and the New York switch is flipped four minutes before the light goes on, while that Chicago switch is flipped three minutes before, then flipping the New York switch is the cause, while flipping the Chicago switch is only a backup. But if all one knows is that the light goes on between one and ten minutes after the special switch is flipped, then (as I have told this hypothetical story) there is no way to say which is the genuine cause and which the backup. This is not an objection to my claim that problems of preemption are to be dealt with by specific knowledge of how causes operate, including particularly knowledge of spatiotemporal relations, because in this case, there is in fact no basis for judging that one of the flippings is the real cause.

[3] I have adapted this way of dealing with the problem of preemption from Menzies (1996, esp. pp. 110–11).

[4] Or (as Christopher Hichcock pointed out to me) your bullet does not startle the deer over a cliff or dislodge a rock that falls on its head or . . . .

## 13.3 Event Fusions

I have maintained that there are causal relations among events in virtue of a certain kind explanatory relation among causally relevant aspects or tropes. When event *c* causes event *e*, some trope that is part of *c* instantiates a property that has a lawful relationship to a property instantiated by a trope that is a part of *e*. Similarly when fact **F** causally explains fact **G**, **F** entails the existence of some trope located at the zone with which **F** is concerned that instantiates a property that has a lawful relationship to a property whose instantiation at the proper location is entailed by **G**. So event *c* can cause event *e* even though *c* and *e* contain parts that are not causally relevant, and fact **A** can causally explain fact **B** even though **A** and **B** mention some irrelevancies. One also needs to say something about the causal relations between *a* and *b* where *a* and *b* are not natural, uncompounded events, but more complicated sorts of things. As we have just seen, an account of causal overdetermination that regards causes of overdetermined events as in some sense disjunctive also places demands on a theory of complex event relata.

My strategy in this book has been to take theories of causation as theories about causal relations among natural events described in terms of their relevant aspects, and then to state how causal relations among other relata derive from the explanatory relations among the tropes they contain. Ideally one should be able to state a set of recursive rules about how causal relations among complex things derive from causal relations among their parts. This is a very difficult thing to do, and it is not obvious that there is any need for a theory that will work for arbitrary complexes.

In the following, when I write phrases including event or trope names and logical operators such as "*a* or *b*" or "*c*&~*d*," I shall be referring to the event or trope that consists of the instantiation at the location of the components of the appropriate logical compound of the properties instantiated by the natural or simple tropes. So "*a* or *b*" refers to the trope consisting of the instantiation of *A* or *B* at the place where *a* and *b* occur, and "*c* & ~*d*" refers to the instantiation of the property *C* and of some property that is instantiated when *D* is not instantiated at the place and time where *c* and *d* occur. I shall also assume that in each of the claims the events named by single letters are distinct from one another.

It is hard to generate any acceptable principles of *decomposition* at all. Consider, for example,

1. If *a*&*b* cause *e*, then *a* causes *e* or *b* causes *e*; if a cause of *e* is a conjunction, one of its conjuncts must cause *e*.
2. If *a* causes *e* or *f*, then *a* causes *e* or *a* causes *f*; if something causes a disjunction, it must cause one of the disjunctions.

These principles might seem plausible, but the first is obviously false, and given the views of probabilistic causation defended in chapter 9, so is the second. Suppose *a* is drinking half a cup of hemlock at (*s*, $[t_1–t_2]$), *b* is drinking half a cup of hemlock at (*s*, $[t_2–t_3]$), *a*&*b* is drinking a whole cup of hemlock at (*s*, $[t_1–t_3]$), and suppose that a half cup, unlike a whole cup, is not a fatal dose. So the first principle of decomposition is false. To generate a counterexample to the second principle, suppose that whether a coin lands heads or tails is genuinely indeterministic but that it is determined that the coin will land heads or tails (and not, for example, on its edge). Then flipping the coin causes it to land heads or to land tails but does not cause either disjunct.

Fortunately, principles of decomposition are not needed. The theories considered in the chapters above tell us how elements are causally related. Now something must be said about how compounds are related. Principles of composition are needed. Here are four that seem relatively uncontroversial:

(C1) If *a* causes *e* and *c* causes *e*, then *a* & *c* cause *e*.
(C2) If *a* causes *e* and *a* causes *f*, then *a* causes *e* & *f*.
(C3) If *a* causes *e* and *a* causes *f*, then *a* causes *e* or *f*.
(C4) If *a* causes *e*, then ~*a* does not cause *e*.

The one question mark hanging over C1–C3 concerns whether their antecedents should require that the regions where the compound events occur be reasonably compact or even convex.[5] But one cannot require that causes and effects never be dispersed or disconnected, or else one has to modify the account of causal overdetermination. If some amalgam of different causal overdeterminers counts as the cause of what is overdetermined, then some causes have parts that are spatially and temporally separated. To maintain that "*a* causes *e* and *c* causes *e*" is not sufficient for "*a*&*c* causes *e*" when *a*&*c* does not occupy a connected region of space and time does not commit one to the general principle that causes and effects are never dispersed events.

C1–C4 do not get one far, and it is hard to go further without expounding falsehoods. But we have got to go further. It can be true that Margaret Thatcher's speech caused Ronald Reagan to fall asleep, even though Thatcher's speech contains spatiotemporal parts (such as those that occurred after Reagan nodded off) that did not cause Reagan to fall asleep. We need some principle that will permit *a* to cause *b* even if not all of *a* causes *b*. One possibility (suggested to me by Christopher Hitchcock) is that a particular trope (such as producing a certain monotone) is instantiated throughout the speech, even though only its instantiation during the minutes before Reagan

[5] If for all points *i* and *j* in a region, the region contains every point on a line segment from *i* to *j*, then the region is convex.

nodded off caused him to go to sleep. The whole speech then caused Reagan to go to sleep in virtue of the whole speech possessing the efficacious trope. This suggestion does not resolve all the difficulties. It can be true that the assassination of Archduke Ferdinand was a cause of World War I, even if it was not a cause of every part and aspect of the war, and even though there may be no trope instantiated everywhere and throughout the war that the assassination caused. We need some principle that permits *a* to cause *b* even if *a* does not cause all of *b*. The following two principles are the best I can do:

(C5) If *a* causes *e*, *b* is not nomically connected to *e*, and *a&b* occupies a connected or convex spatiotemporal region, then *a&b* cause *e*. A cause may contain irrelevant conjuncts and still be a cause of what the relevant conjuncts cause.

(C6) If *a* causes *e*, *a* is not nomically connected to *f*, and *e* or *f* occupies a connected or convex region of space and time, then *a* causes *e* or *f*. An effect may contain irrelevant disjuncts and still be an effect of what caused the relevant disjuncts.

Some sort of requirement of spatiotemporal connectedness or convexity is crucial lest C5 be open to obvious counterexample. Suppose *a* causes *e* and *b* is not nomically connected to *e*, while *b* causes *f* and *a* is not nomically connected to *f*. Striking a match causes it to light but is not causally connected to the door's opening. Turning the door's handle causes it to open and is not nomically connected to the match lighting. From these facts C5 should not enable one to conclude that the door's opening and the match's lighting are effects of the common cause consisting of the fusion of striking the match and turning the door handle. In fact, C5 does not license the conclusion unless striking the match and turning the door handle occupy a connected region of space and time. In a case where they do (one holds the match in the hand holding the door handle and strikes the match with the movement that turns the handle), it is arguably true that the door opening and the match lighting are effects of a common cause.

Though some sort of spatiotemporal connectedness requirement blocks this particular counterexample, I see no guarantee that there will not be others it will not block, and C5 and C6 may require revision. If they should prove to be untenable, then I think one would need to rethink the whole idea of a theory of causation. Without a principle such as C5, one could not say that Thatcher's speech caused Reagan to fall asleep. One would instead have to say that people are speaking loosely when they claim that the speech causes Reagan to fall asleep. Without C6, the claim that the assassination caused the war would have to be regarded as loose talk, too. It is not true that the assassination caused everything about the war. Only some aspects of some spatiotemporal parts of the speech caused Reagan to fall asleep, and the assassination caused only some aspects of some parts of the war. One might explain our causal language by pointing out that we have no convenient way of referring to the cause of falling asleep or to the effects of the assassination. The speech and the war are the only things that are easy to refer to. But if one could not rely on some form of C5 and C6, then it is going to turn out that causal claims are typically false. Striking the match didn't cause it to light. Only certain aspects of certain parts of the striking did. Martha's departure didn't cause Bill's depression. It only caused aspects of parts of Bill's depression. Indeed if one pursues this strategy, it is questionable whether any causal claims will survive except those that relate nothing but the relevant aspects of cause and effect.

These problems cannot be shrugged aside on the grounds that they are problems equally for all theories of token causation, because they constitute a powerful argument against offering any theory of token causation at all. Fortunately, versions of C5 and C6 are defensible.

## 13.4 Conclusion: What Is Causation?

Despite my complaints about Hume's requirements that causes precede their effects and that causation proceed via spatiotemporally contiguous chains of direct causation, the theory of causation defended in this book owes a great deal to Hume. Like Hume this theory tries to conform itself to an empiricist epistemology, even as it recognizes that our conception of causation contains elements that cannot be analyzed in empiricist terms. Hume recognized one such element and called it "necessary connection," while I acknowledge two and call them "nomic connection" (or "causal connection") and "causal priority." Like Hume, I link the notion of a nomic connection to regularities, but offer no analysis of it; and we both offer explanations for why the unanalyzable elements seem so unlike the explication we offer. Hume argued that our idea of necessary connection results from the way our psyche is programmed to react to constant conjunctions. Perceiving one of the conjuncts, we feel our minds turning toward an idea of the other, and we mistakenly take that feeling as representing something extra-mental. Hume's explication appears to satisfy strict empiricist strictures – but only because it fails to distinguish between accidental and lawlike regularities. A strict empiricist cannot draw the needed distinction between accident and law.

Something like Hume's notion of a necessary connection figures in my account as the concept of a nomic or causal connection among events. I take this notion to be a theoretical posit, an idealization of the notion of a probabilistic dependency among kinds of events. Probabilistic dependencies are, however, no more empirically respectable than laws, and the empirical credentials of the notion of a causal connection are shakier still. I offer no psychological theory for why people mistakenly believe that there are nonaccidental connections among events, because I think that this belief is a justifiable theoretical posit, not a mistake at all.

For Hume causal asymmetry is simply a matter of causes preceding their effects in time. There is no such thing as specifically causal priority and nothing beyond the notion of a necessary connection to trouble empiricist scruples. In addition to criticizing Hume's account, I have emphasized that the causal relation has *many* asymmetrical features, and I have argued that a theory of causal asymmetry ought to explain what these features have to do with one another. In attempting to meet this obligation, I have acknowledged a notion of causal priority, which seems as troubling to an empiricist as is the notion of a necessary connection. In addition to believing that there are nomic connections among events, people also believe that causes make their effects happen and not vice versa. This belief is an exaggerated metaphysical pun, which derives from the fact that people can make things happen by means of their causes. This belief, that causes asymmetrically make their effects occur, presupposes the possibility of intervention and the claim that not all the causes of a given event are nomically connected to one another. Independence is the rational kernel to the belief in necessitation. Causal priority is not just an anthropomorphic projection of the notion of agency onto the universe at large. It is constituted by an objective asymmetry of independence.

Figure 13.1 is a schematic summary of the theorems proven in previous chapters. It provides a map of how the different asymmetries of causation are related to one another. The arrows represent logical entailments, and the connection principle, **CC** is left implicit. Most of the entailments require **CC** as a premise.

The view of causal priority that accompanies this picture is not simple. Causal priority is a mind-independent structural feature of reality, but this structural feature gets its significance from human concerns with a particular kind of explanation, and these concerns in turn eventually derive from the practical interests of agents. The systems agents seek to understand are open. If there are closed systems, they are not subject to the same kind of explanation, because they present no line of attack for an abstract intervention, however infeasible it might be. What happens in the systems agents seek to understand is as if made to happen by an agent, and explanation consists in learning the recipes. Recipes are possible only if ingredients can be separately added: Causal explanation requires modularity, and causation itself presupposes independence. Among nomically connected events, those that satisfy the independence condition are the causes, and those that do not are the effects. Although the patterns of independence and connectedness

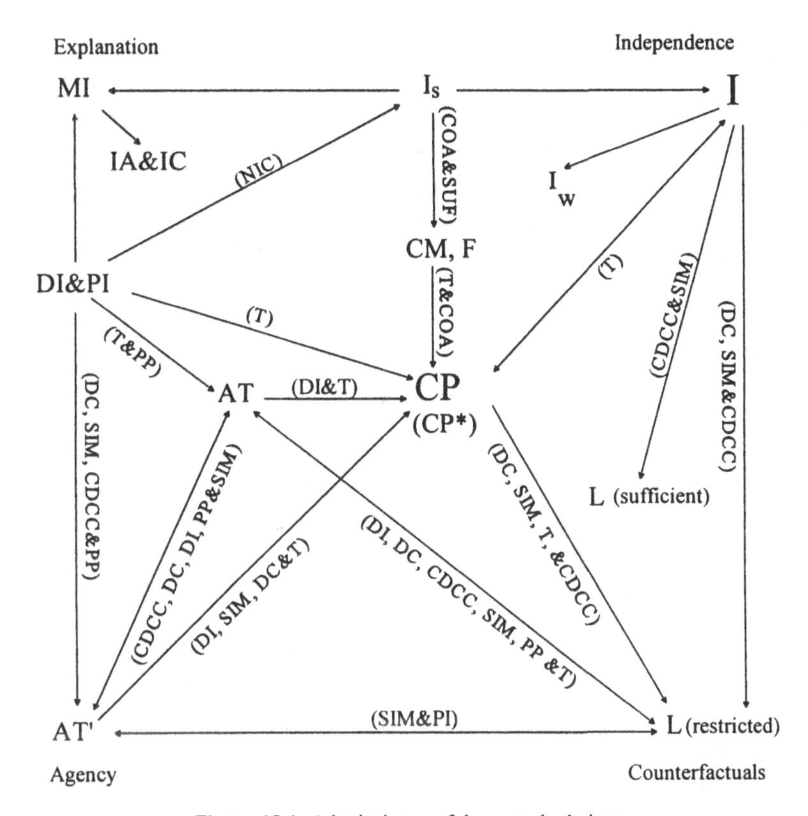


**Figure 13.1:** A logical map of the central relations

that constitute causal priority are mind-independent features of reality, human beings notice such patterns and single them out because of their desire to provide a particular kind of explanation, and the concern to provide this kind of explanation in turn grows out of practical interests in understanding the possibilities for controlling the environment. In closed systems, intervention and control are impossible. Explanation of some sort remains possible, but not specifically *causal* explanation. What happens does not happen in virtue of factors that occur independently. There is no counterfactual dependence on individual factors, no possibility of intervention, and none but a temporal difference between forward and backward in time.

When one has an explanation of the right kind – a causal explanation – then the possibilities of controlling phenomena of the same kind as the explanandum are limited only by the possibilities of intervening to influence the independent factors nomically connected to the explanandum. Independence and its echoes within arguments – modal invariance or at least independent alterability and insensitivity – are essential to this kind of explanation, because without independence there are no openings for interventions and no way of setting the explanandum within the perspective of potential control. The search for causes and the conception of causation derive from explanatory interests, and these in turn derive from our interests as agents.

Whether *a* causes *b* does not, however, depend on human interests or perspective. Recall Menzies and Price's analogy between causation and secondary qualities (1993). If objects were red if and only if they possessed some neatly and intrinsically characterizable surface structure, then the disposition to create a certain experience in us would be identical to something intrinsic, and there would be no need to refer to the disposition except when discussing how human beings react to colors. The dispositional and relational features of the color would play no role in defining what it is. Just such a state of affairs obtains, I maintain, with respect to causation. Our interests and perspectives lead us to look for a certain kind of explanation, but whether such explanations are possible – whether causal relations exist – depends on the existence of an intrinsically and systematically specifiable structure of relations among events. Whether *a* causes *b* depends on whether there are other events causally connected to *b* and independent of *a*.

When human interventions are possible, this condition will be met. The converse does not hold. The possibility of specifically human intervention depends on the extent of human powers, while the existence of causal relations do not. The possibility of an *abstract* intervention (of an intervention that is not necessarily a human action) comes closer to coinciding with the independence that constitutes causal asymmetry. I was able for the most part to make do with a weaker independence condition than is entailed by the possibility of an abstract intervention, but I do not claim to have conclusively established that the weaker rather than the stronger goes to the heart of causation.

The relations between causation and probabilities enable the perspective of an agent to get a grip on the world. Causal independence and causal connection are reflected in probabilistic independence and probabilistic dependency, but causal connection and causal independence are idealizations of probabilistic relations and are not reducible to them. Because of this nonreductive operationalization effects "wiggle" when one "wiggles" their cause, but not vice versa.

Similarly, when *b* is counterfactually dependent on *a* and not vice versa, there will be these other events causally connected to *b* and independent of *a*, and so counterfactual dependence is sufficient for causal dependence. But the converse does not hold because of the possibility of multiple connection. Counterfactual dependence, like robustness, depends on independence among the proximate causes of an event – which is more than is needed for causation or causal explanation. Accounts of causation in terms of robustness and counterfactual dependence are approximations that help one to appreciate the basis and significance of causal asymmetry. Explanation, agency, and counterfactual dependency all point toward the independence condition.

The independence condition could be interpreted as an ambitious metaphysical generalization that breaks down in closed deterministic systems. I maintain on the contrary that when independence breaks down, so does the possibility of giving causal explanations and of identifying causal relations. When there is no room for agency, no asymmetry of overdetermination, and no asymmetry of counterfactual dependence, there is no causal asymmetry and no causation. The notion of tracing out the ways in which independent factors come together to produce an outcome, which is central to a causal explanation, would have no place. One can suppose falsely that interventions are possible even in closed systems, but the supposition is false. Causation presupposes independence.

This interpretation of independence is most poignant in the circumstances in which independence breaks down. In the EPR phenomena discussed in §12.6, the decisive barrier to giving a causal interpretation of the phenomena lay in the fact that the separate spin measurements do not each have a cause that is nomically independent of the other measurements. *Patterns* of spins up and down have causal explanations, but whether (for example) the spin measured by detector 1 in the $x$ direction is up or down does not. The defining feature of causal explanation – the possibility of tracing independent factors to this outcome (or to a determinate probability of this outcome) – is missing. One could say that even though a particular kind of explanation cannot be given, there are still causes here, and the independence condition is false. But what reason could there be for preferring this conclusion to my conclusion that not all species of nomic connection are specifically causal – that, owing to the absence of independence, the measurements, though governed by natural law, are not *caused*? The only reason I can think of would be a commitment to Hume's theory. A Humean might conclude that the particle emission causes the measurements because there are lawlike connections between them, and the former precedes the latter. But everything else we know about causation points in the other direction: The specific measurement result does not counterfactually depend on the purported cause. Intervention to bring about just that result or to prevent it (without simultaneously affecting the other measurement results) is not possible. Robustness and screening-off fail.

Causal asymmetry consists of a pattern of relations among events, which permits a kind of explanation that serves human interests in understanding and control. Causal asymmetry manifests itself in many ways, and the central thread in these manifestations is independence.

# Appendix A

# Alphabetical List of Propositions

**A** *Asymmetry* (p. 80) If *a* causes *b*, then *b* does not cause *a*.

**AOD** *Asymmetry of over-determination* (p.136) If causation is deterministic, then (1) events will be determined by a great many of their natural effects, and (2) events will be not be determined by any of their (natural) causes.

**AT** *Agency theory* (pp. 139, 151) *a* causes *b* if and only if *a* and *b* are distinct, and with respect to some set of (actual) events *E*, if *a* had come about as a result of a direct manipulation with respect to *E*, then that intervention would be a cause of *b*.

**$AT_g$** *Agency theory* (pp. 88, 106) *y* depends on *x* in circumstances *K* if and only if *x* and *y* are distinct and there is some set of variables *V* including *x* and *y* such that in *K* every intervention with respect to *x* given *V* affects *y*.

**AT′** *Counterfactual agency theory* (pp. 141, 154) *a* causes *b* if and only if *a* and *b* are distinct, and if one intervened and prevented *a*, then *b* would not occur.

**CC** *Connection principle* (pp. 59, 75) For all events *a* and *b*, *a* and *b* are causally connected if and only if they are distinct and either *a* causes *b*, *b* causes *a*, or *a* and *b* are effects of a common cause.

**CC′** *Revised connection principle* (p. 252) *a* is nomically connected to *b* if and only if *a* and *b* are distinct and either (i) *a* causes *b*, (ii) *b* causes *a*, (iii) *a* and *b* are effects of a common cause, or (iv) *a* and *b* are mutually dependent.

**$CC_g$** *Type-level connection principle* (p. 108) In circumstances *K* *x* and *y* are causally connected if and only if they are distinct and in *K* either *y* depends on *x*, *x* depends on *y*, or there is some variable *z* that *x* and *y* both depend on.

**CCC** (p. 188; derived from Cartwright) **a** causes **b** if and only if $Pr(B/A\&K_j) > Pr(B/K_j)$ for all state descriptions $K_j$ over the set $\{\mathbf{c}_i\}$ where $\{\mathbf{c}_i\}$ satisfies:

1. If $\mathbf{c}_i$ is in $\{\mathbf{c}_i\}$, then $\mathbf{c}_i$ causes **b** or ~**b**.
2. **a** is not in $\{\mathbf{c}_i\}$.
3. For all **d**, if **d** causes **b** or **d** causes ~**b**, then **d** = **a** or **d** is in $\{\mathbf{c}_i\}$.
4. If $\mathbf{c}_i$ is in $\{\mathbf{c}_i\}$, then **a** is not a cause of $\mathbf{c}_i$.

**CDCC** *Counterfactual dependence implies causal connection* (p. 134) If *a* and *b* are distinct events and *b* counterfactually depends on *a*, then *a* and *b* are causally connected.

**CI** *Pairwise counterfactual independence* (p. 182) If the values of two variables $x$ and $y$ are specified in a derivation, then $y$ is not counterfactually dependent on $x$ and $x$ is not counterfactually dependent on $y$.

**CM** *Causal Markov condition* (p. 211) If $x$ and $y$ are probabilistically dependent conditional on all the direct causes of $x$, then $x$ causes $y$.

**CMD** *Counterfactuals and mutual dependence* (p. 261) If $a$ and $b$ are mutually dependent, then they are counterfactually dependent on one another.

**COA** *Counterfactual operationalizing assumption* (pp. 246, 256) If the variables in some set $S$ were never to vary, $x$ and $y$ would be causally connected if and only if $x$ and $y$ are distinct and in the background circumstances probabilistically dependent conditional on the fixed values of the variables in set $S$.

**CP** *Independence theory of causal priority* (pp. 70, 84) $a$ causes $b$ if and only if $a$ and $b$ are causally connected and everything causally connected to $a$ and distinct from $b$ is causally connected to $b$.

**CP′** *Revised independence theory of causal priority* (pp. 85, 253–4) $a$ causes $b$ if and only if $a$ is nomically connected to $b$, everything nomically connected to $a$ and distinct from $b$ is nomically connected to $b$ and something nomically connected to $b$ is independent of $a$.

**CP$_g$** *Type-level independence theory* (p. 108) $y$ depends on $x$ in circumstances $K$ if and only if in $K$ $x$ and $y$ are causally connected and everything causally connected to $x$ and distinct from $y$ is causally connected to $y$.

**CP*** *Modal independence theory* (pp. 140, 152) $a$ causes $b$ if and only if $a$ and $b$ are causally connected and for all possible events $c$ distinct from $b$, if $c$ were causally connected to $a$, then $c$ would be causally connected to $b$.

**DC** *Deterministic causation – causes as INUS conditions* (pp. 43, 102) If $a$ is a deterministic cause of $b$ in set up $c$ during the time interval $[t, t']$, then given laws of nature **L**, 1. $B(c, t')$ entails and is entailed by $\{A(c, t) \& G(c, [t, t'])$ or $H(c, [t, t'])\}$, but $B(c, t')$ is not entailed by $G(c, [t, t'])$, 2. $B(c, t')$ 3. $A(c, t)$, 4. $G(c, [t, t'])$, and 5. $\neg H(c, [t, t'])$.

**DHI** *Definition of a human intervention* (p. 151) $i$ is a direct human intervention or manipulation that brings about $a$ only (with respect to a set of events $E$ of interest) if and only if (1) $i$ is distinct from every event in $E$, (2) $i$ is a direct cause of $a$, (3) the structure of the causal relations between $a$ and its effects is the same when $a$ is caused by $i$ as when there is no intervention, (4) $i$ has no causal connections to any other events in $E$ except those that follow from its being a direct cause of $a$, and (5) $i$ is a human action.

**DHI$_g$** *Type-level definition of a human intervention* (p. 106) In circumstances $K$, $z$ is a human intervention variable with respect to $x$ only (given a set of variables $V$) if and only if in $K$ (1) $z$ is distinct from all variables in $V$, (2) $z$ directly affects $x$, (3) all other connections between $z$ and members of $V$ follow from $z$ being a direct influence on $x$, and (4) values of $z$ are kinds of human actions.

**DI** *Definition of an abstract intervention* (p. 151) $i$ is a direct intervention or manipulation that brings about $a$ only (with respect to a set of events $E$ of interest) if and only if (1) $i$ is distinct from every event in $E$, (2) $i$ is a direct cause of $a$, (3) the structure of the causal relations between *a* and its effects is the same when *a* is caused by *i* as when there is no intervention, and(4) *i* has no causal connections to any other events in *E* except those that follow from its being a direct cause of *a*.

**$DI_g$** *Type-level definition of an abstract intervention* (p. 106) In circumstances *K*, *z* is an intervention variable with respect to *x* only (given a set of variables *V*) if and only if in *K* (1) *z* is distinct from all variables in *V*, (2) *z* directly affects *x*, and (3) all other connections between *z* and members of *V* follow from *z* being a direct influence on *x*.

**F** *Faithfulness condition* (p. 211) If *x* causes *y*, then *x* and *y* are probabilistically dependent conditional on all the direct causes of *x*.

**G** *Counterfactual generalization view* (p. 102) **a** is a cause of **b** in circumstances *K* if and only if in *K* all events of kind **a** that might occur would cause some event of kind **b** that would bear the right temporal relations to it.

**H** *Humean theory* (p. 47) *a* is a *cause* of *b* if and only if *a* is a direct cause of *b* or there is a unidirectional chain of direct causes running from *a* to *b*.

*a* is a *direct cause* of *b* if and only if *a* and *b* are distinct, *a* is nomically necessary and sufficient in the circumstances for *b*, *a* and *b* are spatially and temporally contiguous, and *a* begins before *b* begins.

**I** *Independence condition* (pp. 64, 81) If *a* causes *b* or *a* and *b* are causally connected only as effects of a common cause, then *b* has a cause that is distinct from *a* and not causally connected to *a*.

**$I_g$** *Type level independence condition* (p. 109) If in circumstances *K*, *x* causes *y* or *x* and *y* are causally connected only as effects of a common cause, then in *K* *y* depends on something that is not causally connected to *x*.

**I*** *Modal independence condition* (p. 153) If *a* causes *b* or *a* and *b* are causally connected only as effects of a common cause, then *b* might have a cause that would be causally independent of *a*.

**$I_s$** *Strong independence condition* (p. 82) Every event *b* that has at least one cause has another cause, *f*, such that for all events *a*

1. If *b* does not cause *a* and there is a causal path that does not go through *f* from *a* to *b* or from some cause of *a* to *b*, then *a* and *f* are causally independent.
2. If *b* causes *a*, then there is no path that does not pass through *b* from *f* to *a* or from any cause of *f* to *a*.

**$I_{gs}$** *Type-level strong independence condition* (p. 109) In circumstances *K*, every non-exogenous variable *y* has some additional cause *z* such that for all variables *x*,

1. If *y* does not cause *x* and there is a causal path that does not go through *z* from *x* to *y* or from a cause of *x* to *y*, then *x* and *z* are causally independent.
2. If *y* causes *x*, then there is no path that does not pass through *y* from *z* to *x* or from any cause of *z* to *x*.

**$I'_{gs}$** *Revised type-level strong independence condition* (p. 260) In circumstances *K*, every nonexogenous variable *y* has an additional cause *z* such that for all variables *x*,

1. If *y* does not cause x and *x* and *y* are not mutually dependent and there is a causal path that does not go through *z* from *x* to *y* or from some cause of *x* to *y*, then *x* and *z* are causally independent.

2. If $y$ causes $x$, then there is no path that does not pass through $y$ from $z$ to $x$ or from any cause of $z$ to $x$.
3. If $x$ and $y$ are mutually dependent, there is no directed or undirected path that does not pass through $y$ between $z$ and $x$ or between any cause of $z$ and $x$.

**$I_s$*** *Strong modal independence condition* (p. 53) Every event $b$ that has at least one cause might have another independent cause, $f$, such that for all events $a$

1. If $a$ were not caused by $b$ and there were a causal path that did not go through $f$ from $a$ to $b$ or from some cause of $a$ to $b$, then $a$ and $f$ would be causally independent.
2. If $b$ were to cause $a$, then there would be no path that would not pass through $b$ from $f$ to $a$ or from any cause of $f$ to $a$.

**$I_w$** *Weak independence condition* (pp. 65, 81) If $a$ causes $b$ and $b$ does not cause $a$, then $b$ has a cause that is causally independent of $a$.

**IA** *Independent alterability* (pp. 167, 182) For every pair of variables, $x$ and $y$, whose values are specified in a derivation, if the value of $x$ were changed by intervention, then value of $y$ would be unchanged.

**IC** *Insensitivity condition* (p. 169) If the value specified in a derivation for any variable were changed by intervention, then the nonspecifying equations in the derivation would continue to hold.

**L** *Lewis's theory* (p. 112) $a$ causes $b$ if and only if $a$ and $b$ are distinct events and if $a$ were not to occur, then $b$ would not occur either.

**L′** *Lewis's theory revised* (p. 261) $a$ causes $b$ if and only if $b$ counterfactually depends on $a$ and $a$ does not counterfactually depend on $b$.

**MI** *Asymmetry of modal invariance* (p. 222) Suppose that $S$ is a complete and correct specification of a causal system involving a set of variables including $x$ and $y$. $x$ causes $y$ if and only if the value of $y$ one calculates when one substitutes a new value of $x$ into $S$ is a correct prediction or a best estimate of what the value of $y$ would be if the value of $x$ were set to its new value via intervention.

**NIC** *Nonintervention causes* (p. 153) If $a$ can be caused by an intervention, then $a$ also has causes that are not interventions.

**$NIC_g$** *Nonintervention causes* (p. 109) If there is an intervention variable $z$ with respect to some variable $x$, then $x$ depends on some variable other than $z$.

**OA** *Operationalizing assumption* (pp. 57, 75) Events or tropes $a$ and $b$ are causally connected if and only if they are distinct and the kinds **a** and **b** or the properties $A$ and $B$ are in the background circumstances probabilistically dependent.

**OBP** *Open bȧck path condition* (p. 83) Every cause $a$ of $b$ that has any causes has at least one cause $d$ such that the only path from $d$ to $b$ is via $a$.

**P** *Prediction condition* (p. 120) The knowledge that $b$ would occur if $a$ were to occur and that an event of kind **a** occurs taken by itself justifies the prediction that an event of kind **b** will occur.

**PDPW** *Probability distributions and possible worlds* (p. 256) The actual probability distribution of $y$ conditional on $x = x^*$ is equal to the probability distribution $y$ would have if the value of $x$ never varied from $x^*$.

**PI** *Possibility of intervention* (p. 151) Given the set of events $E$ of interest, for each event $e$ in $E$, there is a possible situation with the same causal relations between $e$ and its effects in which $e$ and only $e$ comes about via a direct manipulation.

**PI$_g$** *Possibility of intervention* (p. 106) Given the set $V$ of variables and the circumstances $K$ of concern, there is, for each variable $x$ in $V$, some $z$ that is an intervention variable with respect to $x$ only.

**PP$_g$** *Type-level path principle* (p. 107) If $z$ causes $y$, then in a directed graph correctly representing the causal relations, there is a path between the vertex denoting $z$ and the vertex denoting $y$.

**SIM** *Similarity among possible worlds* (pp. 133–4)

1. *Worlds with miracles are not the most similar.* For any event $b$ there are non-$b$ possible worlds without at least one of $b$'s causes that are at least as close to the actual world as are any non-$b$ possible worlds in which all of $b$'s causes occur.

2. *It doesn't matter which cause is responsible.* For any event $b$, if $a$ and $c$ are any two causes of $b$ that are causally and counterfactually independent of one another, there will be non-$b$ possible worlds in which $a$ does not occur and $c$ does occur that are just as close to the actual world as are any non-$b$ possible worlds with $a$ and without $c$, and there will be non-$b$ possible worlds without $a$ and with $c$ that are just as close to the actual world as are any non-$b$ possible worlds without both $a$ and $c$.

3. *The fewer the irrelevant differences in events, the more similar the world.* For any event $b$, consider two possible events $e$ and $f$ that are not causally connected to $b$, where $e$ occurs and $f$ does not. Then there are non-$b$ possible worlds with $e$ that are more similar to the actual world than any non-$b$ possible worlds without $e$, and there are non-$b$ worlds without $f$ that are more similar to the actual world than any non-$b$ worlds with $f$.

4. *The fewer the irrelevant differences in laws, the more similar the world.* For any event $b$ there is a non-$b$ possible world in which all other laws, apart from those relating $b$ to its causes or the causes of $b$ to one another, are the same as the actual world that is more similar to the actual world than is any non-$b$ possible world which differs with respect to some such laws.

**SPI** *Structural presupposition of intervention* (p. 153) Interventions are possible if and only if intervention variables exist.

**SSMD** *Strong symmetry of mutual dependence* (p. 253) Mutual dependence is symmetrical, and if $a$ and $b$ are mutually dependent, then everything nomically connected to $a$ and distinct from $b$ is nomically connected to $b$.

**SUF** *Sufficiency* (p. 183) For all variables $x$ in the set $V$, $V$ contains all direct causes of $x$ that are direct or indirect causes of any other variable in $V$ by paths that do not go through $x$.

**T** *Transitivity* (p. 80) If $a$ causes $b$, $b$ causes $c$, and $a$ and $c$ are distinct, then $a$ causes $c$.

**T$_g$** *Transitivity* (p. 107) If in circumstances $K$, $z$ depends on $y$, $y$ depends on $x$, and z and $x$ are distinct, then $z$ depends on $x$.

# Appendix B

# List of Theorems

**Theorem 4.1** (p. 81) **T** and **CC** entail that if *a* causes *b*, then everything causally connected to *a* and distinct from *b* is causally connected to *b*.

**Theorem 4.2** (p. 81) Given **T** and **CC**, if *a* is distinct from *b* and not causally connected to *b*, then it is not causally connected to any cause of *b*.

**Theorem 4.3** (p. 82) **CC**, **I**, and **T** entail **A**.

**Theorem 4.4** (p. 84) **CC** and **I** imply that if *a* is causally connected to *b* and everything casually connected to *a* and distinct from *b* is causally connected to *b*, then *a* causes *b*.

**Theorem 4.5** (p. 84) **T**, **CC**, and **I** entail **CP**.

**Theorem 4.6** (p. 84) **T**, **CC**, and **I** entail **CP′**.

**Theorem 4.7** (p. 85) **T**, **CC**, **CP**, and **A** (asymmetry) imply **I**.

**Theorem 5.1** (p. 107) $\mathbf{DHI_g}$, $\mathbf{PI_g}$, $\mathbf{T_g}$, and $\mathbf{PP_g}$ imply $\mathbf{AT_g}$.

**Theorem 5.2** (p. 108) $\mathbf{DHI_g}$, $\mathbf{CC_g}$, $\mathbf{T_g}$, and $\mathbf{AT_g}$ entail $\mathbf{CP_g}$.

**Theorem 5.3** (p. 108) $\mathbf{DI_g}$, $\mathbf{PI_g}$, $\mathbf{T_g}$, and $\mathbf{PP_g}$ imply $\mathbf{AT_g}$.

**Theorem 5.4** (p. 108) $\mathbf{DI_g}$, $\mathbf{CC_g}$, $\mathbf{T_g}$, and $\mathbf{AT_g}$ entails $\mathbf{CP_g}$.

**Theorem 5.5** (p. 108) $\mathbf{DI_g}$, $\mathbf{PI_g}$, $\mathbf{CC_g}$, and $\mathbf{T_g}$ entail $\mathbf{CP_g}$.

**Theorem 5.6** (p. 109) Given $\mathbf{CC_g}$ and $\mathbf{NIC_g}$, $\mathbf{I_{gs}}$ is entailed by $\mathbf{DI_g}$ and $\mathbf{PI_g}$.

**Theorem 5.7** (p. 110) $\mathbf{CC_g}$, $\mathbf{PI_g}$, $\mathbf{DI_g}$, and $\mathbf{NIC_g}$ entail $\mathbf{I_g}$.

**Theorem 6.1** (p. 135) **SIM**, **CDCC**, and **I** imply that individual causes will not be counterfactually dependent on individual effects and effects of a common asymmetric cause will not be counterfactually dependent on one another.

**Theorem 6.2** (p. 135) **CDCC**, **SIM I**, and **CC** imply that if *b* counterfactually depends on *a*, then *a* causes *b*.

**Theorem 6.3** (p. 136) If there is no preemption or ordinary overdetermination (overdetermination by conjunctions of natural causes), then **DC** and **I** entail **AOD**.

**Theorem 6.4** (p. 137) Given **CC**, **DC**, **SIM**, and no multiple connections – that is, if *a* had not occurred, no cause of *a* would have been a cause of *b* – if *a* causes *b*, then *b* is counterfactually dependent on *a*.

**Theorem 6.5** (p. 137) **CC**, **I**, **DC**, **SIM**, and **CDCC** entail **L** restricted to circumstances in which there are no multiple connections.

**Theorem 7.1** (p. 152) **DHI** (or **DI**), **PI**, **T**, and **PP** imply **AT**.

**Theorem 7.2** (p. 152) **CC**, **T**, and **I** entail **CP***.

**Theorem 7.3** (p. 153) **CC**, **T**, **DHI** (or **DI**), and **AT** entail **CP***.

**Theorem 7.4** (p. 153) **CC**, **NIC**, **DI**, and **PI** entail $\mathbf{I_s}$*****.

**Theorem 7.5** (p. 154) **DC**, **SIM**, **DI**, **PI**, **CDCC**, and **PP** entail **AT′**.

**Theorem 7.6** (p. 154) **DI**, **CDCC**, **CC**, **T**, **DC**, and **AT′** entail **CP***.

**Theorem 7.7** (p. 155) Given **SIM**, **DI**, **PI**, and the absence of multiple connections, **AT′** entails and is entailed by **L**.

**Theorem 7.8** (p. 155) Given **DI**, **CDCC**, **DC**, **SIM**, **T**, **PP**, and the absence of multiple connections and of preemption or overdetermination, **L** entails and is entailed by **AT**.

**Theorem 7.9** (p. 155) Given **DI**, **CDCC**, **DC**, **SIM**, **T**, **PP**, and the absence of multiple connections or of preemption or overdetermination, **AT′** entails and is entailed by **AT**.

**Theorem 8.1** (p. 182–3) **SIM** and $\mathbf{DI_g}$ imply that if the specified variables in a derivation are independently alterable then they are counterfactually independent.

**Theorem 8.2** (p. 183) Given $\mathbf{CC_g}$, $\mathbf{DI_g}$, and **CDCC**, if the variables whose values are specified in a derivation are not related as cause and effect, then the derivation satisfies **IA**.

**Theorem 8.3** (p. 183) Given **SIM**, $\mathbf{DI_g}$, $\mathbf{T_g}$, $\mathbf{CC_g}$, and **DC**, if a derivation satisfies **IA**, then none of the variables whose values are specified are related as cause and effect.

**Theorem 8.4** (p. 183) $\mathbf{DI_g}$, **SIM**, and **SUF** entail that simple causal derivations satisfy **IA** and **IC**.

**Theorem 8.5** (p. 184) **CDCC** and $\mathbf{DI_g}$ entail that derivations of values of variables from values of their effects or from values of effects of common causes do not satisfy both **IC** and **IA**.

**Theorem 10.1** (p. 217) If $(x)(Ax \Leftrightarrow Cx \& Sx$ or $Zx)$, *C* is probabilistically independent of *S* and *Z*, and Pr(S or Z) > Pr(Z), then Pr(A/C) > Pr(A/~C).

**Theorem 10.2** (p. 217) If (1) $(x)(Ax \Leftrightarrow Cx \& Sx$ or $Zx)$, (2) $(x)(Bx \Leftrightarrow Cx \& Yx$ or $Wx)$, (3) *C* is probabilistically independent of *S*, *Z*, *Y*, and *W*, (4) *S*, *Z*, *Y*, and *W* are probabilistically independent of one another, and (5) all probabilities are intermediate, then *C* and ~*C* screen off *A* and *B*.

**Theorem 10.3** (p. 217–8) **CM** and **F** imply

1. (*Direct causal connection*) For all *x* and *y* in *V*, *x* and *y* are adjacent if and only if they are probabilistically dependent conditional on *every* subset of *V* that does not include them, and

2. (*Direct causation*) For all $x$, $y$ and $z$ in $V$, if $x$ and $y$ and $y$ and $z$ are adjacent, and $x$ and $z$ are not adjacent, then $y$ causally depends on $x$ and $z$ if and only if $x$ and $z$ are probabilistically dependent conditional on every subset of $V$ that contains $y$ but not $x$ or $z$.

**Theorem 11.1** (p. 235) Let $X = \{x_1, \ldots, x_n\}$ be the finite set of $n$ variables represented by the $n$ vertices of some directed acyclic graph $G$ and let $Q$ be a probability distribution over $X$ that satisfies the causal Markov condition – for all $x$, $y$ in $X$, if there is no path in $G$ from $x$ to $y$, then $x$ and $y$ are probabilistically independent conditional on the set of all the parents of $x$, $Px$. Then $Q(x_1, \ldots, x_n) = Q(x_1/Px_1) \cdot Q(x_2/Px_2) \cdot \ldots \cdot Q(x_n/Px_n)$.

**Theorem 12.1** (p. 258) **PDPW** and the assumption that $\mathbf{OA_g}$ is true of all similar possible worlds imply **COA**.

**Theorem 12.2** (p. 258) **SUF**, $\mathbf{CC_g}$, $\mathbf{I_{gs}}$, and **COA** entail **CM** and **F**.

**Theorem 12.3** (p. 258) **CM**, **F**, $\mathbf{CC_g}$, $\mathbf{T_g}$ and **COA** imply $\mathbf{CP_g}$

**Theorem 12.4** (p. 259) **SUF**, $\mathbf{CC_g}$, $\mathbf{I_{gs}}$, and **COA** imply that $x$ and $y$ are adjacent if and only if they are probabilistically dependent conditional on every subset of the sufficient set of variables $V$ that does not contain $x$ or $y$.

**Theorem 12.5** (p. 260) **T**, **CC′**, **I** and **SSMD** entail **CP′**.

**Theorem 12.6** (p. 260) **T**, **CC′**, **SSMD**, and **CP′** imply $\mathbf{I_w}$.

**Theorem 12.7** (p. 260) Given $\mathbf{CC'_g}$, and $\mathbf{NIC_g}$, $\mathbf{I'_{gs}}$ is entailed by $\mathbf{DI'_g}$ and $\mathbf{PI_g}$.

**Theorem 12.8** (p. 261) **CC′**, **I**, **DC**, **SIM**, **CMD**, and **CDCC** entail **L′** restricted to circumstances in which there are no multiple connections.

**Theorem 12.9** (p. 262) **SUF**, $\mathbf{CC'_g}$, $\mathbf{I'_{gs}}$, and **COA** entail that $x$ causes $y$ if and only if $x$ and $y$ are probabilistically dependent conditional on all the direct causes of $x$ and $x$ and $y$ are probabilistically independent conditional on all the direct causes of $y$.

# References

Achinstein, Peter. 1983. *The Nature of Explanation*. Oxford: Oxford University Press.

Anderson, John. 1938. "The Problem of Causality." *Australasian Journal of Psychology and Philosophy* 16: 127–42.

Anscombe, Elizabeth. 1971. *Causality and Determinism*. Cambridge: Cambridge University Press.

Arntzenius, Frank. 1990. "Physics and Common Causes." *Synthese* 82: 77–96.

_____. 1993. "The Common Cause Principle." In Hull, D., M. Forbes, and K. Okruhlik, eds. *PSA 1992*, vol. 2. East Lansing: Philosophy of Science Association, pp. 227–37.

Aronson, Jerrold. 1971a. "On the Grammar of Cause." *Synthese* 22: 414–30.

_____. 1971b. "The Legacy of Hume's Analysis of Causation." *Studies in the History and Philosophy of Science* 2: 135–57.

Barnes, Eric. 1992. "Explanatory Unification and the Problem of Asymmetry." *Philosophy of Science* 59: 558–71.

Beauchamp, Thomas and Alexander Rosenberg. 1981. *Hume and the Problem of Causation*. Oxford: Oxford University Press.

Bell, John. 1964. "On the Einstein-Podolsky-Rosen Paradox." *Physics* 1: 195–200.

_____. 1966. "On the Problem of Hidden Variables in Quantum Mechanics." *Reviews of Modern Physics* 38: 447–52.

Bennett, Jonathan. 1984. "Counterfactuals and Temporal Direction." *Philosophical Review* 93: 57–91.

_____. 1988. *Events and Their Names*. Indianapolis: Hackett Publishing.

Berofsky, Bernard. 1971. *Determinism*. Princeton: Princeton University Press.

_____. 1977. "Review of *The Cement of the Universe: A Study of Causation*." *Journal of Philosophy* 74: 103–18.

Braddon-Mitchell, David. 1993. "The Microstructural Causation Hypothesis." *Erkenntnis* 39: 257–83.

Bradie, Michael. 1984. "Recent Work on Criteria for Event Identity, 1967–1979." *Philosophy Research Archives*, pp. 29–77.

Brand, Myles. 1977. "Identity Conditions for Events." *American Philosophical Quarterly* 14: 329–37.

_____. 1980. "Simultaneous Causation." In van Inwagen (1980), pp. 137-53.

Bridgman, P. 1938. "Operational Analysis." *Philosophy of Science* 5: 114–31.

Bromberger, Sylvain. 1966. "Why Questions." In R. Colodny, ed. *Mind and Cosmos: Essays in Contemporary Science and Philosophy*. Pittsburgh: University of Pittsburgh Press, pp. 86–111.

Brown, Bryson. 1992. "Defending Backwards Causation." *Canadian Journal of Philosophy* 22: 429–44.

Byerly, Henry. 1979. "Substantial Causes and Nomic Determination." *Philosophy of Science* 46: 57–81.

_____. 1990. "Causes and Laws: The Asymmetry Puzzle." In A. Fine, M. Forbes, and L. Wessels, eds. *PSA 1990*, vol. 1. East Lansing: Philosophy of Science Association, pp. 545–55.

Carroll, John. 1988. "General Causation." In A. Fine and J. Leplin, eds. *PSA 1988*, vol. 1. East Lansing: Philosophy of Science Association, pp. 311–17.

_____. 1991. "Property-Level Causation?" *Philosophical Studies* 63: 245–70.

Cartwright, Nancy. 1979. "Causal Laws and Effective Strategies." *Noûs* 13. Rpt. in Cartwright 1983, pp. 21–43.

_____. 1983. *How the Laws of Physics Lie*. Oxford: Oxford University Press.

_____. 1989. *Nature's Capacities and Their Measurement*. Oxford: Clarendon Press.

_____. 1995. "Probabilities and Experiments." *Journal of Econometrics* 67: 47–59.

Castañeda, Hector-Neri. 1980. "Causes, Energy and Constant Conjunctions." In van Inwagen (1980), pp. 81-108.

_____. 1984. "Causes, Causity and Energy." *Midwest Studies in Philosophy* 9: 17–28.

Clendinnen, F. John. 1992. "Nomic Dependence and Causation." *Philosophy of Science* 59: 341–60.

Collingwood, R. G. 1940. *An Essay on Metaphysics*. Oxford: Clarendon Press.

Cushing, James and Ernan McMullin, eds. 1989. *The Philosophical Consequences of Quantum Theory*. Notre Dame: Notre Dame University Press.

Davidson, Donald. 1980. *Essays on Actions and Events*. Oxford: Clarendon Press.

_____. 1985. "Reply to Quine on Events." In LePore and McLaughlin (1985), pp. 172–6.

Davis, Wayne. 1980. "Swain's Counterfactual Analysis of Causation." *Philosophical Studies* 38: 169–76.

Dieks, D. 1981. "A Note on Causation and the Flow of Energy." *Erkenntnis* 16: 103–8.

Dowe, Phil. 1992a. "Process Causality and Asymmetry." *Erkenntnis* 37: 179–96.

_____. 1992b. "Wesley Salmon's Process Theory of Causality and the Conserved Quantity Theory." *Philosophy of Science* 59: 195–216.

_____. 1995. "Causality and Conserved Quantities: A Reply to Salmon." *Philosophy of Science* 62: 321–33.

_____. 1996. "Backwards Causation and the Direction of Causal Processes." *Mind* 105: 227–48.

Dretske, Fred. 1977. "Referring to Events." *Midwest Studies in Philosophy* 2: 90–9.

Dretske, Fred and Aaron Snyder. 1972. "Causal Irregularity." *Philosophy of Science* 39: 69–71.

Dretske, Fred and Berent Enç. 1984. "Causal Theories of Knowledge." *Midwest Studies in Philosophy* 9: 517–28.

Dummett, Michael. 1964. "Bringing About the Past." *Philosophical Review* 73: 338–59.

Dupré, John. 1984. "Probabilistic Causality Emancipated." *Midwest Studies in Philosophy* 9: 169–75.

_____. 1993. *The Disorder of Things: Metaphysical Foundations of the Disunity of Science*. Cambridge, MA: Harvard University Press.

Earman, John. 1976. "Causation: A Matter of Life and Death." *Journal of Philosophy* 73: 5–25.

Eells, Ellery. 1987. "Probabilistic Causality: Reply to John Dupré." *Philosophy of Science* 54: 105–14.

_____. 1991. *Probabilistic Causality*. Cambridge: Cambridge University Press.

Eells, Ellery and Elliott Sober. 1983. "Probabilistic Causality and the Question of Transitivity." *Philosophy of Science* 50: 35–57.

Ehring, Douglas. 1982. "Causal Asymmetry." *Journal of Philosophy* 79: 761–74.

_____. 1997. *Causation and Persistence: A Theory of Causation*. Oxford: Oxford University Press.

Einstein, A., B. Podolsky, and N. Rosen. 1935. "Can a Quantum Mechanical Description of Physical Reality be Considered Complete?" *Physical Review* 47: 777–80.

Elby, Andrew. 1992. "Should We Explain the EPR Correlations Causally?" *Philosophy of Science* 59: 16–25.

Engle, R., D. Hendry, and J. Richard. 1983. "Exogeneity." *Econometrica* 51: 277–304.

Fair, David. 1979. "Causation and the Flow of Energy." *Erkenntnis* 14: 219–50.

Faust, C. H. and T. H. Johnson, ed. 1935. *Jonathan Edwards*. New York: ??.

Festa, Roberto. 1993. *Optimum Inductive Methods*. Dordrecht: Kluwer.

Fine, Arthur. 1987. *The Shaky Game*. Chicago: University of Chicago Press.

Forster, Malcolm. 1988. "Sober's Principle of Common Cause and the Problem of Comparing Incomplete Hypotheses." *Philosophy of Science* 55: 538–59.

_____. 1996a. "The GHZ Version of Bell's Argument." Unpublished manuscript.

_____. 1996b. "The Symmetry of Probabilities and the Direction of Causal Arrows." unpublished manuscript.

Foster, J. A. 1975. "Testing the Cement: An Examination of Mackie on Causation." *Inquiry* 18: 487–98.

Frankel, Lois. 1986. "Mutual Causation, Simultaneity and Event Description." *Philosophical Studies* 49: 361–72.

Freudlich, Y. 1977. "The Causation Recipe." *Dialogue* 16: 472–84.

Friedman, Michael. 1974. "Explanation and Scientific Understanding." *Journal of Philosophy* 71: 5–19.

Gasking, Douglas. 1955. "Causation and Recipes." *Mind* 64: 479–87.

Glymour, Clark. 1978. "Two Flagpoles Are More Paradoxical Than One." *Philosophy of Science* 45: 118–19.

_____. 1980. "Explanations, Tests, Unity and Necessity." *Noûs* 14: 31–52.

Glymour, Clark, Richard Scheines, Peter Spirtes, and Kevin Kelly. 1987. *Discovering Causal Structure*. New York: Academic Press, 1987.

Goggans, Phillip. 1992. "Do the Closest Possible Worlds Contain Miracles?" *Pacific Philosophical Quarterly* 73: 137–49.

Goldman, Alvin. 1970. *A Theory of Human Action*. Princeton: Princeton University Press.

Good, I. J. 1961. "A Causal Calculus, I–II." *British Journal for the Philosophy of Science*. 11: 305–18, 12: 43–51.

Goosens, William. 1979. "Causal Chains and Counterfactuals." *Journal of Philosophy* 76: 489–95.

Gorovitz, Samuel. 1965. "Causal Judgments and Causal Explanations." *Journal of Philosophy* 62: 695–711.

Granger, C.W.J. 1969. "Investigating Causal Relations by Econometric Models and Cross-Spectral Methods." *Econometrica* 37: 424–38.

Haavelmo, Trygve. 1944. "The Probability Approach to Econometrics." *Econometrica* 12 (Supplement): iii–118.

Hart, H. L. A. and A. M. Honoré. 1959. *Causation in the Law*. Oxford: Clarendon Press.

Hausman, Daniel. 1982. "Causal and Explanatory Asymmetry."In T. Nickles and P. Asquith, eds. *PSA 1982*, vol. 1. East Lansing: Philosophy of Science Association, pp. 43–54.

_____. 1984. "Causal Priority." Noûs 18: 261–79.

_____. 1986. "Causation and Experimentation." *American Philosophical Quarterly* 23: 143–54.

_____. 1989a. "The Insufficiency of Nomological Explanation." *Philosophical Quarterly* 39: 22–35.

_____. 1989b. "Ceteris Paribus Clauses and Causality in Economics." In A. Fine and J. Leplin, eds. *PSA 1988*, vol. 2. East Lansing: Philosophy of Science Association, pp. 308–17.

_____. 1990. "Supply and Demand Explanations and their Ceteris Paribus Clauses." *Review of Political Economy* 2: 168–86.

_____. 1992. "Thresholds, Transitivity, Overdetermination, and Events." *Analysis* 52: 159–63.

_____. 1993a. "Linking Causal and Explanatory Asymmetries." *Philosophy of Science* 60: 435–51.

_____. 1993b. "Why Don't Causes Explain Their Effects?" *Synthese* 94: 227–44.

_____. 1996. "Causation and Counterfactual Dependence Reconsidered." *Noûs* 30: 55–74.

_____. 1998a. "Causation, Agency, and Independence." *Philosophy of Science* 63 (Supplement).

_____. 1998b. "Deterministic Causation of Probabilities." *Communication and Cognition*.

Healey, Richard. 1983. "Temporal and Causal Asymmetry." In Richard Swinburne, ed. *Space, Time and Causality*. Dordrecht: Reidel, pp. 79–105.

_____. 1992. "Discussion: Causation, Robustness, and EPR." *Philosophy of Science* 59: 282–92.

Hempel, Carl. 1965. *Aspects of Scientific Explanation*. New York: Free Press.

Hesslow, G. 1976. "Two Notes on the Probabilistic Approach to Causality." *Philosophy of Science* 34: 290–2.

_____. 1981. "Causality and Determinism." *Philosophy of Science* 48: 591–605.

Hitchcock, Christopher. 1993. "A Generalized Probabilistic Theory of Causal Relevance." *Synthese* 97: 335–64.

_____. 1995a. "The Mishap at Reichenbach Falls: Singular vs. General Causation." *Philosophical Studies* 73: 257–91.

_____. 1995b. "Discussion: Salmon on Explanatory Relevance." *Philosophy of Science* 62: 304–20.

_____. 1996a. "Farewell to Binary Causation." *Canadian Journal of Philosophy* 26: 267–82.

_____. 1996b. "The Mechanist and the Snail." *Philosophical Studies* 84: 91–105.

_____. 1996c. "The Role of Contrast in Causal and Explanatory Claims." *Synthese* 107: 395–419.

_____. 1998. "Causal Knowledge: That Great Guide of Human Life." *Communication and Cognition.*

Honderich, Ted. 1982. "Causes and *if p, even if x, still q*." *Philosophy* 57: 291–317.

Hoover, Kevin. 1990. "The Logic of Causal Inference." *Economics and Philosophy* 6: 207–34.

_____. 1991. "The Causal Direction between Money and Prices." *Journal of Monetary Economics* 27: 381–423.

_____. 1993. "Causality and Temporal Order in Macroeconomics or Why Even Economists Don't Know How to Get Causes from Probabilities." *British Journal for the Philosophy of Science* 44: 693–710.

_____. 1994. "Econometrics as Observation: the Lucas Critique and the Nature of Econometric Inference." *Journal of Economic Methodology* 1: 65–80.

_____. 1995. "Comments on Cartwright and Woodward: Causation, Estimation, and Statistics." In Little (1995), pp. 75–89.

Hoover, Kevin and Stephen Perez. 1994. "Post Hoc Ergo Once More: An Evaluation of 'Does Monetary Policy Matter?' in the Spirit of James Tobin." *Journal of Monetary Economics* 34: 47–73.

Hoover, Kevin and Steven Sheffrin. 1992. "Causation, Spending, and Taxes: Sand in the Sandbox or Tax Collector for the Welfare State?" *American Economic Review* 82: 225–48.

Horgan, Terence. 1978. "The Case Against Events." *Philosophical Review* 87: 28–47.

_____. 1980. "Humean Causation and Kim's Theory of Events." *Canadian Journal of Philosophy* 10: 663–79.

Horwich, Paul. 1987. *Asymmetries in Time*. Cambridge, MA: MIT Press.

Hume, David. 1738. *A Treatise of Human Nature*. Rpt. Oxford: Clarendon Press, 1966.

_____. 1748. *An Inquiry Concerning Human Understanding*, Rpt. Indianapolis: Bobbs-Merrill, 1955.

Humphreys, Paul. 1989. *The Chances of Explanation*. Princeton: Princeton University Press.

Irzik, Gürol. 1996. "Can Causes Be Reduced to Correlations?" *British Journal for the Philosophy of Science* 47: 249–70.

Irzik, Gürol and Eric Meyer. 1987. "Causal Modeling: New Directions for Statistical Explanation." *Philosophy of Science*. 54: 495–514.

Jeffrey, Richard. 1971. "Statistical Explanation vs. Statistical Inference." In Salmon, ed. 1971, pp. 19–28.

Jobe, Evan. 1976. "A Puzzle Concerning DN Explanation." *Philosophy of Science* 43: 542–9.

_____. 1985. "Explanation, Causality and Counterfactuals." *Philosophy of Science* 52: 357–89.

Kendall, M. 1948. *The Advanced Theory of Statistics*. London: Charles Griffin and Co.

Kiiveri, H. and T. Speed. 1982. "Structural Analysis of Multivariate Data: A Review." In S. Leinhardt, ed. *Sociological Methodology*. San Francisco: Jossey-Bass.

Kim, Jaegwon. 1969. "Events and their Descriptions: Some Considerations." In N. Rescher, ed. *Essays in Honor of Carl G. Hempel*. Dordrecht: Reidel, pp. 198–215.

_____. 1973. "Causation, Nomic Subsumption and the Concept of Event." *Journal of Philosophy* 70: 217–36.

_____. 1980. "Events as Property Exemplifications." In M. Brand and D. Walton, eds. *Action Theory*. Dordrecht: Reidel, pp. 159–77.

_____. 1984. "Epiphenomenal and Supervenient Causation." *Midwest Studies in Philosophy* 9: 257–70.

Kistler, Max. unpublished. "Causes as Events and Facts."

Kitcher, Philip. 1981. "Explanatory Unification." *Philosophy of Science* 48: 507–31.

_____. 1989. "Explanatory Unification and the Causal Structure of the World." In Kitcher and Salmon (1989), pp. 410–505.

Kitcher, Philip and Wesley Salmon, eds. 1989. *Minnesota Studies in the Philosophy of Science*. vol. 13 *Scientific Explanation*. Minneapolis: University of Minnesota Press.

Kyburg, Henry. 1965. "Comments." *Philosophy of Science* 32: 147–51.

Lauwers, Steven. 1978. "A Reexamination of Causal Irregularity." *Philosophy of Science* 45: 471–3.

Lemmon, E. 1967. "Comments." In N. Rescher, ed. *The Logic of Decision and Action*. Pittsburgh: University of Pittsburgh Press, pp. 96–103.

LePore, E. and B. McLaughlin, eds. 1985. *Actions and Events: Perspectives on the Philosophy of Donald Davidson*. Oxford: Blackwell.

Lewis, David. 1972. "Psychophysical and Theoretical Identification." *Australasian Journal of Philosophy* 67: 642–63.

_____. 1973a. "Causation." *Journal of Philosophy* 70: 556–67.

_____. 1973b. *Counterfactuals*. Cambridge: Harvard University Press.

_____. 1976. "Probabilities of Conditionals and Conditional Probabilities." *Philosophical Review* 85: 297–315. Rpt. in Lewis 1986b, pp. 133-56. Rpt. in and cited from Jackson, Frank, ed. *Conditionals*. Oxford: Oxford University Press, pp. 76–102.

_____. 1979. "Counterfactual Dependence and Time's Arrow." *Noûs* 13: 455–76.

_____. 1986a. "Causal Explanation." In Lewis 1986b, pp. 214–40.

_____. 1986b. *Philosophical Papers*. vol. 2. New York: Oxford University Press.

_____. 1986c. "Postscripts to 'Counterfactual Dependence and Time's Arrow.'" In Lewis 1986b, pp. 52–67.

_____. 1986d. "Postscripts to 'Causation.'" In Lewis 1986b, pp. 173–218.

Little, Daniel, ed. 1995. *The Reliability of Economic Models*. Dordrecht: Kluwer.

Lombard, Lawrence. 1986. *Events: A Metaphysical Study*. London: Routledge.

McDermott, Michael. 1995. "Lewis on Causal Difference." *Australasian Journal of Philosophy* 73: 129–39.

Mackie, John. 1966. "The Direction of Causation." *Philosophical Review* 75: 441–66.

_____. 1979. "Mind, Brain, and Causation." *Midwest Studies in Philosophy* 4: 19–29.

_____. 1980. *The Cement of the Universe*. Oxford: Oxford University Press.

McLaughlin, James. 1925–6. "Proximate Cause." *Harvard Law Review* 39: 149–99.

Meek, Christopher and Clark Glymour 1994. "Conditioning and Intervening." *British Journal for Philosophy of Science* 45: 1001–21.

Mellor, D. H. 1995. *The Facts of Causation*. London: Routledge.

Menzies, Peter. 1989a. "Probabilistic Causation and Causal Processes: A Critique of Lewis." *Philosophy of Science* 56: 642–63.

_____. 1989b. "A Unified Account of Causal Relata." *Australasian Journal of Philosophy* 67: 59–83.

_____. 1996. "Probabilistic Causation and the Pre-emption Problem." *Mind* 105: 85–117.

Menzies, Peter and Huw Price. 1993. "Causation as a Secondary Quality." *British Journal for Philosophy of Science* 44: 187–203.

Mermin, N. David. 1990. "Quantum Mysteries Revisited." *American Journal of Physics* 58: 731–4.

Miller, Richard. 1987. *Fact and Method*. Princeton: Princeton University Press.

Orcutt, Guy. 1952. "Actions, Consequences and Causal Relations." *Review of Economics and Statistics* 34: 305–13.

Otte, Richard. 1981. "A Critique of Suppes' Theory of Probabilistic Causality." *Synthese* 48: 167–89.

_____. 1985. "Probabilistic Causality and Simpson's Paradox." *Philosophy of Science* 52: 110–25.

_____. 1986. "Reichenbach, Causation, and Explanation." In Fine, Arthur and Peter Machamer., eds. *PSA 1986*, vol. 1. East Lansing: Philosophy of Science Association, pp. 59–65.

Papineau, David. 1985a. "Causal Asymmetry." *British Journal for the Philosophy of Science* 36: 273–89.

_____. 1985b. "Probabilities and Causes." *Journal of Philosophy* 82: 57–74.

_____. 1986. "Causal Factors, Causal Interference, Causal Explanation." *Aristotelian Society Supplementary Volume* 50: 115–36.

_____. 1989. "Pure, Mixed and Spurious Probabilities and the Significance for a Reductionist Theory of Causation." In Kitcher and Salmon (1989), pp. 307–48.

_____. 1990. "Causes and Mixed Probabilities." *International Studies in the Philosophy of Science* 4: 79–88.

_____. 1991. "Correlations and Causes." *British Journal for the Philosophy of Science* 42: 397–412.

_____. 1993. "Can We Reduce Causal Direction to Probabilities?" In Hull, D., M. Forbes, and K. Okruhlik, eds. *PSA 1992*, vol. 2. East Lansing: Philosophy of Science Association, pp. 238–52.

Pearl, Judea. 1993. "Comment: Graphical Models, Causality, and Intervention." *Statistical Science* 8: 266–9.

_____. 1995. "Causal Diagrams for Empirical Research." *Biometrika* 82: 669–88.

Pearl, Judea and Thomas Verma. 1994. "A Theory of Inferred Causation." In D. Prawitz, B. Skyrms, and D. Westerståhl, eds. *Logic, Methodology and Philosophy of Science, IX*. Amstersdam: Elsevier, 1994, pp. 789–811.

Pollock, John. 1976. *Subjunctive Reasoning*. Dordrecht: Reidel.

_____. 1984. "Nomic Probability." *Midwest Studies in Philosophy* 9: 177–204.

Price, Huw. 1991. "Agency and Probabilistic Causality." *British Journal for Philosophy of Science* 42: 157–76.

_____. 1992. "Agency and Causal Asymmetry." *Mind* 101: 501–20.

_____. 1993. "The Direction of Causation: Ramsey's Ultimate Contingency." In Hull, D., M. Forbes, and K. Okruhlik, eds. *PSA 1992*, vol. 2. East Lansing: Philosophy of Science Association, pp. 253–67.

_____. 1996. *Time's Arrow and Archimedes' Point: New Directions for the Physics of Time*. Oxford: Oxford University Press.

Putnam, Hilary. 1962. "The Analytic and the Synthetic." In Herbert Feigl and Grover Maxwell, eds. *Minnesota Studies in the Philosophy of Science*. vol. 3. Minneapolis: University of Minnesota Press, pp. 350–97.

Quine, W. 1953. "Two Dogmas of Empiricism." In *From a Logical Point of View*. Cambridge, MA: Harvard University Press, pp. 20–46.

_____. 1960. *Word and Object*. Cambridge: MA: MIT Press.

_____. 1985. "Events and Reification." In LePore and McLaughlin (1985), pp. 162–71.

Railton, Peter. 1980. "Explaining Explanation: A Realist Account of Scientific Explanation." Dissertation, Princeton University.

Ray, Greg. 1992. "Probabilistic Causality Reexamined." *Erkenntnis* 36: 219–44.

Redhead, Michael. 1987. *Incompleteness, Nonlocality, and Realism: A Prolegomenon to the Philosophy of Quantum Mechanics*. Oxford: Oxford University Press.

Reichenbach, Hans. 1956. *The Direction of Time*. Berkeley: University of California Press.

Rosen, Deborah. 1978. "In Defense of a Probabilistic Theory of Causality." *Philosophy of Science* 45: 604–13.

_____. 1982. "A Critique of Deterministic Causality." *Philosophical Forum* 14: 101–30.

Russell, B. 1913. "On the Notion of Cause." *Proceedings of the Aristotelian Society* 13: 1–26.

Salmon, Wesley. 1971. "Statistical Explanation." In Salmon (1971), pp. 29–87.

_____. 1978. "Why Ask, 'Why?' An Inquiry Concerning Scientific Explanation." *Proceedings and Addresses of the American Philosophical Association* 51: 683–705.

_____. 1980. "Probabilistic Causality." *Pacific Philosophical Quarterly* 61: 50–74.

_____. 1985. *Scientific Explanation and the Causal Structure of the World*. Princeton: Princeton University Press.

_____. 1994. "Causality without Counterfactuals." *Philosophy of Science* 61: 297–312.

Salmon, Wesley, ed. *Statistical Explanation and Statistical Relevance*. Pittsburgh: University of Pittsburgh Press, 1971.

Sanford, David. 1976. "The Direction of Causation and the Direction of Conditionship." *Journal of Philosophy* 73: 193–207.

_____. 1984. "The Direction of Causation and the Direction of Time." *Midwest Studies in Philosophy* 9: 53–75.

_____. 1985. "Causal Dependence and Multiplicity." *Philosophy* 60: 215–30.

Sen, Amartya. 1970. *Collective Choice and Social Welfare*. San Francisco: Holden-Day.

Simon, Herbert. 1953. "Causal Ordering and Identifiability." In W. Hood and T. Koopmans, eds. *Studies in Econometric Method*. New York: John Wiley & Sons, pp. 49–74. Rpt. and cited from Simon (1977), pp. 53–80.

_____. *Models of Discovery and Other Topics in the Methods of Science*. Dordrecht: Reidel, 1977.

Simon, Herbert and Nicholas Rescher. 1966. "Cause and Counterfactual." *Philosophy of Science* 33: 323–40. Rpt. and cited from Simon (1977), pp. 107–31.

Simpson, C. 1951. "The Interpretation of Interaction in Contingency Tables." *Journal of the Royal Statistical Society* Series B 13: 238–41.

Skyrms, Brian. 1980. *Causal Necessity: A Pragmatic Investigation of the Necessity of Laws.* New Haven: Yale University Press.

_____. 1984. "EPR: Lessons for Metaphysics." *Midwest Studies in Philosophy* 9: 245–55.

Sober, Eliott. 1985. "Two Concepts of Cause." In Peter Asquith and Philip Kitcher, eds. *PSA 1984*, vol. 2. East Lansing: Philosophy of Science Association, pp. 405–24.

_____. 1986. "Causal Factors, Causal Inference, Causal Explanation." *Aristotelian Society Supplementary Volume* 50: 97–113.

_____. 1987. "Discussion: Parsimony, Likelihood, and the Principle of the Common Cause." *Philosophy of Science* 54: 465–9.

_____. 1988. "The Principle of the Common Cause." In Fetzer, James, ed. 1988. *Probability and Causality: Essays in Honor of Wesley Salmon.* Dordrecht: Reidel, pp. 211–28.

Sober, Elliott and Martin Barrett. 1992. "Conjunctive Forks and Temporally Asymmetric Inference." *Australasian Journal of Philosophy* 70: 1–23.

Spirtes, Peter, Clark Glymour, and Richard Scheines. 1993. *Causation, Prediction, and Search.* New York: Springer-Verlag.

Spohn, Wolfgang. 1983a. "Deterministic and Probabilistic Reasons and Causes." *Erkenntnis* 19: 371–96.

_____. 1983b. "Probabilistic Causality: from Hume via Suppes to Granger." In M. Galvotti and G. Gambetta, eds. *Causalitá e Modelli Probabilistici.* Bologna: Clueb, pp. 69–87.

_____. 1990. "Direct and Indirect Causes." *Topoi* 9: 125–45.

_____. 1991. "A Reason for Explanation: Explanations Provide Stable Reasons." In Wolfgang Spohn, et al. eds. *Existence and Explanation.* Dordrecht: Kluwer Publishing, pp. 165–96.

_____. 1994. "On Reichenbach's Principle of the Common Cause." In Salmon, W. and G. Wolters, eds. *Logic, Language, and the Structure of Scientific Theories.* Pittsburgh: University of Pittsburgh Press, pp. 215–39.

Stegmueller, W. 1973. *Personnelle und Statistische Wahrscheinlichkeit.* Berlin: Springer-Verlag.

Suppes, Patrick. 1970. *A Probabilistic Theory of Causality.* Amsterdam: North-Holland.

_____. 1984. "Conflicting Intuitions about Causality." *Midwest Studies in Philosophy* 9: 151–68.

Swain, Marshall. 1978. "A Counterfactual Analysis of Event Causation." *Philosophical Studies* 34: 1–19.

Taylor, Richard. 1964. "Can a Cause Precede its Effect?" *The Monist* 48: 136–42.

_____. 1966. *Action and Purpose.* Englewood Cliffs, NJ: Prentice-Hall.

Teller, Paul. 1989. "Relativity, Relational Holism, and the Bell Inequalities." In Cushing and McMullin (1989), pp. 208–23.

Thomson, Judith. 1977. *Acts and Other Events.* Ithaca: Cornell University Press.

_____. 1985. "Causal Priority: A Comment." *Noûs* 19: 249–53.

Van Fraassen, Bas. 1980. *The Scientific Image.* Oxford: Oxford University Press.

_____. 1982. "The Charybdis of Realism: Epistemological Implications of Bell's Inequality." *Synthese* 52: 25–38.

Van Inwagen, Peter, ed. 1980. *Time and Cause*. Dordrecht: Reidel.

Verma, Thomas. 1992. "Graphical Aspects of Causal Models." Dissertation, University of California at Los Angeles.

Verma, Thomas and Judea Pearl. 1990. "Equivalence and Synthesis of Causal Models." In Bonnisone, P., M. Henrion, L. Kanal, and J. Lemmer, eds. *Uncertainty in Artificial Intelligence*. Cambridge, MA: Elsevier, pp. 220–27.

von Wright, Georg. 1975. *Causality and Determinism*. New York: Columbia University Press.

Williams, Donald. 1953. "The Elements of Being." *Review of Metaphysics* 7: 3–18, 171–92.

Wold, Hermann. 1954. "Causality and Econometrics." *Econometrica* 22: 162–77.

Woodward, James. 1979. "Scientific Explanation." *British Journal for the Philosophy of Science* 30: 41–67.

_____. 1984. "Explanatory Asymmetries." *Philosophy of Science* 51: 421–42.

_____. 1989. "The Causal Mechanical Model of Explanation." In Kitcher and Salmon (1989), pp. 357–83.

_____. 1993. "Capacities and Invariance." In Earman, J. A. Janis, G. Massey, and N. Rescher, eds. *Philosophical Problems of the Internal and External Worlds: Essays Concerning the Philosophy of Adolph Grünbaum*. Pittsburgh: University of Pittsburgh Press, pp. 283–328.

_____. 1995. "Causation and Explanation in Econometrics." In Little (1995), pp. 9–61.

_____. 1997. "Causal Models, Probabilities, and Invariance." In McKim, V. and S. Turner, eds. *Causality in Crisis? Statistical Methods and the Search for Causal Knowledge in the Social Sciences.* Notre Dame: University of Notre Dame Press, pp. 265–317.

_____. 1998. "Explanation, Invariance and Intervention." *Philosophy of Science* 64.

_____. forthcoming a. "Causal Independence and Faithfulness."

_____. unpublished. *Explanation, Invariance and Intervention.*

Yule, G. 1903. "Notes on the Theory of Association of Attributes in Statistics." *Biometrika* 2: 121–34.

_____. 1926. "Why Do We Sometimes Get Nonsensical Relations between Time-Series? – A Study in Sampling and the Nature of Time Series." *Journal of the Royal Statistical Society* 89: 1–64.

# Index